The Globalization of Wood:

Supply, Processes, Products, and Markets

Papers presented at a conference sponsored
by the Forest Products Society and the
Oregon Forest Resources Institute
November 1-3, 1993
Portland, Oregon

Proceedings No. 7319

Forest Products Society
2801 Marshall Court
Madison, WI 53705-2295 USA
phone: 608-231-1361
fax: 608-231-2152

ISBN 0-935018-64-6

Printed in the United States of America.

9403-500

Preface

It was a great pleasure that I accepted the task of co-chairing the conference *The Globalization of Wood: Supply, Processes, Products, and Markets*" which led to the production of this proceedings. There are three key reasons that I feel honored to have been associated with the planning and development of this conference.

The first reason stems from the fact that the *Globalization of Wood* conference evolved from an international conference, organized by the Forest Products Society and chaired by Dr. Jim Bowyer, titled *Wood Product Demand and the Environment* which was held in Vancouver, British Columbia in November of 1991. To co-chair and help organize a conference of the same caliber of the previous meeting was truly a challenge, given the importance of the material presented in Vancouver. To be considered in the same vein, *The Globalization of Wood* would have to maintain the global scope, the preeminent quality of presentations, and the high level of what I call "future think" established by the previous conference. I hope readers of this proceedings will agree that the high caliber of presentations established in the Vancouver conference has been maintained.

The second reason was the opportunity to associate with some of the most perceptive minds in the field of wood and material use from a global perspective. This group of people included not only those presenting papers at this conference but also the session co-chairs and moderators. It was truly an honor to be associated with such an illustrious, knowledgeable, and friendly group of practitioners and scientists.

The third reason was the opportunity to work with and help fulfill the dreams of not only a good friend but a scientist of consequence — Dr. Steven A. Sinclair. Steve passed away during the planning of this conference, and as a former student, good friend, and co-chair, I am particularly pleased to dedicate this proceedings to Steve. The quality and importance of much of the material presented in the following pages would have made Steve proud and pleased. I want to dedicate this proceedings to the memory of a seminal figure in the development of the field of forest products marketing in North America.

Dr. David H. Cohen
Associate Professor
Department of Wood Science
Faculty of Forestry
University of British Columbia
Vancouver, British Columbia, Canada

About the Sponsors

Forest Products Society

The Forest Products Society is an international non-profit educational association founded in 1947 to provide a common forum for all segments of the forest products industry from standing tree to finished product. The Society encourages the development and application of new technologies and provides a link for the interchange of information between industry and research through the collection and dissemination of the latest technical information. Membership is open to all interested individuals and organizations.

For further information on membership, write or call:
Forest Products Society
2801 Marshall Court
Madison, WI 53705-2295, USA
Telephone (608) 231-1361
Fax (608) 231-2152

Oregon Forest Resources Institute

The Oregon Forest Resources Institute (OFRI) was created by the Oregon legislature in 1991 to promote public understanding of forest products and practices. The Institute is governed by a board of directors elected by those who are taxed to support it — the harvesters of timber from federal, state, county, and private lands within the state. The board also includes a labor representative and ex-officio members representing various public interests.

The Institute's goals are:

- educate the public about the region's forests;
- to communicate how forests can be managed for society's benefit;
- to promote understanding of the relationship between the global environment, science-based forestry, and enlightened wood products manufacturing.

For information contact:
OFRI
1440 S.W. 3rd Ave.
Suite 305
Portland, OR 97201, USA
Telephone (503) 229-6718
Fax (503) 229-5823

CONFERENCE PLANNING COMMITTEE

Co-Chairs

Dr. David H. Cohen
Associate Professor
Department of Wood Science
Faculty of Forestry
University of British Columbia
Vancouver, British Columbia, Canada

Dr. Steven A. Sinclair (deceased)
Formerly, Professor
Center for Forest Products Marketing
Virginia Tech
Blacksburg, Virginia, USA

Committee Members

Arthur Brauner
Executive Vice President
Forest Products Society
Madison, Wisconsin, USA

Dr. Robert L. Ethington
Professor
Forest Products Lab.
Oregon State University
Corvallis, Oregon, USA

Bernard E. Fuller
Vice President
Resource Information Systems, Inc.
Bedford, Massachusetts, USA

Leslie A. Lehmann
Executive Director
Oregon Forest Resources Institute
Portland, Oregon, USA

Floyd Vike
Executive Vice President, Marketing
Willamette Industries, Inc.
Albany, Oregon, USA

Ronald W. Willis
Export Markets Manager
Southern Forest Products Association
Kenner, Louisiana, USA

Session Cochairs and Moderators

Dr. Jim L. Bowyer
Professor and Head
Department of Forest Products
University of Minnesota
St. Paul, Minnesota, USA
Session I

Dr. David H. Cohen
Assistant Professor
Forest Products Marketing
University of British Columbia
Vancouver, British Columbia, Canada
Session V

Dr. Ivan L. Eastin
Assistant Professor
College of Forest Resources
University of Washington
Seattle, Washington, USA
Session V

Bernard E. Fuller
Vice President
Resource Information Systems, Inc.
Bedford, Massachusetts, USA
Session I

Dr. Thomas C. Maness
Assistant Professor
Department of Wood Science
University of British Columbia
Vancouver, British Columbia, Canada
Session III

Dr. Thomas E. McLain
Professor and Head
Department of Forest Products
Oregon State University
Corvallis, Oregon, USA
Session IV

Dr. David V. Plackett
Manager-Composites and Treated Wood Products
Forintek Canada Corporation
Vancouver, British Columbia, Canada
Session III

Joe K. Sample
Vice President — Marketing
Temple-Inland Forest Products
Diboll, Texas, USA
Session II

Dr. Albert T. Schuler
Manager — Market Research
Norbord Industries, Inc.
Toronto, Ontario, Canada
Session II

Dr. Paul M. Smith
Assistant Professor
School of Forest Resources
Pennsylvania State University
University Park, Pennsylvania, USA
Session IV

Table of Contents

Luncheon Address and Technical Forum Presentations

SESSION I:

Global Fiber Supply and Demand — The Engine of Change

The Globalization of Wood: Supply, Processes, Products, and Markets

Wood production and the post-industrial forest

Alexander Mather

Abstract

The aim of this paper is to consider the extent to which a shift to the "post-industrial" paradigm of forest management will accelerate existing spatial trends in the wood industry. The extent to which globalization already exists is reviewed, and the characteristics of the post-industrial paradigm are outlined. It is concluded that spatial trends evident in the wood industry in recent decades are likely to be strengthened as this paradigm is adopted. Bulk fiber production, based on forest plantations, may increasingly become concentrated in the South (i.e., the developing world), where comparative advantage may exist in both biophysical and socio-economic terms. The implications, for both South and North, are considered.

Introduction

The theme of this conference is the globalization of wood. This paper will first consider the extent to which the wood industry is already globalized. I will review changes in the perceptions of forests that have been underway in some parts of the world in recent years; these represent the advent of the "post-industrial" forest. Thirdly, I shall attempt to relate the advent of the post-industrial forest to spatial trends and trends in globalization, and to speculate about possible future patterns. My main interest lies in the forest itself, rather than in downstream industry, but clearly trends in the use and management of the forest will have downstream consequences.

To what extent is the wood industry globalized?

We sometimes assume that we now live in a global village and that most, if not all, forms of human activity involve interaction at the global scale. To use the words of the eminent American conservationist, Raymond Dasmann, we are now "biosphere" people, whereas we once were "ecosystem" people. By this he means that once we drew our resources from local areas immediately around us; now we assemble them from around the world. The question of whether we draw our resources from near or far is indeed an important one: as self-centered and self-serving individuals we all too easily ignore the effects of our resource demands when they are met from the other side of the world. In the meantime, however, let us consider the extent to which the wood industry is at present globalized.

The author is Reader, Dept. of Geography, Univ. of Aberdeen, Aberdeen, Scotland.

Measures of globalization

Numerous different measures or yardsticks could be used to indicate the extent of globalization. A simple one, though imperfect and crude, is the proportion of production entering international trade. A related characteristic is the flow pattern of products: are they truly global, or are they little more than regional? We might also look at flow patterns of capital investment, and at the actors in the industry. To what extent do they play on a global, as opposed to a national or regional, stage?

The availability of data to allow us to use these criteria is variable. Perhaps the simplest yardstick — that of international trade — is also the most accessible.

Proportions of wood products entering international trade

The proportion of the products of the wood industry that entered international trade during the 1980s was approximately 17 percent (1). The proportion varies from product to product, but in comparison with other sectors this proportion is not unusually high. The initial impression, then, is that the industry is not globalized to an unusually large extent.

Wood in its raw form is a bulky, low-unit value material, and it is not surprising that it does not loom large in trade, even with modern developments in cheap, bulk transport. As might be expected, the proportion of production entering international trade increases in general terms with the level of processing and value added (Table 1). For industrial roundwood, it is around 7 percent, while for paper it is approaching 25 percent (4).

Trends in relative trade volumes

The proportions may still be relatively small, but have they been rising? For the period of the 1960s to 1980s, the proportional share of exports increased for every forestry-industry sector (12). During the 1980s, they increased for most products (Table 1). For value-added products such as sawn wood and paper there have been clear and steady increases in relative terms as well as in absolute terms. For other products such as sawlogs and industrial roundwood in general, there has either been little change or a downward trend. This is perhaps to be expected in the light of specific measures such as log-export bans or restrictions, as well as of more general factors such as the development of wood-processing industries in new areas. To some extent, complementary trends are evident: e.g., the relative volume of sawn wood entering trade has increased, but that of non-coniferous sawlogs has decreased. Again, it would seem that globalization is limited in extent.

Global or merely international?

Developed and developing worlds: Trends in relative trade volumes are imperfect indicators of globalization. They may simply reflect regional trade between neighboring countries, rather than the growth of global networks of interaction. As another indicator we can look at the relative proportions of wood products generated by the developed and developing worlds, and their contributions to trade.

The developing world's share of production rose steadily during the 1980s, the only (partial) exception being in the case of sawlogs (Table 2). Trends in the relative pattern of production, however, are not necessarily the same as those in the relative pattern of trade. Regional shifts in productive capacity are not necessarily reflected directly in the global pattern of trade, as over 80 percent of global production is directed at domestic markets. Much of the growth in developing-world production in particular is geared to home markets.

Trade has, for the most part, traditionally involved the developed world, although some countries have for centuries imported wood products from their empires in what is now the developing world. About 10 percent of industrial forestry production in developing nations enters international trade, compared to 20 percent from the developed world. Nevertheless, developing-country exports have been growing faster than those of developed countries. The share of exports originating from the developed world decreased from 85 percent in 1963 to 78 percent in 1980 (7). Increasing proportions of developing-world exports go to other developing countries (1). In other words, the pattern of interaction has become more complex in recent decades.

Patterns of concentration. — The extent to which trade is dispersed or concentrated is another indicator of the extent to which globalization exists. One measure is the proportion of trade accounted for by the three leading exporters and importers of the various wood products. As Table 3 suggests, strong concentration exists in some of these patterns. Concentration is as prominent a feature amongst exporters from the developing world as it has traditionally been from the developed world. Three developing countries (Brazil, Malaysia, and Indonesia) have accounted for about half of the exports from that part of the world. A related point is the extent to which trade is concentrated regionally around specific

Table 1. — Proportion of production entering international trade (percentage exports by weight or volume).

	1970	1980	1990
Industrial roundwood	7.4	7.9	7.1
Sawnwood	13.9	17.7	18.3
Wood-based panels	14.0	16.1	25.0
Wood pulp	16.6	16.8	16.2
Paper and paperboard	18.3	20.6	23.2
Sawlogs (total)	8.3	8.0	6.8
Sawlogs (C)	4.4	4.5	4.9
Sawlogs (NC)	18.7	16.0	11.7

Table 2. — The developing world's share of production (percentage by volume or weight).

	1970	1980	1990
Industrial roundwood	16.0	23.3	23.7
Sawn wood	12.2	20.0	23.2
Wood-based panels	9.0	13.8	20.5
Wood pulp	3.4	6.1	6.9
Paper and paperboard	7.8	11.5	16.0

foci of imports and exports. Several distinct trade flows and areas can be recognized, rather than a truly global network. Three flows account for half of the world trade in forest products: within western Europe, from northern to western Europe, and from North America to western Europe (7).

Multinational corporations. — Yet another indicator of globalization is the involvement of multinational corporations and the extent to which they function in terms of production and marketing. In qualitative terms, there is ample evidence of the involvement of corporations based in North America, Europe, or Japan in the production of wood as well as in the processing and marketing of wood products. It may be the case that increasing proportions of international trade are channeled through these corporations. But we need to be cautious about reaching definite conclusions about globalization trends. Some of these links are long established, dating from the colonial era, and are not the manifestation of completely new trends. On the other hand, there is increasing evidence of cross-links between parts of the developed world (e.g., involving European and Japanese interests in North America), as well as between the developed and developing worlds, in both directions. American corporations, for example, have invested in forests and in processing facilities in areas such as Southeast Asia and South America, as well as in Canada. They have also invested in facilities in Europe, while some European investment has occurred in the United States (5). Recent purchases of forestland in New Zealand have involved corporations based in developing countries such as Indonesia and Malaysia, as well as in developed countries such as the United States and Japan (Table 4). This is just one manifestation of the internationalization of the forest industry in New Zealand (15), a country with a small population and domestic market but with a potential comparative advantage for fiber production. The growth of these complex patterns of interaction, linking corporations in different parts of the world, is probably the best indicator of globalization, though it is difficult to chart and even more difficult to interpret. This issue should feature prominently on our research agendas if we are to obtain a better understanding of the world wood industry.

We should note, however, that while some of these corporations operate on a truly global stage (e.g., one based in Britain is involved with forest plantations in Chile, New Zealand, and Thailand), others have a more limited sphere, such as around the Pacific Basin. We should also note that some multinational corporations may be involved in globalization at certain levels but not at others (e.g., either in production or in manufacturing). Fully globalized integration clearly does not necessarily accompany the involvement of multinationals.

We can conclude, then, that the wood industry is at present globalized only to a limited degree. It can scarcely be said that there is a single world market for most forest products. Globalization is impeded by the inertia of traditional trading links, by long-term agreements, and by a reluctance to venture towards new supplies or qualities unless existing ones are seriously threatened (25). On the other hand, the process of globalization, though in general slow, has not always been steady. There have been phases when new patterns have developed rapidly. Examples include the growth of tropical hardwood exports after World War II (13), the growth of Japanese imports from the 1960s (timber self-sufficiency dropped from around 70% then to less than 30% by the end of the 1980s (26,22)), and the growth of plywood exports from Indonesia and other southeast Asian countries and of pulp exports from Brazil. Such episodes have certainly been associated with internationalization, if not with full globalization. Perhaps the years ahead will witness further comparable episodes or pulses, possibly on a grander and more genuinely global scale.

In short, the degree of global interaction will probably continue to increase in the years ahead, perhaps

Table 3. — Concentration in trade patterns (percentage accounted by three leading exporters and importers, 1990).

	Exporters	Importers
Chips	72	86
Sawlogs (C)	78	74
Sawlogs (NC)	73	56
Wood pulp	51	42
Paper and paperboard	40	36

Table 4. — Foreign acquisitions of New Zealand forests, 1990-92.[a]

Purchaser	Base	Area (ha)
Ernslow One	Malaysia/Singapore	23,801
Juken-Nissko	Japan	43,531
Wenita	Hong Kong/China	20,521
Winstone Pulp International	Indonesia	8,331
Oji Paper-Sanyo Kokusaken Pulp	Japan	30,232
ITT Rayonier	USA	97,453

[a] Source: Rowling, W. 1993. Forestry ownership — some implications of change. New Zealand Forestry (May). p. 3.

at a faster rate than in past decades. Why? Part of the reason may follow from the fact that at present we are in the midst of a transition, or rather several different transitions.

Forest transitions

The forest management transition

The transition in forest management that we are currently in has been likened by Roger Sedjo to that from hunting to farming (28). To a large degree, we until now depended on natural, old-growth forests rather than "farmed" or managed forests. This is perhaps less true in parts of Europe, where managed and planted forests have long been extensive, than in other parts of the world, but it is a matter of time before the transitional stage is reached globally. What has happened in the past is that production was extended as demand increased: England looked first to Norway and Scotland, then to the Baltic and North America. In the same way Japan has drawn supplies from a succession of southeast Asian areas more recently. Sooner or later, however, intensification takes place: ultimately there are no new areas to which production can be extended. This process can take the form of plantation production: forests established by human hands then begin to supply the wood formerly provided by natural forests. These plantations are well known for their high productivities, at least in terms of wood sought by industry, compared with natural forests. But while natural forests are sometimes regarded as the "free gifts of nature," whose costs are no more than those of harvesting and transport, plantations have price tags. As long as the "free gifts of nature" were available in the form of cheap wood from accessible natural forests, the cost of wood from plantations was unattractive, except when cost of transport was less than from the natural forests. It seems that conditions have changed during the present century, and especially during the last few decades. Plantation establishment has now become an almost global phenomenon, as well as accelerating in rate in tropical latitudes (Table 5). The trend is set to continue, and perhaps to increase further (17).

Table 5. — Areas of forest plantations in the tropics and hotter sub-tropics.[a]

	1965	1980	1990
		(1,000 ha)	
Africa	1,378	2,724	3,773
Asia (including S. China)	4,421	13,046	29,245
Latin America	798	4,934	9,256
27°N-27°S	6,667	20,973	42,694

[a] Source: Evans, J. 1992. Plantation forestry in the tropics. Clarendon, Oxford, England.

Some very extensive areas of natural forest still survive, but most of them are likely to remain indefinitely as potential rather than actual resources for the wood industry. In some of them, such as the Soviet Far East, the biological potential is huge, but so also are the economic and social problems confronting development (32). The construction of the Baikal-Amur railway has made it possible to begin the exploitation of a huge forest area of some 54 million hectares, with a tradable stock of 2,000 million m^3 (31). In general, however, most of the potential areas are in continental interiors so inaccessible that the cost of transport would mean that the wood could scarcely compete in price with that from accessible plantations. Such areas may also offer species as yet unfamiliar or unfavored. The historical pattern of substitution of new species for known and previously accepted species may continue to some extent, but the attraction of uniformity of supplies available from plantations may discourage research and development in the use of new species from previously unutilized natural forests. Furthermore, many natural forests are gaining "protected area" status, and in general logging in these forests is meeting with growing opposition, as need hardly be said in this setting. Overall, it is difficult to avoid the conclusion that the scope for the opening up of major new regions of supply from natural or old-growth forests is limited.

The switch from natural to planted forests will not happen overnight, but the trend seems inexorable. And it introduces questions of comparative advantage in a new way. Plantations can be located in areas that are optimal in terms of productivity, accessibility to markets, and of costs of labor and management. New patterns become not only possible but also likely, and these new patterns may be more dispersed and more global than the ones they replace. In short, comparative advantage operates in different ways in relation to locations of plantations and of old-growth forests.

The forest area transition

This transition away from reliance on old-growth forests towards more intensively managed forests or plantations is one clear feature of the present century. Another is the forest area transition, or the change from shrinking to expanding forests that has occurred in most of Europe (19,21). In parts of Europe in the 19th century, deforestation was proceeding as rapidly as it is in parts of the developing world today. A turnround was achieved — rapidly and dramatically in the case of countries such as France, Hungary, and Italy — and the forest is now expanding in area. On this evidence, it is tempting to speculate that the same will happen in the developing world, and that the present declines in forest areas there will be halted and eventually reversed. Remaining areas of natural forest are receiving greater protection, fiscal and other policies destructive to the forest are being modified, a switch away from fuelwood as the primary energy source will presumably eventually occur, and of course the establishment of plantations will itself be a factor.

One factor that contributed to the forest area transition in Europe was the slowing of population growth rates, and hence of growth in demand for food. This was combined with increasing yields of food crops, and a switch away from reliance on fuelwood as the primary source of energy. It was also linked to changing values and perceptions of the forest, as a resource base which could offer many benefits of which wood was only one. This leads us to the concept of the post-industrial forest.

The post-industrial forest

Three sequential models of forest resource use can be recognized, which for convenience we shall call the pre-industrial, industrial, and post-industrial stages (Table 6). One distinguishing variable is the diversity of goods and services sought from the forest. An associated variable is the degree of primacy afforded to wood production as a management goal. A further variable is the perception of modes of forest management that are acceptable to the public at large.

There is clear evidence in parts of Europe that a shift is occurring away from the industrial forest and towards the post-industrial forest. This shift is exemplified by what has happened in Britain during the 1980s, and especially during the latter part of the decade (20). Note that this shift of perception and of acceptability applies to plantations, and not only to natural forests. Note also that it applies to plantations that for the most part have been established on bare land considered by many to be degraded. (The shift began earlier, of course, in natural forests. Timber has been cut from these as recently as 10 to 20 years ago, but now they are protected.) Nor does it seem that the shift is peculiar to Britain or Europe, although I shall concentrate on that part of the world. In relation to Canada, for example, it is suggested by Pearse (1993) that the current public perception of forest management is that interests of timber production ride roughshod over forest values (27). Perhaps this hints at an eventual transition there also.

Some of the evidence for this shift in Britain is summarized in Table 7. A point to emphasize is that it has occurred abruptly or rapidly over a period of a few years since the mid-1980s. The monocultural Sitka spruce plantation was by 1990 the symbol of what was perceived to be wrong with forestry: as recently as the 1970s such plantations were being welcomed by environmental gurus such as Fraser Darling as environmental improvements.

To some extent, the shift can be described as "post-material," in that the primacy or relative importance afforded to wood production has dwindled. But "post-material" is inadequate as a descriptor, for two reasons. First, it is not wholly accurate. "Material" or wood production is still sought, but there are signs in countries such as Britain and Spain of shifts to preferences for hardwoods rather than bulk fiber production (10). A similar preference for quality over quantity of wood in Japan is evidenced by Marchak (1992) in her comparative study involving that country and British Columbia (18). The other reason for the inadequacy of "post-material" as a descriptor is concerned with how the forest is managed, as well as with the goals of management and the relative priority afforded to wood production.

"Post-industrial" as employed here is not synonymous with "multiple use." It relates to popular perceptions and expectations of planted forests in private ownership, as well as those of public old-

Table 6. — Post-industrial forest characteristics.

Stage	Resource characteristics	Resource regime
Pre-industrial	Diversity of products	Common-property
Industrial	Wood primary or sole product	Concession/private
Post-industrial	Increasing range of goods and services	Private-regulated

Table 7. — Indicators of shift to post-industrial forest in Britain.

- Removal of tax advantages that were main driving force for afforestation up to 1988.
- New planting-grant scheme, with stringent environmental conditions; wood production need not be the management objective
- Grants conditional on adherence to guidelines and codes of practice
- Wide differentials in grants, in favor of broadleaves
- "Presumption against" commercial coniferous planting in English uplands
- Strengthening influence of planning system on afforestation
- Creation of community and urban-fringe forests
- New estimates of recreational value of state forests

growth forests. It signifies a questioning of ownership rights: in effect, it contends that there is a public or common-property interest in private forests, even if these forests have been created for the specific purpose of timber production for private profit. This is one facet of the emerging wider issue of forest ownership and sovereignty (33); at the larger scale there is the question of the relationship between national and global interests in the fate of the tropical forest.

In short, there has been a marked shift away from perceptions of new forests as potential sources of wood (and especially bulk fiber), and towards perceptions of new forests as attractive environments, especially for recreation. There is little doubt that this shift has been encouraged by what are now seen as some of the excesses of industrial forestry and its environmentally unfriendly behavior, as recently as the second half of the 1980s. Accompanying the shift has been a decline in new planting rates in Britain to half their level of 5 to 10 years ago, despite steep increases in levels of planting grants. Institutional investors such as pension funds, which a decade ago were actively involved in industrial afforestation, now show little interest in planting in Britain. Some have sold their forests. Some have become involved in international investments in the United States and elsewhere. Capital has become more mobile, and can quickly perceive comparative advantage and disadvantage at a scale which is increasingly more global than national.

This shift to the post-industrial forest is perhaps unusually abrupt in the case of Britain, but there are also symptoms of it, sometimes in less acute form, in many other European countries. These symptoms may include the setting aside of tracts of forest for purposes of environmental conservation, and increasing constraints on forms of forest management that are perceived as environmentally unfriendly. For example, artificial drainage and the replacement of native stands with faster growing exotics can attract opposition. Similarly, the private ownership of forests is increasingly being motivated by factors such as environmental conservation, residential amenity (e.g., in association with hobby farms) and recreation. Timber production, in the view of many such owners, is a secondary consideration: the forest is valued more as an amenity than as a source of profit or raw material in the form of bulk fiber.

Intensive plantation production is sometimes seen as environmentally beneficial, as a means of relieving pressures to supply wood demands from old-growth forests. But the environmental acceptability of intensive production geared exclusively to bulk-fiber production is itself increasingly being questioned. Intensive agricultural production is now widely associated in Britain and some other European countries with adverse environmental impacts. There are plenty of signs that forestry is viewed in the same light.

The post-industrial forest and wood industry patterns

The shift towards the post-industrial forest in parts of Europe and perhaps elsewhere in the developed world will probably reinforce a trend that is already in existence. This trend is reflected in differential growth patterns, and consists essentially of a southward adjustment, in both wood production and in processing. It can be exemplified at different scales, and in a variety of ways. Within Europe, the production of industrial roundwood and pulp has been increasing faster in countries such as Spain and Portugal than in the traditional forest-products countries of Scandinavia (Table 8). I suspect that a similar trend may be evident in the American South. On a global scale, the growth of wood production has been much greater in recent years in countries such as Brazil, Chile, and New Zealand than in traditional forest countries such as Canada and the former Soviet Union (Table 8).

On a global scale (and perhaps also on a continental scale), this southward shift can be expected to continue and probably to accelerate. Some of the contributing reasons can be regarded as "pull" factors, reflecting the attractions and comparative advantages of new areas. Others are "push" factors, reflecting impediments to growth in production in some of the existing areas.

1. Higher productivities can be achieved in low-latitude plantations than in many parts of North America and Europe (2,11,29). Comparative advantage may exist both in the biophysical and the socio-economic environments. Costs of land and labor are often lower, or at least no higher, in the South than in the North (hereafter, unless

Table 8. — Industrial roundwood (production indices 1970=100).

	1970	1980	1990
Canada	100	131	127
Sweden	100	79	91
Finland	100	115	103
Soviet Union	100	93	95
Spain	100	159	214
Brazil	100	259	305
Chile	100	177	261
New Zealand	100	121	146

otherwise qualified, South and North relate respectively to the developing and developed worlds). Already by the mid-1980s it was being stated that the worldwide potential competitiveness of the new, low-cost raw material sources of the less developed countries has gradually started to reshape the global structure of the forest industry (12).

2. The shift towards the post-industrial forest will probably continue in the North, and the traditional industrial forest will become less acceptable there. In European countries such as Britain, multiple objectives and managerial constraints (e.g., on choice of species and silvicultural practices) will limit the extent to which wood production will increase, and/or result in additional costs. If wood production can be pursued more single-mindedly in parts of the South (i.e., the developing world) and if fewer constraints are placed on intensive management for wood production, then international capital will soon identify the opportunity, as indeed it already has done to some extent. A related point is that environmental controls on wood-using industries may be perceived by some as weaker in the South, further lowering costs there relative to the North. In other words, both the "industrial-flight" and the "pollution-haven" hypotheses may apply (16), at both the timber-growing and timber-processing stages.
3. The third reason is that demand for many wood products is growing faster in the South than in the North. In general terms, there is an inverse relation between the pattern of population growth rates and that of the wood industry. By and large, the areas of fastest population growth are areas in which the wood industry (from the forest to the processing plant) is poorly developed. A further complication is that paper-making in the North is increasingly likely to see the substitution of recycled fiber for raw wood. (New locational factors will operate, or at least changes in the relative weights of locational factors.)
4. More speculatively, future prospects for the boreal forest are uncertain in the light of possible climatic change. Furthermore, shortfalls in regeneration in logged areas in some of the more accessible parts of the forest are reported to exist in both Canada and in the former Soviet Union (19). It may be an exaggeration to claim the state of the boreal forest will become as big an issue in the years ahead as tropical deforestation has been in recent times. Nevertheless, it can probably be concluded that huge increases in wood production from the boreal forest are unlikely in the foreseeable future, and that increases in demand are more likely to be satisfied from the South.

In short, what may develop is a pattern in which bulk wood or fiber production, and the associated processing industries, will grow more rapidly in the South than in the North. This is not an entirely new trend, but it is argued that it is likely to accelerate, because of both "pull" factors and "push" factors. The former include the possible advent of the forest area and forest management transitions there, and the potential for establishing large-scale, high-productivity plantations in which management for maximum fiber production is unhindered by other "post-industrial" considerations. Conversely, the "post-industrial" shift in the North is a "push" factor.

On the analogy of manufacturing industry, mass production of bulk products, or "Fordism," will perhaps begin to characterize the South. In the North, "post-Fordism" may apply in the post-industrial forest: small-scale production of diverse but specialized hardwoods and their products may increasingly be sought, as indeed it already is in countries such as Britain and Spain (6,10). Indeed, if supplies of hardwood logs become more difficult to obtain from tropical countries, either because of resource exhaustion or "green" embargoes, then markets for home-produced hardwoods may well grow rapidly. (Supplies of high-grade tropical hardwoods from plantations are likely to be relatively small over the next few decades (9)). In other words, while globalization and a southward shift may characterize the bulk wood-products industry, the opposite trend may apply in specialized sectors. At the same time, the wood-products industry in the North will probably have little option but to follow the path of "added-value" as pioneered by countries such as Finland. Over the last 30 years it has undergone a shift from a role as major exporter of relatively low-value products such as roundwood and plywood to that of exporter of higher value products such as fine paper and forest industry machinery.

One scenario for the future is that of "Japanization" expanded to the global scale. Production forestry and bulk-fiber production becomes concentrated in the South, in areas of optimal biophysical productivity and in areas where intensive and single-minded industrial forest management is possible. Meanwhile, in the North, post-industrial forest management predominates, perhaps with the production of high-quality specialized wood being sought alongside recreational and environmental benefits.

An alternative to this scenario of polarization is one where "post-industrial" management prevails worldwide: where management for wood production proceeds at an acceptable intensity, and where objectives involving environmental and other services are also achieved at the same time. This paradigm resembles in some respects the features of the "complex plantations" advocated by some commentators: these include an intimate association with other land uses, direct involvement with local people, and more complex patterns of species composition and plantation structure (11). As the advocates recognize, a major conceptual change or paradigm shift is required if "complex plantation" forestry is to be adopted.

Environmental variations and comparative advantages will ensure that uniformity is never achieved, but the contrasts between North and South will be more muted and less stark than in the first scenario. Perhaps the wood industry would also be more stable, in terms of location and in other characteristics.

Perhaps the trends in the years ahead will lie somewhere between these two scenarios, and between the countervailing pressures of globalization and autarchy. Globalization in the sense of the exchange of all wood products at a truly global scale is unlikely to develop: for raw and semi-processed products the trend is more likely to be in the opposite direction. At the same time autarchy is unachievable for many countries — not only for those in arid areas but also for those that are densely populated and have high levels of demand for wood products.

Trends and uncertainties

At present we are in an age that history may come to deem decisive. On the one hand, there are signs that trade is becoming freer. On the other hand, we are seeing the first tentative steps towards the global application of uniform environmental standards. If the latter trend lags behind the former, the shift to the South will probably be accelerated.

There are many uncertainties about the future trends of which we have been discussing, and about their strengths and rates. One is the local acceptability of plantations of the industrial forest model, and of the associated industries. Our first scenario, of an industry relying heavily on intensive timber plantations in the South, implies a continuation and strengthening of the tradition of colonial and post-colonial forestry. Local perceptions of this tradition are aptly summarized in the words of the Indian writers Gadgil and Gula (8): "the unpopularity of scientific forest management with the mass of the Indian population is a powerful indicator of its social inappropriateness" (8). Concern with both the socio-economic and environmental impacts of such forestry have been widespread, being reported from settings as diverse as Scotland, Portugal, India, and Chile (for Chile, see (14)). Will successive countries in the South begin to impose effective curbs and constraints on the wood industry, and to manifest shifts towards the post-industrial forest in their own territories? Will the industry then demonstrate its own variation of "industrial-flight" and "pollution-haven" behavior, and move sequentially from area to area and from country to country? Will it simply repeat, albeit in modified form, the pattern of the past, when it exploited natural forests?

The answer will probably lie in the balance between the attitudes of countries in the South and the behavior or performance of the multinational and domestic corporations and investors. It seems that awareness of the "ecological shadows" cast by countries such as Japan (24) and Britain is growing. Resource activities unwanted in the North will not always be welcomed in the South. The other factor is how the new plantations are established and managed: how serious are their social and environmental impacts and those of the downstream industries, and what steps are taken to ensure that the receiving areas can enjoy a positive balance between the benefits and costs arising from the growth of a local wood industry? The Rio Forestry Principle that "forest resources and forestlands should be sustainably managed to meet the social, economic, ecological, cultural, and spiritual human needs of present and future generations" (30) sounds like advocacy of a shift towards the post-industrial forest worldwide. How this principle will be observed remains to be seen. Perhaps the day of true globalization will arrive only when the wood industry is based on forest plantations established, as it were, on a level playing field: when management practices and objectives are standardized and only environmental conditions are variable.

Endnote

Forests, whether natural or planted, can be perceived as symbols of environmental purity or of environmental devastation. In a recent paper, Clark Binkley refers to McKibben's "snarl of the chain saw" as symbolizing the "end of nature," though he rejects McKibben's hypothesis that nature has ended (2,23). There can be little dispute that negative perceptions abound, however strongly the forestry profession may consider that they are unfair. Whether the perceptions are positive or negative depends on how forests are managed and for what purposes, as well as on changing popular attitudes and outlooks. Perhaps the task of the wood industry is to devise forms of management of plantations that will provide not only wood but also a range of benefits that satisfy popular aspirations in the areas in which they are located, or at least does not employ management methods that attract strong opposition. Sustainability of development and of production has its ecological and economic dimensions, but there is also a political one. Popular opposition to an operation or a form of management can build up to a level that makes it unsustainable. One option is then to move to a new area where opposition is weaker, and ideally also one where productivity is higher. As has been suggested, this may happen in the years ahead, and a southward shift may accelerate. However, relocation cannot continue indefinitely. Comparative advantage may endure in terms of productivity, but it is unlikely to do so permanently in terms of laxity or stringency of environmental controls. Sooner or later, they are likely to become both more stringent and more uniform, even if a "window of opportunity" is perceived to exist in parts of the South in the short- and medium-term future. Adoption of high environmental standards and of at least some of the characteristics of post-industrial forest management before they are strictly enforced in successive countries in the South could help to ensure that the wood industry is welcomed and valued by their governments and peoples. It would be ironic if a realm of activity that devised the concept of sustained yield management were to fail to adapt to this concept of sustainability, and if the plantation era were to be characterized by continual sequential shifts to a succession of new areas, as was the era of wood production from the natural forest.

Literature cited

1. Alexandratos, N. (ed.) 1988. World agriculture: toward 2000 — an FAO study. Belhaven Press, London, England.
2. Binkley, C. 1992. Forestry after the end of nature. J. Forestry 90(10):33-37.
3. Evans, J. 1992. Plantation forestry in the tropics. 2nd ed. Clarendon, Oxford, England.
4. FAO Yearbooks of Forest Products, Rome, Italy. [Source of all statistical data except where otherwise indicated.]
5. Flora, D.F. 1988. USA: trade channels for its forest products. *In:* International trade in forest products. A. Nagy, ed. AB Academic Publishers, Bicester, England. pp. 171-181.
6. Forestry Commission. 1993. Growing broadleaves for timber. HMSO, London, England.
7. Francescon, A. and A. Nagy. 1988. Historical analysis: international trade in forest products. *In:* International trade in forest products. A. Nagy, ed. AB Academic Publishers, Bicester, England. pp. 197-225.
8. Gadgil, M. and R. Guha. 1993. This fissured land: an ecological history of India. Oxford Univ. Press, Delhi, India.
9. Grainger, A. 1988. Future supplies of high-grade tropical hardwoods from intensive plantations. J. World Forest Resource Management 3:15-29.
10. Groome, H. 1993. Afforestation policy and practice in Spain. Afforestation: policies, planning and progress. A. Mather, ed. Belhaven Press, London, England. pp. 72-91.
11. Kanowski, P.J. and P.S. Savill. 1992. Forest plantations: towards sustainable practice. *In:* Plantation politics: forest plantations in development. C. Sargent and S. Bass, eds. Earthscan, London, England. pp. 121-155.
12. Kiljonen, K. 1986. Growth of Third World forestry: possible impact on Finland. Silva Fennica 20(3):159-179.
13. Laarman, J.G. 1988. Export of tropical hardwoods in the twentieth century. *In:* World deforestation in the twentieth century. J.F. Richards and R.P. Tucker, eds. Duke Univ. Press, Durham, N.C. pp. 148-163.
14. Lara, A. and T.T. Veblen. 1993. Forest plantations in Chile: a successful model? *In:* Afforestation: policies, planning and progress. A. Mather, ed. Belhaven Press, London, England. pp. 118-139.
15. Le Heron, R. 1988. The internationalisation of New Zealand forestry companies and the social reappraisal of New Zealand's exotic forest resource. Environment and Planning A 20:489-516.
16. Leonard, H.J. 1988. Pollution and the struggle for world product. Cambridge Univ. Press, Cambridge, England.
17. Lyon, K.S. and R.A. Sedjo. 1992. Comparative advantage in timber supply: lessons from history and the timber supply model. *In:* Emerging issues in forest policy. P.N. Nemetz, ed. Univ. of British Columbia Press, Vancouver, B.C., Canada. pp. 171-186.
18. Marchak, M.P. 1992. Global markets in forest products: sociological impacts on Kyoto Prefecture and British Columbia interior forest regions. *In:* Emerging issues in forest policy. P.N. Nemetz, ed. Univ. of British Columbia Press, Vancouver, B.C., Canada. pp. 339-369.
19. Mather, A.S. 1990. Global forest resources. Belhaven Press, London, England.
20. ________. 1991. Pressures on British forest policy: prologue to the post-industrial forest? Area 23(3):245-253.
21. ________. 1992. The forest transition. Area 24(4):367-379.
22. Maull, H.W. 1992. Japan's global environmental policies. *In:* International politics of the environment. A. Hurrell and B. Kingsbury, eds. Clarendon, Oxford, England. pp. 354-372.
23. McKibben, W. 1990. The end of nature. Anchor Doubleday, N.Y.
24. McNeill, J., P. Winsemius, and T. Yakushiji. 1991. Beyond interdependence: the meshing of the world's economy and the Earth's economy. Oxford Univ. Press, N.Y.
25. Nagy, A. and A. Anderson. 1988. Introduction. *In:* International trade in forest products. A. Nagy, ed. AB Academic Publishers, Bicester, England. pp. 1-16.
26. Nomura, I. 1988. Japan: the timber trade and its problems. *In:* International trade in forest products. A. Nagy, ed. AB Academic Publishers, Bicester, England. pp. 115-133.

27. Pearse, P.H. 1993. Forest tenure, management incentives and the search for sustainable development policies. *In:* Forestry and the environment: economic perspectives. W.L. Adamowicz, W. White, and W.E. Phillips, eds. CABI, Wallingford, England. pp. 77-96.
28. Sedjo, R.A. 1987. Forest resources of the world: forests in transition. *In:* Global forest sector: an analytical perspective. M. Kallio et al., eds. John Wiley, Chichester, England. pp. 7-31.
29. ________. 1992. Can tropical forest management be economic? *In:* Issues in forest policy. P.N. Nemetz, ed. Univ. of British Columbia Press, Vancouver, B.C., Canada. pp. 505-518.
30. United Nations Conference on Environment and Development. 1992. [It is explicitly stated that the principles should apply to all types of forests, both natural and planted. A convenient source of the full text is in: K. Ramakrishna and G.M. Woodwell, eds. 1992. World forests for the future: their use and conservation. Yale Univ. Press, New Haven, Conn. pp. 141-148.]
31. Volkov, V.O. 1988. USSR: trends and prospects in the forest products trade. *In:* International trade in forest products. A. Nagy, ed. AB Academic Publishers, Bicester, England. pp. 183-196.
32. Waggener, T.R. and C.A. Backman. 1992. United States-Soviet roles in Pacific Rim softwood log trade: the Japanese market as the competitive link. *In:* Changing Pacific forests: historical perspectives on the forest economy of the Pacific Basin. J. Dargavel and R. Tucker, eds. Forest History Soc., Durham, N.C. pp. 1-18.
33. Woodwell, G.M. 1993. Forests: what are they for? *In:* World forests for the future: their use and conservation. K. Ramakrishna and G.M. Woodwell, eds. Yale Univ. Press, New Haven, Conn. pp. 1-20.

The Globalization of Wood: Supply, Processes, Products, and Markets

The world's need for wood

W.R.J. (Wink) Sutton

Abstract

Globally we use a lot of wood both for fuel and industrial purposes. That consumption is driven by population. As the global population continues to increase, total wood demand is also predicted to increase. Globally we are now near (or even beyond) the harvest level of natural forests. The increasing supply deficiency can only partly be met from maturing plantations. As forest management becomes more intensive and the move to plantations hastens, there will be increasing recognition that forestry is very capital intensive. There are certain to be increasing problems attracting the necessary long-term investment. Substitution is not a realistic option because a major global move to wood substitutes would significantly increase energy demand. An effective (and democratic) means of slowing down population growth is to improve living standards. But that would greatly increase wood demand. The frightening global dilemma is that it will become increasingly difficult for the world's forests to satisfy the increasing demand for wood. That production objective will be even more difficult if society desires that our forests also satisfy a multitude of other forest values. Yet if we do not satisfy that demand for wood there will be catastrophic consequences.

Introduction

The scope of this paper is global. The time horizon is very long — decades, not years. The paper considers all wood use, including fuelwood. The inclusion of fuelwood is because most fuelwood comes from forests and because there are increasing problems defining the boundary between what is industrial wood and what is fuelwood. We used to call off-cuts "waste" — now we call them "residues." With the increasing interest in bio-energy as renewable energy source, there are certain to be even more problems in actually defining the boundary between industrial wood and fuelwood.

The author is Senior Policy Advisor, Industry, Trade, and Economics, Canadian Forest Serv., Victoria, B.C., Canada.

Historical and current wood use

There are few estimates, let alone reliable statistics, of global wood consumption prior to the U.N.'s Food and Agricultural Organization (FAO) effort beginning in the late 1940s. Since 1950 the global wood consumption has gone from around 1.5 billion m^3 (52 billion $ft.^3$) to 3.5 billion m^3 (120 billion $ft.^3$) in 1991 (Fig. 1).

Over that time the official global estimates consistently show that we use about equal amounts of fuel and industrial wood.

The principal driver for wood consumption is population. The global population has increased from 2.5 billion in 1950 to 5.2 billion by 1991 (Fig. 1).

Many wood uses have been lost to substitutes. We have also improved the efficiency of conversion as well as increased the use of residues from solid wood processing. But we have also created new uses for wood. The net effect has been that the global per capita consumption of all wood has actually increased. It has gone from less than 0.6 m^3 per person per year in 1950 to nearly 0.7 m^3 per person per year in 1991.

The conversion of global and national totals to personal levels of consumption provides an easily understandable measure of the importance of wood.

A volume of 3.5 billion m^3 used annually by 5.2 billion people works out at 0.67 m^3 (24 $ft.^3$) per person per year. That is equivalent to a cube of wood with edges equal to just under 0.9 meter (or about 35 in.). If we average 0.67 m^3 per person over a year, that represents a daily wood consumption of 1.8 liters (just under 0.5 U.S. gallons or just over 3 Imperial pints per person per day). Since it is rare to express wood volumes (which are of solid form) in liters, gallons, or pints (values which are normally associated with liquids), a more appropriate

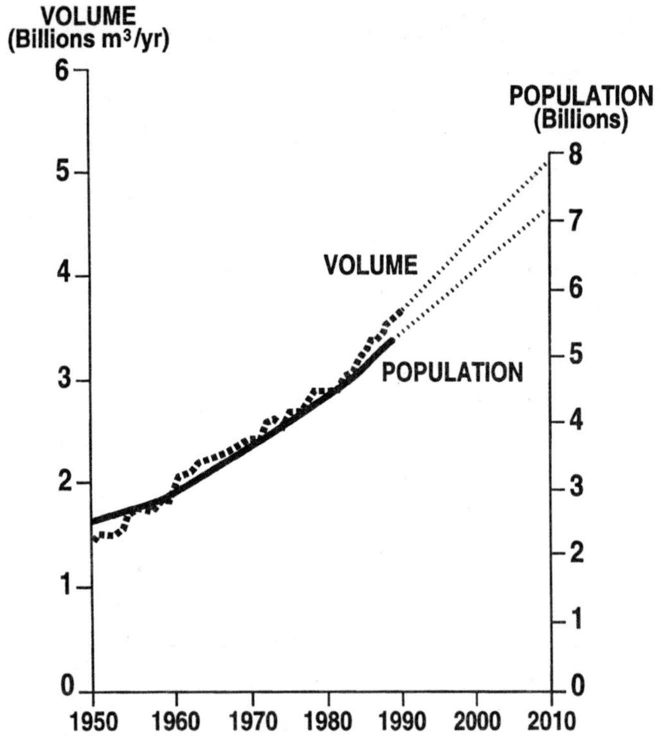

Figure 1. — Total world wood harvest and world population (with projections to 2010).

measure is to demonstrate the actual amount. I use this approach to considerable effect in my presentations.

I begin by taking an even thickness wooden disc cut from a log with an underbark diameter of 60 cm (24 in.).[1] That disc is then cut into equal sized wedges or segments. It can make 8 × 45° wedges, 10 × 36°, or, 12 × 30° wedges. To represent the average daily consumption of wood that is used globally each wedge should have a volume of 1.8 liters. To have that volume, the 60-cm (24-in.) disc would need to have a thickness of:

- 8 cm (3 in.) when cut into 12 × 30° wedges
- 6.5 cm (2.5 in.) when cut into 10 × 36° wedges
- 5 cm (2.0 in.) when cut into 8 × 45° wedges

For every person in the 5.4 billion global population, that wedge represents the average daily consumption of both fuelwood and industrial wood.

There are, of course, major variations in consumption levels. The United States has one of the highest per capita wood consumption rates in the world. Bowyer (4) has estimated that the annual consumption of wood in the United States is 80 ft.^{3} (or 2.3 m^3) per person per year. That represents a consumption of 3.4 times the global average. On a daily basis, that consumption translates as 1.65 U.S. gallons (1.4 Imperial gallons or 6.3 liters per person per day).

To illustrate the U.S. level of personal daily consumption using the same underbark diameter 60-cm (24-in.) wooden disc, that disc should have a thickness of:

- 26.5 cm (10.5 in.) when cut into 12 × 30° wedges
- 22 cm (8.75 in.) when cut into 10 × 36° wedges
- 17.5 cm (7 in.) when cut into 8 × 45° wedges

This all demonstrates just how very important wood is as a raw material, especially in developed economies. Wood demand seems insatiable.

Surprisingly, little attention has been devoted globally to wood production and consumption. Especially in the media, other forest values like species and biodiversity protection, concerns over clearcuts, preserving old growth, etc. have received much attention. Yet wood remains one of our most important raw materials.

Future wood demand

Global wood consumption in the past has been driven by an increasing global population. It is unlikely that this will change.

There are few forecasts of future wood demand. An exception was in 1991 when the FAO (3) forecast through to the year 2010. This is given in Table 1 together with the global population and wood production levels for 1991. Between the years 1991 and 2010 the global annual wood demand is predicted to increase from 3.43 to 5.07 billion m^3. That means that over the next 19 years the global wood demand is predicted to increase every year by an average of 86 million m^3. To put that increase in perspective, that global annual increase is slightly more than the current total harvest of British Columbia (74 million m^3). That means that in each of the next 19 years we will need to find globally the equivalent of an extra British Columbia just to satisfy the increasing wood demand. This total in-

Table 1. — Global wood demand and population.

	Current (1991)	Projected (2010)	Increase
Wood demand	- - - - - - -	(billion m^3)	- - - - - - -
Industrial wood	1.6	2.7	+1.1
Fuelwood	1.8	2.4	+0.6
All wood	3.4	5.1	+1.7
Population (in billions)	5.3	7.2	+1.9

Source: (3).

[1] For convenience all of the following daily consumption calculations have been rounded. Hence, care should be taken when multiplying up these daily average values.

cludes fuelwood. If we consider only industrial wood then we need to find the equivalent of a current British Columbia every 2 years.

The future demand for wood could be even higher than the FAO has predicted. There are two reasons for this:

1. *The improving global economy, especially in Asia.* As will be illustrated later in this paper, improving personal wealth greatly increases the demand for wood. As more economies improve, especially those in Asia, so will the demand for wood.
2. *A possible move to more bio-energy.* To meet their obligations under the Climate Change Convention, countries are committed to maintain, or even reduce, emissions of fossil carbon into the atmosphere. This will stimulate an increased interest in renewable energy sources. Bio-energy using wood is an obvious choice. This seems certain to increase the demand for wood as a fuel, even in countries which no longer use much wood as an energy source.

This leads us to the question of future supply.

Future wood supply

It is possible that some of the increasing demand for wood will be satisfied from fast growing plantations on very short rotations — like 6 to 10 years. However, such wood will be only suitable for pulping or fuelwood. But many solid wood end uses require larger and more mature logs. Very few, if any, quality saw or peeler logs can be grown (even with fast growing species) on rotations of less than 20 years. This means that any tree required over the next 20 to 30 years for solid wood end uses is already established and growing. Prediction of availability should therefore present fewer problems then prediction of demand.

Where then are those trees and when will they be harvested to meet the future demand for wood? There are two possible sources:

1. *Traditional supply areas based on natural forests.* These natural forests may have been harvested in the past and then regenerated, or have yet to be harvested.
2. *Plantations of either indigenous or introduced tree species.* These are areas where trees have been actually planted. They may or may not have been in areas which were previously forested immediately prior to planting.

Natural forest supply

Some of the increased wood needs will be satisfied from traditional natural forests. There are possibilities of small increases in wood harvest from several European countries and in Eastern Europe. Siberia is another possibility but there are many doubts about how much of this vast forest region can be economically harvested. There may be future environmental opposition to the further utilization of the Siberian resource.

In the United States and Canada there is increasing opposition to forest utilization, particularly in the West. It seems likely that harvest levels from these natural forests will be further reduced in the future.

Asia, especially Southeast Asia, is facing major environmental conflicts over tropical forest use. Similar conflicts will continue over further utilization of the forest of the Amazon basin. Harvest levels will decline, not increase.

Africa and other areas of the world have limited scope of increasing production.

On balance there appears to be a limited scope globally to increase wood harvest levels from traditional sources. Wood harvest levels from natural forests might even decline significantly if other forest values (like biodiversity, protection of endangered species) are considered by society to be more important and to be incompatible with wood harvesting. This could well happen. Because some consider clearcutting to be ecologically unsound, Greenpeace International is rumored to be planning in 1994 to campaign against all clearcutting harvesting systems globally.

Plantation supply

Globally there are about 100 million hectares of plantations. At least 85 million hectares are slow-growing species in the northern hemisphere. Many of these plantations are still under 30 to 35 years of age. Although these plantations will provide increasing volumes of wood, especially in Europe and the southeast United States, the increased volume of supply will not be sufficient to make up for the predicted supply deficit (a deficit resulting from the increased wood demand and declining harvest from the traditional natural forest sources).

As they mature, the fast growing plantations of species like radiata pine in New Zealand and Chile will add to the global wood supply. However, especially for sawlogs, the total volume available from all countries will not be great. Certainly not more than 20 to 25 million m^3 of sawlogs can be expected above 1990 levels by the year 2010.

Future supply — general

Taking a relatively long-term global perspective, it is difficult to see where any significant increases in volume can come from. The supply position will be worst for the bigger and better quality saw and peeler logs.

As fewer utilizable natural forests remain, we can only maintain and increase future wood harvest either by regenerating the natural forest which has been harvested or by establishing more plantations.

In the harvested natural forest it will be necessary to ensure regeneration. The regeneration must not only be adequate but also it must be achieved quickly. As the level of management increases in these regenerating natural forests, so increases the capital expenditure. The total capital required to adequately regenerate harvested natural forest can be similar to (or even in excess of) the capital required to establish plantations.

This highlights a very important issue — how do we treat that investment? Few appreciate the absolute size and relevance of this investment. It is, however, an issue of major importance and will be addressed in the next section.

Forestry is very capital intensive

At first consideration, forestry (either as plantations or as regenerated natural forest) appears to be labor intensive. This is because of the high labor requirements for site preparation, planting, releasing, and early tending. However, a characteristic of a labor-intensive industry is that the labor input is quickly converted into a readily saleable product. That does not happen in forestry. The input of labor into forestry is not converted back into actual money until the tree is finally harvested, but harvesting will not be for decades after the labor input. Those costs compounded over the decades means that forestry is very capital intensive. Evidence of just how important this is comes from a country which established its first plantation nearly 140 years ago. Since then that country has experienced nearly five rotations of sawlogs and now gets over 95 percent of its wood from plantations. That country now exports (with a fair proportion as logs) more than it uses internally. That country is New Zealand.

Plantation forestry in New Zealand is a true business. This is because plantation forests are now mostly privately owned, there are no subsidies for forestry, and the plantation business is treated by most companies as an independent profit center. Every forestry expenditure therefore is treated as an investment in the same way that it is in any other industry. Plantation forestry can only attract further investment if it earns a competitive rate of return. Growth rates are high in New Zealand and rotations are less than half those of many traditional wood supply areas (25 to 30 compared with 50 to 70 years). However, New Zealand plantations are still very capital intensive. Many evaluations in New Zealand have demonstrated that the capital invested in plantations is greater than the combined investment required for the harvesting, transporting, sawing, board manufacturing, and pulp and paper production of the wood that will be harvested from the plantations.

Where rotations are longer the capital investment is even greater.

To illustrate this, let us compare $1 invested at 3 percent and 7 percent on rotations of 30 and 70 years (Table 2). Let us now compare a $1 investment in a 30-year radiata pine plantation with a similar investment in northern hemisphere forests grown on a 70-year rotation. This is what we might expect on the coast of the Pacific Northwest or British Columbia. While costs and returns will not be the same for both regions, differences will not be great. Any difference in actual values will be swamped by the cost of capital.

Assuming that costs and returns are more or less similar in both regions, then the major difference is the cost of capital resulting from the length of the rotation. An investment of $1 in the New Zealand planting accrues, at the 7 percent compound interest, to 7.6 times its original value at the end of the 30 years. If the rotation had been extended to 70 years then, for a similar accrued value, the interest rate (because of the longer rotation) would only be a fraction over 3 percent ($1 at 7% for 30 years accrues to $7.60; $1 at 3% for 70 years accrues to $7.90) To achieve a similar return at 7 percent, but with a longer rotation, the invested $1 would have to have resulted in a 15-fold increase in value.[2]

A 15-fold improvement in economic efficiency with the longer rotation cannot be achieved. It could even be that the growing costs of efficient plantations could now be lower than regenerating natural forest in the Pacific Northwest or British Columbia. Because of good management, plantation returns per unit

Table 2. — Compounded value over a rotation of $1 invested at forest establishment.

Rotation	Value appreciation at compounding rates	
	3%	7%
30 years	$2.40	$7.60
70 years	$7.90	$114.00

[2] The reasoning being that $1 compounded at 7 percent for 30 years becomes $7.60; that same $1 compounded at 7 percent over 70 years becomes $114.00. So, a 70-year accrual at 7 percent is $114/7.6, or 15 times, that of the accrual at 7 percent for 30 years.

area are now high and costs are lower. On the other hand, public pressure is forcing more costly practices in managed natural forests, practices which may also reduce returns. This leads us to a major problem, a problem very few are even aware of and even fewer appreciate its significance.

In an increasingly global economy, why should anyone invest in the forests of the Pacific Northwest or British Columbia (or even slower growing forests) where returns will be low (and the risk high)? At best they may get a 3 percent return. Alternatively, they can invest in fast-growing southern hemisphere plantations and achieve at least a 7 percent return. But without investment in what are relatively fast-growing natural forests, then future harvest levels must decline. Industry must also decline. Investments will be even less attractive where growth rates are lower and/or rotations are longer.

We are not discussing small sums of money. Because hundreds of millions of hectares of forest are needed at a direct cost of several thousand dollars a hectare, the total money required globally will be in the hundreds of billions of dollars — even one or two trillion dollars. Governments are most unlikely to invest for such long time periods even if they had the money, which they do not have. Individuals and companies do not have that surplus investment capital either. The only sector with the necessary finance is that of the institutional investors. But why should they invest in long-rotation forestry (where time horizons are long, risks are high, and returns are low) when they can just as easily invest in short-rotation plantations (where returns will be very much higher, etc.).

The global dilemma is frightening and the long-term consequences are very serious. Without a large investment in sound management, future wood yields in the traditional forestry regions will be lower, the forestry industry will be smaller, and there will be fewer wood products produced.

This raises the question of wood substitutes.

The substitution of wood

Wood is astonishingly versatile. It is probably the most versatile raw material mankind has ever known. Wood has been estimated to have been used in more than 10,000 different ways. Yet there are product substitutes for almost every one of those numerous wood end uses. This implies that if we were forced to, we could continue to all enjoy our present high standards of living and not require any wood at all. Why not then phase out the whole forest product industry and simply let all wood use go to substitutes? This would certainly get us out of further conflicts over forest utilization as well as the issues of finding capital.

Unfortunately it is not that simple.

We shall now consider the substitution of fuelwood and industrial wood separately.

Fuelwood

About half of the world's wood consumption is used as fuelwood. It is very likely that wood is still the primary energy source for the majority of the global population. But there are, of course, many fuelwood substitutes — peat, animal dung, oil, natural gas, and electricity (from sources such as nuclear, hydro, wind, waves, solar, and fossil fuels). The problem is that every one of these energy substitutes has both economic and environmental costs of its own. Most of those communities that rely primarily on wood as an energy source are rarely in a position to make the investment to produce energy and/or to pay for the energy from an alternative source. Another aspect is that any significant increased use of fossil fuels will hasten the rate of exploitation of that non-renewable source. The use of fossil fuels also means the atmospheric release of fossil carbon. For most countries an increased use of fossil fuels would put them in contravention of the recent Climate Change Convention. That convention commits countries to at least stabilize, if not reduce, future fossil carbon emissions. Advanced economies, which currently use only limited volumes of wood for fuel, are now increasingly looking at wood as an energy source (bio-energy) because it is environmentally benign — using wood recycles atmospheric carbon rather than releasing more fossil carbon.

Substitution of fuelwood simply increases the demand for other energy supplies. A major move to fuelwood substitutes poses serious environmental as well as economic consequences for the planet.

A move to more bio-energy could actually increase the demand for wood. This would come at a time when global wood harvest levels are at best static.

Industrial wood

It is generally believed that we use most of our industrial wood for the making of paper. This suggests that paper recycling and/or the use of an agriculture fiber (like hemp) will greatly reduce our global need for wood. This is not so. At least two-thirds of the industrial wood used globally is either sawn or peeled. Paper is generally made from either logs that are unsuitable for sawing or peeling, or from residues arising from sawing or peeling. Because few forests are harvested primarily for pulping, practices such as recycling paper or finding a paper substitute can only have minimal influence in reducing wood harvest levels.

The issue of wood substitution is primarily about the substitution of solid wood (lumber or plywood).

When we consider substitutes for solid wood what are we really talking about? The most common wood substitutes are steel, cement (concrete), aluminum, plastics, and ceramics (brick, porcelain, etc.). The total global production in millions of tonnes of these products (except ceramics) is given in Table 3. Not included are fiber substitutes like hemp, bagasse, cotton, and other agricultural fibers.

It may come as a surprise to many that on a weight/volume basis we currently use more wood globally then all the wood substitutes combined. Because of the sheer size of the global wood market, any significant wood substitution is going to require major changes in the global economy.

Of course, for North America where the consumption levels for wood and wood substitutes are already 3 to 4 times the global average, the implications are even more serious.

Of the many studies that have looked at solid wood substitution, one of the most recent, relevant, and comprehensive was written by Koch (5). Although Koch was primarily looking at the energy implications of using wood substitutes in U.S. residential construction, his work has major global implications.

Koch analyzed the energy implications of replacing building wood products with substitutes. Using just a sample of Koch's calculations of Btu (or equivalents) for each product comparison (Table 4), we find that:

- to replace wood studs by steel requires 9 times as much energy
- to replace wooden flooring systems with those of concrete requires 21 times as much energy
- to replace plywood sidings with brick veneer requires 30 times as much energy

Applying Koch's general findings, we come to a startling conclusion. If we substitute solid wood in construction we will use 10, 20, or even 30 times as much energy as we would if we had used wood in the first place! This same conclusion on energy consumption seems likely to apply to all solid wood substitution. We use so much wood globally that any significant wood substitution must have enormous implications for global energy usage. Without a major breakthrough in the production of cheap, renewable, and environmentally friendly energy, it is difficult to see where all this additional energy can come from.

Wood substitution — general

Without that major energy production breakthrough the long-term global outlook is grim. This is especially so as many countries are now committed to stabilizing or reducing their use of fossil fuel, as almost all hydroelectric dams face greater and greater environmental opposition, and as nuclear power production seems less and less likely to be able to deliver on its earlier promise.

There is no doubt that large-scale substitution of both fuelwood and industrial wood is theoretically possible. The problem is where, and how, are we going to find the energy (up to 30 times as much in some cases) that will be required for a major substitution of wood products. This is not to mention any of the other downsides of substitutes: more mining, more pollution from industrial production, and fossil carbon release in the production of some products, especially cement. We also have a problem with disposal at the end of wood substitutes' life cycle.

It is not possible to consider large-scale substitution of wood without a major breakthrough on energy production. The possibility of such a breakthrough seems increasingly unlikely. For the present we must conclude that we cannot replace significant volumes of the global wood demand without major consequences for both the global economy and the global environment.

There is another interpretation of Dr. Koch's findings. If we want to reduce our global energy con-

Table 3. — Global production of wood and wood substitutes (1990-92 values).

Product	Annual production	Daily per capita consumption
	(million tonnes)	(g)
Cement (1991)	1,150	585
Steel (1991)	734	370
Plastics (1992)	104	55
Aluminum (1990)	18	9
Wood (1991)	3,429 million m^3	1,800 ml

Table 4. — Energy usage by wood and by wood subsititutes.

	Btu (oil equivalent) for:		Ratio of
Substitution process	1 tonne of wood product	Replacement by wood substitute	wood product to wood substitute
Wood studs by steel	2.91	26.67	9
Wood floor by concrete	4.14	86.31	21
Plywood siding by brick veneer	6.0	175.2	30

Source: (5).

sumption but are reluctant to reduce living standards, then Koch's work suggests a solution. All we need to do is to use wood instead of steel, plastics, aluminum, cement, and ceramics. Such a move would greatly reduce energy requirements. Even that energy (as with all the other energy we need) could come from using fuelwood. Carried to a logical conclusion we could, if we had enough wood, have the whole global economy based on using either wood directly, or using the energy derived from wood. Such a move would be very environmentally friendly. It would also be sustainable for as long as the sun continues to shine.

The only problem of course is "where are we going to find the wood?" Such a move would also require the public and the politicians to reconsider attitudes on forest use and our whole current value system.

Possible solutions

Reduced consumption is an obvious option. However, it is difficult to see how the advanced economies, like North America and Europe, can reduce consumption without major economic and social consequences.

As we have already discussed, and without a major energy production breakthrough, neither recycling nor the increased use of wood substitutes are realistic. We must look for other solutions.

The basic driving force in future wood consumption (other than increasing economic well-being) is population growth. It is difficult to discuss population growth without running into all kinds of religious, cultural, and ethical conflicts. We have to find a democratic solution that can slow down population growth but which does not intrude into these conflict areas.

One solution which is socially just, morally acceptable, and largely devoid of human suffering is that of improving living standards.

When people are poor and life is a daily struggle, children are often seen as an asset. Children can provide labor and/or earn money. More importantly, perhaps, children guarantee that someone will be around to look after the present parent generation in its old age. Also, it is unrealistic to expect poor people to be concerned about their environment.

With higher standards of living, large families are often seen as a liability. Not surprisingly, those in well-off economies tend to have smaller families (and therefore have lower rates of population increase).

But do we have any evidence that improving living standards actually reduces the growth in population? We do. A study of the Republic of South Korea through the 1980s provides an interesting insight. (South Korea has a mostly homogeneous population with minimal immigration and emigration.)

Through the first half of the 1980s, the population of South Korea was increasing by an average of 1.6 percent per year. South Korea's economy in real terms was also growing. However, the growth was not as spectacular as in the second half of the decade. By 1990 the economy (GDP) was growing at an impressive annual rate of 9 percent. In 1990 the annual GDP per person reached US $5,400. In monetary value South Koreans were then wealthier than the Portuguese and more than twice as wealthy as the average Malaysian, Mexican, or Argentinean. That dramatic increase in wealth was associated with a significant fall in the rate of population growth.

The population growth in 1990 had fallen (from 1.6 percent 5 years earlier) to less than 1 percent. While this growth rate is still higher than for many advanced economies, it is low compared with the global average (over 1.7 percent per year) (Table 5). As far as I can determine, the South Koreans do not exercise any compulsory form of population control.

A comparison with North Korea is most revealing. In North Korea in 1980 the population growth was running at 2.4 percent per annum (compared with 1.6 percent in South Korea). By 1990 the North Korean population growth had slowed to 1.9 percent per annum (but South Korea had dropped to just under 1 percent). Statistics on annual GDP per capita suggest that North Korea has much lower income levels than South Korea. However, no meaningful comparisons can be made because there are real problems in obtaining reliable economic statistics from North Korea. There are problems also in realistically converting their nonconvertible currency.

South Korean increase in wealth was associated with a decline in the rate of population growth. However, that wealth increase was also associated with major increases in industrial wood consumption.

Table 5. — Republic of South Korea, 1980 to 1990.

	1980	1985	1990
Population (in millions)	38.2	41.2	42.8
Annual rate of population increase (%)	1.60	1.61	0.98
GDP per capita (in US$ per person per year)			
Actual	1310	2065	5400
Inflation adjusted to 1990 dollars	2290	2510	5400
Per capita domestic industrial wood consumption (m^3 per person per year)	0.27	0.28	0.41

Sources: (1,2).

Between 1985 and 1990, the South Korean per capita consumption of industrial wood increased 46 percent (from 0.28 m^3 to 0.41 m^3 per person per year, respectively). South Korea is not regarded as a traditional wood-using nation, but by 1990 had begun to use more industrial wood than the global average consumption (0.34 m^3 per person per year).

Ideally we should make comparisons with the much poorer economy of North Korea. Unfortunately, North Korea does not provide any information on either its harvest levels or its industrial production.

While association does not necessarily imply cause and effect, there is circumstantial evidence that improving living standards (and hence wealth) reduces population growth. We also have other circumstantial evidence that shows that if we improve personal wealth we also increase the demand for industrial wood.

Conclusions — a frightening global dilemma

To summarize this brief overview:

- Globally, we use a great deal of wood. After water, wood is probably our most important raw material. The forecasts are for very significant increases in future wood demand.
- That increased wood demand cannot be met from the remaining natural forests because there is increasing pressure for these forests to satisfy many other demands. There is only limited potential for the increasing supply deficit to be supplied from plantations.
- Without a very major breakthrough in energy production, we cannot realistically consider any significant substitution of wood (either fuelwood or industrial wood). If we used even more wood we could reduce our global energy needs and still maintain current standards of living.
- Forestry is not only a very long-term investment but also a very capital intensive one. There will be major increasing problems attracting investment capital into forestry. If we are to have any hope of increasing the level of wood harvest while also maintaining, or even enhancing, the other values then hundreds of billions of dollars (and maybe several trillion) of investment are required globally. From where will this money come? How can we make these long-term (and probably low yielding) investments attractive to those institutional investors which might invest?
- The driving force behind the increasing global wood demand is population growth. The most democratic means of slowing down population growth is to improve living standards. But raising living standards significantly increases wood demand.

All of which creates a real "Catch-22" situation:

> Where do we get the wood to improve living standards which slows down the population growth which drives the demand for wood...

This suggests that the global society now has some very difficult choices to make. It may well decide that the protection of the environment, biodiversity, preserving old growth, preventing the extinction of endangered species, etc. is now so important that forest harvesting must be reduced. But we must appreciate that such a decision will have major consequences for the global economy and the global environment. Unless we find, and find quickly, an alternative energy source which is cheap, renewable and environmentally friendly there is no possibility of moving globally to wood substitutes. If we have less wood and cannot move to wood substitutes we will have problems of maintaining, let alone improving, living standards. If we do not improve global living standards then we probably have little chance of slowing down population growth. Continued exponential growth of the global human population will eventually have catastrophic effects not only for human existence and survival but also for the entire global environment.

A solution to the dilemma is to use more wood, not less. This would reduce other energy usage and could improve living standards (and so help slow down population growth). But using more wood must mean more forest harvesting. More harvesting is seen by many of the public as being detrimental to other major forest values. These values are considered to be so important that many see reduced global wood harvest levels as the only option.

Somehow we have to find better ways of focusing attention on this frightening and crucial global dilemma. This will be a very difficult task.

Literature cited

1. Anonymous. 1950 to 1993. FAO Yearbooks: Forest Products. Food & Agri. Org. Forestry Serv., No. 26.
2. ________. 1980 to 1993. Britannica Books of the Year (several editions). Encyclopaedia Britannica.
3. ________. 1991. 1989:2010 — Wood & Wood Products. Food & Agri. Org. 39 pp.
4. Bowyer, J.L. 1992. Responsible Environmentalism — Forests, People, Raw Material Needs and Environmental Protection. Hamilton Roddis Memorial Lecture Series No. 1. Dept. of Forestry, College of Agri. and Life Sci., Univ. of Wis. 16 pp.
5. Koch, P. 1992. Wood versus non-wood materials in U.S. residential construction: some energy related global implications. Forest Prod. J. 42(5):31-42.

Impacts of changing global resources on product development and markets

Lynn O. Michaelis

Abstract

The character of the raw material available to the forest products industry has been and will continue to change dramatically — in part because of the inventory of standing timber, but also because of emerging political priorities and constraints. At the same time, the markets that forest products serve are also changing. The relative price of alternate materials, demographic changes, new technologies, and emerging economies all suggest the possibilities of different product forms in the decades ahead.

This paper will explore two separate but very important issues facing the forest products industry. The first issue is the importance of wood fiber to a world with a growing population. A political climate has developed that may favor other materials. However, there are reasons to promote wood fiber products as the preferred material. Wood can be made available in a manner that will help, not hurt, the environment.

The second issue is to identify what the market "needs" of the future might be. Those needs are changing both because of the raw material available to the industry and because of the evolving markets that our products serve. Several examples of changes that have already occurred or are currently underway will be discussed.

The paper will conclude that some elements of change are inevitable and socially desirable. A successful forest products business must recognize and adapt to these changes. Other changes, however, especially those that inhibit optimal use of the land, could lead to expensive and potentially undesirable social outcomes. These choices could not only be costly in direct financial terms, but they may not even achieve the desired environmental goal. Everyone associated with the industry, whether in government, academia, or business, must work together to find solutions that meet broad social goals. At the same time, we must use the power of the marketplace to achieve the best solutions.

The author is Chief Economist, Weyerhaeuser Co., Tacoma, Wash.

Introduction

Clearly, the timing of this conference and the issues to be considered come at a critical time for the forest products industry. The recent "solution" or compromise proposed by the Clinton Administration effectively removes the federally owned forestland from the production of wood products. Less than 20 percent of West Coast timberland will be left in timber production with this proposal. At the same time, harvests from public lands in the Inland area are down over 30 percent and are likely to fall further. Restrictions in coastal British Columbia and wetland issues in the South could reduce production further in those production regions. It appears that under the best of circumstances, North American lumber production in 1994 is likely to be 58 billion board feet (BBF), about the same as in 1992 and 10 percent below 1988 levels, even if prices rise another 20 to 30 percent.

The situation gets much more complex when we look at resource supplies globally. There are regions where the resource is expanding for some species because of plantation activities. In other areas, the production of native species will be curtailed dramatically because of declining inventories.

To illustrate how markets will adjust to changing raw materials, I will recall an experience I had in Japan 10 years ago. At the time I was the product manager for Weyerhaeuser's particleboard business. As part of a fact-finding mission to assess the potential for expanding our sales of finished products to Japan, we visited furniture manufacturers, particleboard mills, and plywood operations. It was a very instructive visit in determining what is possible in foreign markets and how people in other countries think about a changing raw material situation.

The first major finding was how the Japanese assess a changing raw material situation and adjust accordingly. Since they have been running out of raw materials for a thousand years, the Japanese have become very adept at shifting technology or raw material. We visited the corporate offices of Dantani, a major plywood producer, whose primary raw material was Indonesian hardwood logs. After the ceremonial tea and small talk, we asked them what they would do if there was a total ban on Indonesian logs. The 76-year-old patriarch and chairman of the board had said nothing up to this point. He then lifted his head and, at great length, discussed how they had considered a wide range of alternative materials, including logs from Papua, New Guinea, and even softwood veneers from the United States.

He then said something profound: "At a price, we will very quickly shift to another raw material." To an economist this was not terribly shocking. But some of the U.S. mill managers had been so locked into softwood, they had not even considered the range of options, much less had they explored the potential cost differences as extensively as the Japanese. The Japanese had clearly considered many contingencies in the event of a ban.

The second major finding had to do with the nature of a market itself. Once again when we left for Japan there was a strong belief that we could easily penetrate the Japanese market with U.S. particleboard either because of changing raw material cost or because of the expected appreciation of the yen. What we found was a much more complex situation. The furniture and cabinet industry had adopted European technologies. The market required boards cut to metric standards and a density profile similar to German particleboard because of the hinges used in the cabinets. To be competitive, it was imperative to deliver small orders on a 1- to 2-day notice. Finally, the Japanese had a 15 percent duty on the delivered-to-dock price. On the other hand, our Springfield plant has a 5- by 24-foot press and uses fir residuals. To be economic, the panels needed to be shipped in bulk and wrapped to protect them from the high humidity of a sea voyage. The combination of all these factors made it impossible for us to compete effectively in this market. The bottom line was that the Japanese could out-compete us by paying 20 to 30 percent more for the sawdust and shavings, ship them to Japan, and then produce their own particleboard.

There was a slightly humorous example of trying to overcome the cost barrier, while taking full advantage of mill economics. We took an order from a Japanese speaker manufacturer for 5- by 24-foot sheets that we offered at a very attractive price. Everything went fine until we tried to get the packages off the dock in Tokyo. Their fork lifts could not handle the weight, and their trucks and streets were too small to carry sheets that large. The end result was that we had to cut the sheets to smaller sizes with hand saws and then hand-load them into the trucks. We lost a lot of money, but we learned a great deal about the nuances of a foreign market.

Thus, as we consider the global marketplace, it is clear that markets will adjust to relative prices and each market will adjust in a unique way. As a result, it is imperative that we avoid glib generalizing statements driven from large, simplistic econometric models. Second, it is important that we discuss the factors that are causing the relative price shift. If the reason is arbitrary policy restrictions, the world could move to a suboptimal solution, in terms of market prices and environmental impact.

A brief historical perspective

Unlike high-technology industries where change is dramatic and fast, the forest products markets and technologies are relatively mature and change slowly. Occasionally, participants in the industry look at short time intervals and are lulled into a false sense of security. In fact, they may use those short periods to deny that change is occurring, believing that strategies pursued during the last 10 years will work for the next 10 years.

The U.S. structural panels business is one of the best examples of how changing raw material supplies/prices, combined with changing technology, can change the nature of a market. I have experienced this change first hand. When I joined Weyerhaeuser in 1973, the structural panels business was western plywood. Southern plywood production was only 20 percent of the industry and was viewed as an inferior product. The product manager at that time refused to believe that the market would accept southern plywood as a substitute for Douglas-fir plywood. But lower prices and improving quality, given the enhanced small-log peeling technologies and drying capability, led the market to embrace the product.

In the late 1970s, rising timber prices caused the industry to look closely at waferboard technology that had been around for 20 years. This enabled the industry to use a weed species, like alder or aspen, to make structural panels. Once again, the plywood people refused to believe that the market would accept this new product because of appearance, weight, slickness, and difficulty in nailing. Improved processes shifted the industry quickly to oriented strands and different resins that improved the new board's performance. The entry strategy of Weyerhaeuser as a new oriented strandboard (OSB) producer was to focus on performance characteristics, such as modulus of rupture, etc., rather than on appearance grading. Today, OSB is viewed as a product superior to plywood in many applications, since there are no core voids and the product performance is reliable. With nailing guns, the nailing barrier has also disappeared.

With the decline in western timber from public lands, West Coast plywood has dropped from 9 billion square feet (BSF) in 1988 (down from 13 BSF in 1973) to only 4 BSF in 1993. Southern plywood is now near 14 BSF. OSB is growing rapidly and will approach 9 to 10 BSF in the next few years.

There are a host of other examples: using particleboard for plywood in furniture, medium-density particleboard (MDF) for hardwood lumber, hardwood veneers as a core in plywood, engineered products for wide-dimension lumber, and more. On the paper side, we see recycled fiber displacing pulpwood, hardwood pulp used instead of northern softwood, and potential technologies that need no paper at all.

There is another trend that will dramatically affect the demand for forest products in the future: demographics. The aging population of the industrialized world implies changes in housing needs and other consumptive patterns. Newspaper readership, for instance, varies considerably depending on age. Younger people are much more likely to rely on television for news versus the newspaper. Younger people are also more comfortable with computers and are more willing to rely on electronic media for other information.

Our history is rich in examples where our industry and markets have adapted to changing raw material and product characteristics. But this is always at a price — whether in lower direct cost or improved quality or lower overall cost of operation. Dramatic cost changes in the old product or material can greatly accelerate the substitution process.

Asking what a market needs, how it currently works, and why?

The forest products industry is engaged in meeting fundamental human needs. People need shelter, and that requires building materials. People need to have things to sit on and places to store food, and that leads to furniture and cabinetry. People need food, and that requires packaging to move it from farms to kitchen tables. People need to communicate, since that is one of our fundamental differences with other species, so they need mediums to convey and store that information, such as newsprint and paper.

There will always be a normative debate about how much housing one needs or whether paper is wasted with all those Sunday inserts. There can also be normative debates about what is the most desirable material to use in order to fulfill some of these basic needs. Is plastic, steel, or wood better in a broad social sense for meeting our housing or furniture needs? It is in this arena that politics can enter to distort or enhance the relative economics of materials used to the marketplace.

For the last several years our management has been looking beyond current or immediate customer needs to determine what the final consumer or user wants. For many forest products, the immediate customer is simply an intermediary for the final customer. Although a sawmill may sell to a retailer, the ultimate consumer is the homeowner. The same is true for a newsprint mill or a papermill. Looking to the final customer is critical in understanding how any market, whether it is the United States, Germany, or Korea, will adjust to changing raw materials, technologies, or demographics.

In the case of wood products, the material is used in either structural or appearance applications. When used in a structural application, the primary market requirement relates to strength, ease of use, or durability. The builder needs to put up a house at a relatively low cost. How sensitive he is to quality will vary depending on the price level of the house. The ultimate homeowner may be concerned about other factors, such as the ability of the roof to hold a heavy snow or how quiet the floor is when walked on, or even more fundamental things like not punching through the floor when walking with high heels and the house staying warm in winter. How the builder goes about constructing the house to assure these basic needs is of little concern to the homeowner. Similarly, a builder is willing to consider alternative materials or designs if he can meet the buyer's needs at a lower cost. This is one reason why builders have shifted to roof trusses and OSB, for example. In addition, builders in the United States are currently taking a serious look at steel studs to replace wood in nonload-bearing applications.

But using a U.S. assumption on construction or practices on a global scale clearly does not work. Each country has adapted to the raw materials that have been available historically and have used forest products when they were available for certain applications. China, for instance, builds most residential units with brick. Log sales to China have been for building docks or other infrastructure projects. Lumber is too expensive for the average home buyer. Even softwood pulps may be too expensive; hence they will shift to pulp manufactured in Indonesia.

For appearance applications, such as cabinets and furniture, improvements in finishing techniques and high-speed production lines have led producers to materials with more consistent surfaces and uniform density profiles. When furniture manufacturers shifted to these production techniques, the preferred raw material became particleboard and MDF instead of lumber or plywood. Once again, the interrelation of changing price, quality, and technology led to changes in the way furniture was made. The consumer adjusted to the style and appearance changes, given the price difference.

Global demand for appearance applications will increasingly be met by composite products. These technologies can use a wide range of raw materials, including recycled materials. Market designs and styles will eventually accommodate the type of product produced with local raw materials.

A similar situation exists on the paper side of our business. Paper is merely a vehicle for conveying or storing information. Needs differ considerably from those of a consumer or a household. Businesses require a reliable, accurate, inexpensive means for conveying, storing, and retrieving information. Consumers want information sources that are inexpensive, disposable, transportable, and entertaining. Paper has advantages in many of these applications. But new electronic technologies have the potential to displace paper's historic role. Even so, this may not mean less paper demand. Cheaper printers could lead to a boom in household printers. Even the CD-ROM could create more paper demand rather than less, if people shift from text books to laser printer copies of selected sections.

As we look at potential paper demand in the developing world, we cannot use the same growth pattern as in the industrial world. Some analysts have argued that as per capita income rises, so will paper demand — the same as it did in the industrialized world. That kind of analysis does not fully understand the power of current technologies to avoid the cumbersome way our information systems worked as our income grew. It is already evident that these emerging countries will skip over to the new technology. For instance, Singapore did not put down copper wire for its telephone system. It went directly to cellular and fiberoptic cables, thus avoiding all the infrastructure cost of the old system.

Assessing global market conditions in light of changing raw material supplies

The information laid out up to this point can help us discuss where the markets will go from here. Even without the recent sharp contraction of western timber or the problems in the tropical rainforest, the trends of the last 50 years would continue. The characteristic of western second-growth wood is different from that of old-growth timber. The new plantation woods in the southern United States, Chile, or New Zealand do not have the same strength characteristics as loblolly or ponderosa pine of 10 years ago. The wood has more knots, which affects its value in plywood and in appearance applications.

Just the changing characteristics of the wood would keep the trend toward engineering the fiber in place. Composite products (particleboard, OSB, or MDF) will displace lumber or plywood. This will be particularly true in developing countries that, because of income, will need to use lower cost materials. Engineered lumber will displace wide-dimension lumber for strength, but also for other performance reasons. One problem these products may face is fire code issues.

On the paper side, the trend to use recycled fiber will continue. With plantations in Indonesia, Chile, and Brazil, much of the next decade's pulping needs will be met from these sources. The current program in Indonesia to get 600,000 acres of plantation in the *Acacia* species will significantly affect the supply situation for pulp and paper over the next 20 years. If they achieve their target, a back-of-the-envelope calculation suggests that these 600,000 acres could sustain 15 to 17 scale pulpmills. Acacia grows rapidly (30 m^3/acre per year) and can be managed on a 7-year rotation.

The demand for products will be affected by other technological developments. Super-absorbents will reduce the demand for fluff pulp. The electronic revolution and the wireless age will affect paper demand — if not the growth, then certainly the composition. The growth in laser printers and fax machines will be offset by declines in directory and printer demands.

On the plus side, paper demand is likely to grow because of the explosion in information. Technology is making it cheaper for anyone to design and print an advertising flyer. I was struck recently by a new real estate gimmick: little boxes with paper flyers under the "For Sale" sign of a house. Population growth and rising incomes will still tend to increase paper demand.

But one other factor could so fundamentally alter the supply of near-term fiber that, instead of wood-for-wood substitution through technology, we get nonwood-for-wood substitution that is not socially desirable. If policies limit the supply of wood fiber or effectively destroy the incentive to grow new forests, then both short-term and long-term supplies of raw material will change, driving the price of wood fibers up.

There seems to be a strong bias developed by the more radical environmental groups that it will be a good thing if we stop harvesting trees entirely. There is a large social cost if that extreme is selected. Wood is a wonderful carbon sink and young vibrant forests absorb the sun's energy. There are many benefits derived from maintaining a vibrant, growing forest system around the globe.

If you restrict the use of wood, there are other costs. Since wood has been the cheapest way to meet market needs, direct economic costs must be absorbed in the near term. The best example of this is what is happening with lumber prices in the United States. As Table 1 illustrates, there is presently a hypothetical "shortage" of lumber in North America. The shortage has been caused by effectively removing millions of acres of the world's most productive softwood land from the production stream. This has driven prices up 50 to 70 percent. This in turn is leading builders to look at plastics and steel substitutes.

The work of Peter Koch[1] has made it clear this is not desirable. He has pointed out how much more energy is required to produce the alternative materials. In the process, the shift would increase the level of CO_2 emissions dramatically.

So the question remains: how do we achieve a globally optimal solution? — A solution where we set aside tracts of land to preserve habitat and native forest characteristics, but also encourage the management and growth of wood fiber to meet the lifestyle needs of a growing world population. One solution is to intensively manage the most productive lands in temperate zones. Evidence shows that private land management has the ability to do a better job than public ownership — if there is an incentive to do so. This has been demonstrated in the United States. New Zealand also believes in this solution and is encouraging private investment in trees and is exiting public ownership.

But to encourage private ownership to intensively manage these forestlands, there needs to be some stability to the legal structure and parameters for an eventual return on the investment. The future is risky enough for a 35- to 70-year investment without the problems of an unstable political environment. The last 10 years have done just the opposite. This isn't to say that some of the changes were unwarranted. As we learn more about our environment, the private landowner must adapt better practices as we learn what the trade-offs are. But when there are dramatic and sometimes capricious changes, the private owner needs to be compensated for the change. If not, intensive management will be abandoned. The industry will meet state minimums, which again, would have a serious long-term consequence.

An average acre in the Northwest could produce 40 to 60 percent more sawtimber per acre through intensive management. This includes a herbicide treatment, two commercial thinnings, and a treatment of fertilizer. To illustrate the size of the difference, consider 10 million acres of Northwest forestlands. If managed intensively, the acreage could produce roughly an additional 3 BBF of lumber when compared to the capacity of the same unmanaged acreage. That is nearly 10 percent of current U.S. lumber production. Besides, it would be in

Table 1. — The North American lumber shortage.

	1988	1992	1994[a]
		(BBF)	
Lumber demand	64	58	64
Lumber production	**64**	**58**	**58**
U.S. total	38	35	32
West	24	19	16
Canada	26	24	26
B.C. Coast	5	4	4
Eastern Canada	10	9	11

[a] Demand forecast assumes 1992 usage rates.

[1] Koch, P. 1992. Wood versus nonwood materials in U.S. residential construction: some energy-related global implications. Forest Prod. J. 42(5):31-42.

one of the most desirable construction grade materials — Douglas-fir.

If one were to then conclude that we should just mandate tougher management standards, the industrial forest manager would try to sell the land to alternative uses. Clearly, this would reduce habitat and dramatically reduce wood fiber supply even further. Once again, a potential CO_2 sink would be lost.

What we are seeing is the major growth in plantations in New Zealand, Chile, Indonesia, Brazil, and Venezuela. At a recent conference on forest products, I suggested, half in jest, that if you wanted to be competitive in the North American forest products industry you would build a plant in Venezuela. There are very good reasons why Louisiana Pacific has chosen this course.

World raw material needs are growing along with population and real income. Encouraging plantations in other parts of world, while restricting some of the best and most productive lands in the United States, just doesn't make good sense. This course is forcing people to use less desirable materials or to harvest trees from more sensitive parts of the world. The choices are being made based on emotional and erroneous perceptions about the forest and its capacity to meet multiple goals.

We now know, for instance, that forests can be managed for spotted owls and for timber. I hope that the professionals in this meeting will help to communicate the role of forests in meeting the ultimate customer needs and our ability to do it a way that is better for the global environment. Failure to meet this growing social challenge to use one of our most natural and sustainable competitive advantages will negatively affect our standard of living and will potentially make the world a less desirable place to live.

Adapting to global supply constraints

or: Timber famine and six reasons why it won't occur

Clark S. Binkley

Abstract

As global population and incomes increases, so too will the demand for wood increase. At the same time, the supply of timber from many traditional sources is declining, and the increased supply from such "emerging" regions as New Zealand and Chile is probably not adequate to offset the reductions elsewhere in the world. Increased demand will confront constrained supply, setting the stage for "timber famine" or, in economic terms, increases in timber prices. These prospective price increases will induce changes in both timber supply and timber demand that will tend to offset the effects of increased scarcity. This paper focuses on changes in timber demand.

For an economist, price increases produce substitution and income effects. In this case, substitution effects include 1) substitution of domestically grown wood by imports; 2) substitution of capital and labor for timber in wood processing; 3) substitution of technology for wood in wood products; 4) substitution of nonwood products for wood products; and 5) substitution of high wood-consumption life-styles for ones that require less wood. Even after all of this substitution occurs, higher timber and product prices will reduce consumer welfare. Economists call these losses in consumer welfare "income effects."

Each of these adjustments carries economic and environmental costs. The net effect, however, will be a significant abatement of the latent demand for wood inferred from population and income data alone. Despite these adjustments, timber demand is apt to increase faster than supply, and some real price increases can still be anticipated. These increases will reduce consumer welfare but will also create opportunities for profitable investment in timber production, in process improvements, in new product development, and in the R&D that supports these new businesses.

The author is Dean, Faculty of Forestry, Univ. of British Columbia, Vancouver, B.C., Canada.

Introduction: the ingredients for timber famine?

> "If we accept the larger estimate of annual growth and apply it to the present rate of consumption, the result shows a probable duration of our supplies of lumber of not more than 33 years...[It] is certain that the United States has already crossed the verge of a timber famine."
>
> *Gifford Pinchot, 1907*

Despite a doubling of population and a seven-fold increase in economic activity, the United States has somehow avoided Pinchot's dire prediction. But now humans have pushed into nearly every corner of the world, and our high income levels have fueled massive increases in the consumption of wood products. The productivity of our forests seems to be threatened by regional air pollution and global climatic change. Demands for the important environmental services that forests provide increasingly constrain their capacity to produce industrial products. Are we now on the verge of a real timber famine?

As global populations increase, so too will the latent demand for wood. By applying estimates of the current per-capita consumption of industrial roundwood to widely accepted forecasts of population growth, Sutton (20) concluded that the world will need about 75 million m^3 of additional industrial timber supply each year for the next two decades. As a point of reference, British Columbia — one of the world's largest suppliers of softwood timber — has an annual allowable cut of about this same amount.

Sutton's forecasts ignore the increases in consumption associated with increases in income. For many forest products, the income elasticity of demand is greater than one, so an increase in income has more than a proportionate increase in consumption (21). For example, when restated using purchasing power parity exchange rates, the average income level in the People's Republic of China is now about $2,500, with the high population coastal areas having a level of income about twice that. If the rapid economic development there continues, within two decades a large fraction of that country will possess a standard of living equivalent to that in Japan, Germany, or the United States today — and conceivably the same high levels of wood-products consumption.

Because of its importance to long-term industrial planning, many researchers have studied the prospective demand for wood products. Table 1 shows the estimates of industrial timber consumption in the year 2000 from seven such studies. Not surprisingly, there is considerable range in the estimates. The median growth rate reported by these studies (1.7% per year) suggests that timber consumption will rise by about 30 million m^3 per year. Sutton's forecasts lie on the high end of the range, but even the median estimate implies that substantially more wood will be required every year. Will timber supply expand accordingly?

Perhaps not. At the same time that the underlying factors driving demand — population, income, and economic activity — head upward, economic and environmental factors constrain timber supply in many parts of the world. Table 2 shows one estimate of future timber supply in the major regions of the Pacific Basin.

Table 1. — Forecasts of industrial timber consumption, 2000.

Study	Forecast consumption	Growth rate from 1985
	(billion m^3)	(% per year)
Food & Agriculture Organization (1982)		
High demand scenario	2.6	4.0
Low demand scenario	2.3	3.2
Food & Agriculture Organization (1979)	1.8	1.6
SRI (1979)	1.9	1.9
World Bank (1978)	2.8	4.5
Kallio, Dykstra, and Binkley (1987)	1.8	1.6
Sedjo and Lyons (1990)		
Base case scenario	1.7	1.2
High demand scenario	1.8	1.6
Cardellichio et al. (1988)	1.8	1.7
Median projection for 2000	**1.8**	**1.7**

Both British Columbia and the U.S. Pacific Northwest have increasingly restrictive environmental regulations that are likely to force significant reductions in harvest levels. In British Columbia, the current round of timber supply reviews is likely to conclude that current harvest levels are 20 to 25 percent above the levels that can be sustained in perpetuity. However, these timber supply reviews reflect neither the provincial government's stated objective of doubling the areas of parks and preserves nor some of the more restrictive environmental constraints contained in the provincial Forest Practice Code that has recently been released for public comment. Completing the Protected Areas Strategy could conceivably reduce harvest levels by an additional 6 percent, although the actual impact is apt to be less. Recent analyses of the constraints contained in the proposed Forest Practices Code (e.g. reduction in size of clearcuts, mandatory green-up periods before adjacent cutblocks can be harvested, networks of old-growth reserves within the working forest, cover constraints to retain at all times a significant area of old-growth forests within the working forest) found significant reductions in all areas studied, ranging from 29 percent in the coastal Tsitika and interior Tangiers drainages to 65 percent on coastal Hardwick Island (11,12).

Some Indonesian analysts forecast that, as a consequence of forest depletion and expanded domestic consumption, Indonesia could become a net importer of timber by 2000 (18). Poor infrastructure, institutional uncertainty, and depleted forests limit supply increases from the Russian Far East (3). Timber supply will increase in New Zealand and Chile, but these increases are not sufficient to fully offset reductions elsewhere.

In short, we seem to have the basic ingredients in the recipe for timber famine of the kind Gifford Pinchot heralded 85 years ago. But instead of timber famine, we are apt to see a host of offsetting adaptive responses.

Table 2. — Timber harvests from Pacific Rim supply regions (1).

Supply regions	1990	1995	Change
	- - - - - - -	(million m^3)	- - - - - - -
U.S. Pacific Northwest	81.0	65.0	(16.0)
British Columbia	78.3	65.0	(13.3)
Russia	100.0	97.0	(3.0)
Malaysia and Indonesia	66.0	61.3	(4.7)
New Zealand	13.0	14.0	1.0
Chile	16.5	18.2	1.7
Total	354.8	320.5	(34.3)

Adaptive responses to timber scarcity

For an economist, timber scarcity means rising timber prices. Rising timber prices induce responses in both supply and demand. Considerable attention has been paid to supply-side responses (16). These include shifts in the extensive margin as places previously too remote or rough to log become economically attractive, increases in management intensity to grow more wood per unit area, reduction in timber rotations, and increases in plantation area. These effects can be powerful. For example, plantations established to absorb carbon are one possibility for incremental supply that many analysts commonly ignore. Such plantations are a cost-effective method of CO_2 abatement (17). Haynes et al. (6) has estimated that a 6-million-acre afforestation program would provide enough timber supply to drive prices down to "stabilize at a low level" by 2040.

This paper focuses on the demand-side responses that are not well understood. Economists decompose price-induced reductions in demand into substitution effects and income effects.

Five kinds of substitution effects will tend to limit the impact of timber scarcity:

1. Changes in trade flows to substitute timber from places where it is relatively abundant for timber from places where it is comparatively scarce.
2. Greater use of capital and labor in processing facilities to substitute these inputs for timber and to economize on wood consumption.
3. Incorporation of higher levels of knowledge per unit of product to substitute technology for wood while meeting customer needs with less fiber.
4. Increased use of aluminum, plastics, and steel to substitute nonwood products for wood products.
5. Cultural changes to substitute new ways of living for old ones requiring more wood.

Even once the full latitude for substitution has occurred, consumers are likely to still be worse off than they were when timber was more abundant. This residual loss of consumer welfare is called an income effect: Higher prices reduce the income consumers have to spend on other goods and services. Because the demand for some wood products is fairly inelastic, the income effects associated with higher timber prices may be significant.

Let us examine each of these substitution and income effects in greater detail.

Changes in trade flows

Every since deforestation in the Mediterranean forced the Greek navy to import cedar for its triremes, timber trade has moved wood from places of relative abundance to places of relative scarcity. Wood products remain one of the world's most heavily traded goods.

By making timber from one region available to consumers in another, trade offsets the effects of local scarcity. Indeed, some environmentalists decry trade precisely because it permits importing countries to "live beyond their means." Conversely, restrictions on trade exacerbate local imbalances between supply and demand. For example, the propensity to restrict log exports when local timber supplies decline is nearly universal. This misguided policy reduces the value of local forests and thereby thwarts some logical responses to increased scarcity such as improved processing efficiency or reduced demand (22). Such restrictions also drive up consumer prices, exacerbating the global welfare losses associated with rising timber prices. Non-tariff barriers to trade such as the current limitations on the import of unpasteurized Canadian lumber into Europe produce similar pernicious effects.

Perez-Garcia (13) analyzed the impacts on forest products trade on the likely reductions in timber supply in the U.S. Pacific Northwest (20 million m^3) and in British Columbia (13.3 million m^3) due to environmental restrictions. Figure 1 shows that 65.8 percent of the reduction in harvests in these northwest coastal regions of North America are sim-

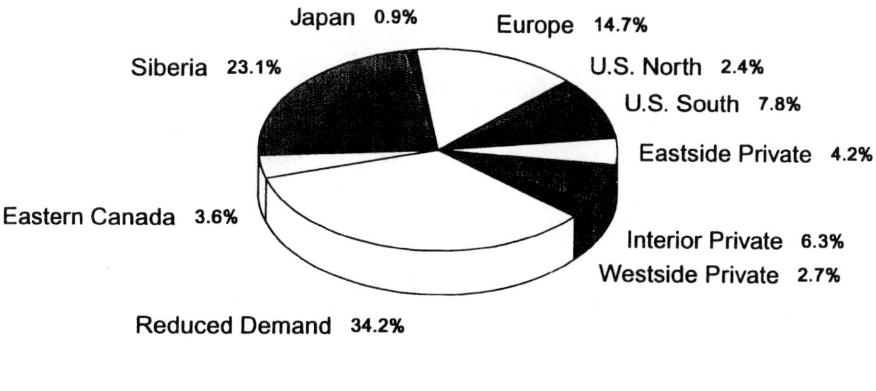

Figure 1. — Changes in regional production due to reduced northwest coastal North American timber harvests (13).

ply redistributed among other regions. Because most of the world's forests carry lower timber volumes than do those of the Northwest, the *area* harvested annually would actually increase by 12 to 61 percent, depending on the precise assumptions used. As a consequence, it is not clear that — when viewed from a global perspective — harvest restrictions in the U.S. Pacific Northwest and in British Columbia produce net positive environmental effects for the world as a whole.

Improved processing efficiency

Gifford Pinchot's predictions were not realized, partly of because steady improvements in the technical efficiency of wood processing. Using standard econometric techniques, the rate of technical improvement is difficult to measure, but the trend probably lies in the range of 1 to 2 percent per year for many wood products manufacturing activities (7).

Two examples drawn from sawmills in the British Columbia interior provide some insight into the nature of technological progress. Figure 2 shows that a typical British Columbia interior sawmill has increased lumber recovery by 12.8 percent as a result of capital investments taken over the last few years. Even at current factor and output prices, these investments all carried high internal rates of return, so the firm readily chose to make them. With increased timber prices, the incentive to invest in such technology would be even greater. Put another way, ordinary market forces make it economical to offset a significant reduction in log supply.

Such anecdotes about individual mills tell little about overall market impacts. To reflect on these market effects, Figure 3 shows the lumber-recovery factor for northern interior sawmills in British Columbia between 1986 and 1992. Note first that the interquartile range varies between 7.6 percent and

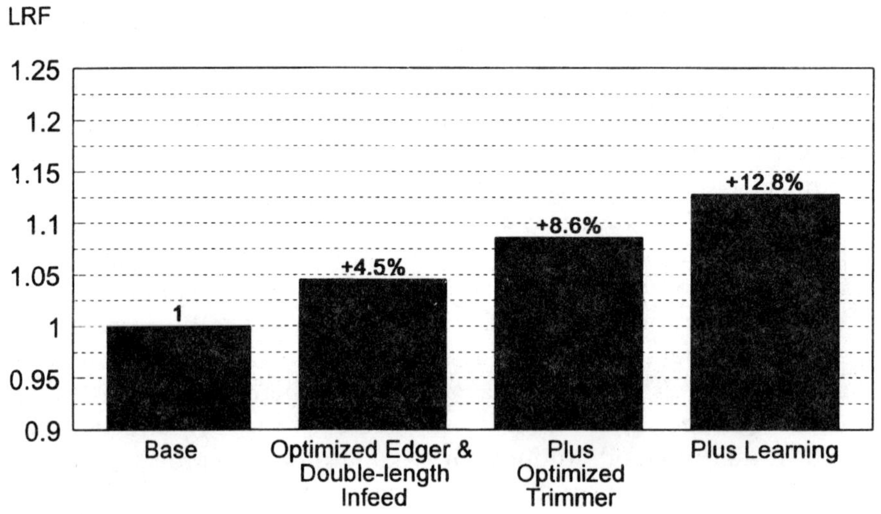

Figure 2. — Capital-time substitution in an interior B.C. sawmill (LRF is scaled so the pre-investment level equals 1.00).

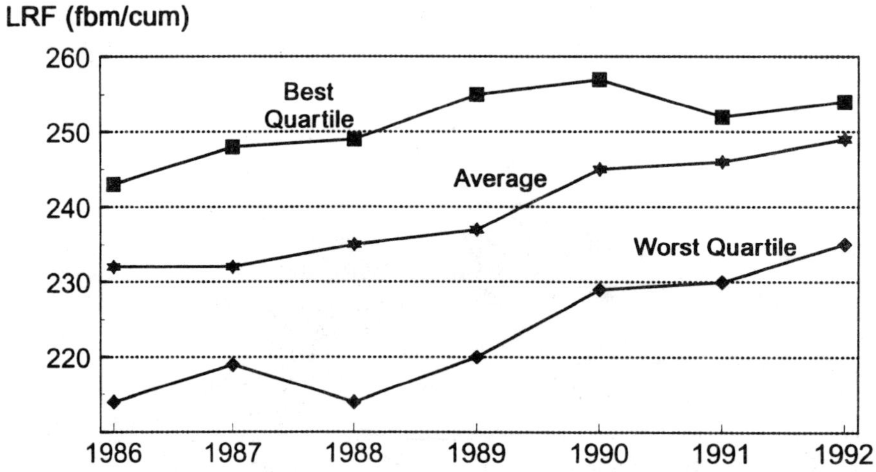

Figure 3. — Trends in lumber recovery for interior B.C. sawmills.

12.5 percent of the average. As a consequence, there is considerable room for industry-wide improvement in technical efficiency if the worst mills came up to the standard set by the best ones. Second, interior sawmills face similar factor prices and produce similar products (mainly SPF dimension lumber for U.S. markets). Lumber recovery is therefore a good proxy for production costs. Since prices are set by the marginal production cost of the least efficient mill, all of the mills that are more efficient generate profits (or what economists call "quasi-rents") equal to the difference between their own production costs and those of the least efficient mill. The data in Figure 3 suggest that adoption of the technology that is most efficient at one point in time continues to produce quasi-rents for at least 7 years. These quasi-rents have persisted despite the fact that the spread in lumber recovery between the least and most efficient mills has narrowed over time. These persistent excess returns to technology provide a powerful economic incentive to invest in more efficient processing technology, and thereby to reduce timber demand.

Technologically advanced wood products

Increases in timber prices provide an incentive to develop new products that provide similar performance characteristics to the old ones but use less wood. Such developments substitute technology for wood. Examples of such engineered wood products include laminated veneer lumber, Trus-Joist/MacMillan's Parallam, and older products such as oriented strandboard (OSB).

The case of OSB is instructive. In many applications OSB can be substituted for plywood, and consumers choose between the two products largely on the basis of comparative costs. The relative cost of the two products depends heavily on the price of peeler logs, the most important single cost of producing plywood. Figure 4 shows the growth of OSB's share of the U.S. panel market as the price for peeler logs rose. Even though the technology to produce OSB became commercial in the 1960s, only the timber shock of the 1970s created an adequate incentive for consumers to adopt OSB. As scarcity drives timber prices up, there is probably considerable capacity for new, higher technology wood products to substitute for old ones.

Wood-nonwood substitution

Increases in timber prices will tend to drive up the prices of products manufactured from timber. These increases in product prices provide an incentive for consumers to use formerly more costly nonwood products with similar performance characteristics in place of their wooden counterparts.

Table 3 shows that there is considerable room for price-induced substitution between wood and nonwood products. For example, a 1 percent increase in the price of softwood lumber will increase the consumption of structural steel by 0.32 percent. Indeed, the recent run-up in SPF lumber prices caused some homebuilders to look for alternative materials. *The Economist* (19 June 1993, p. 68) noted this effect, and reported that "a typical home built [as suggested by steel producers] would contain between four and five tons of hot-dipped galvanized steel doing the work of about 20 tons of timber." As scarcity

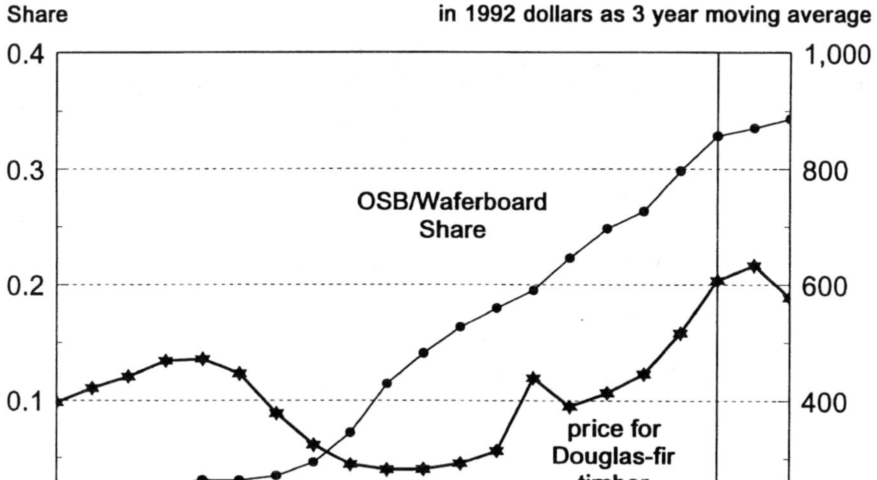

Figure 4. — Increase in OSB/waferboard share of U.S. panel markets (RISI, personal communication).

pushes up the prices of wood products, consumers will buy products manufactured from other materials to satisfy the demands formerly met by wood. This process of substitution reduces the impact of timber scarcity on consumers.

Of course, the nonwood substitutes for wood products may produce deleterious side effects that warrant policy attention. For example, cement production is second only to fossil fuel consumption as a source of CO_2 emissions in Canada. Using the elasticities of substitution reported in Table 3, Prins (15) estimated the amount of other materials that would be substituted for SPF lumber produced in Canada if the price of SPF doubled. He then computed the amount of additional energy that would be required to manufacture these other materials, and estimated that energy consumption would increase by the equivalent of 3.4 million barrels of oil/year to offset the reduction in Canadian lumber production alone. Similarly, Perez-Garcia (13) estimated that wood/nonwood substitution associated with the 33.3 million m^3 timber harvest reduction anticipated in the northwest coastal region of North America would release 18 million tons of additional CO_2 into the atmosphere each year.

One policy to control pollution calls for internalizing the cost of environmental damage into the price of the products that cause it. This could be accomplished, for example, through a carbon tax, through mechanisms such as tradable effluent permits, or through regulations. Timber-supply restrictions occasioned by tightened environmental constraints internalize the costs of these environmental concerns into the price of wood products. If the environmental costs of the substitute nonwood products were incorporated into the price of those products, then wood products would compete more effectively against them. Wood-products manufacturers might therefore be well advised to demand — in the interest of both equity and environmental objectives — that the prices of *all* products should carry the burden of the environmental damage caused by the product.

Cultural changes

The Knight-Ridder newspaper company is experimenting with electronic tablets to replace newspapers (*The Economist,* 10 July 1993, p. 77). Adoption of this technology will require a change in the culture of newspaper reading. But even without high technology, consumption of wood products varies dramatically among countries and among individual households within a country. It is apparently possible to maintain a high standard of living with widely differing levels of wood-products consumption.

Figure 5 shows the per capita apparent consumption (i.e., production plus imports minus exports) of newsprint and printing and writing papers in six industrialized countries. Each of these countries enjoys a very high material standard of living. Yet people in such places as Canada apparently consume over three times as much newsprint as do others

Table 3. — Elasticities of substitution (9,10,15).

Wood product	Non-wood product	Elasticity
Wood windows	Aluminum windows	1.7
Softwood lumber	Bricks	0.51 to 0.79
Softwood lumber	Cement	0.15
Softwood lumber	Structural steel	0.32

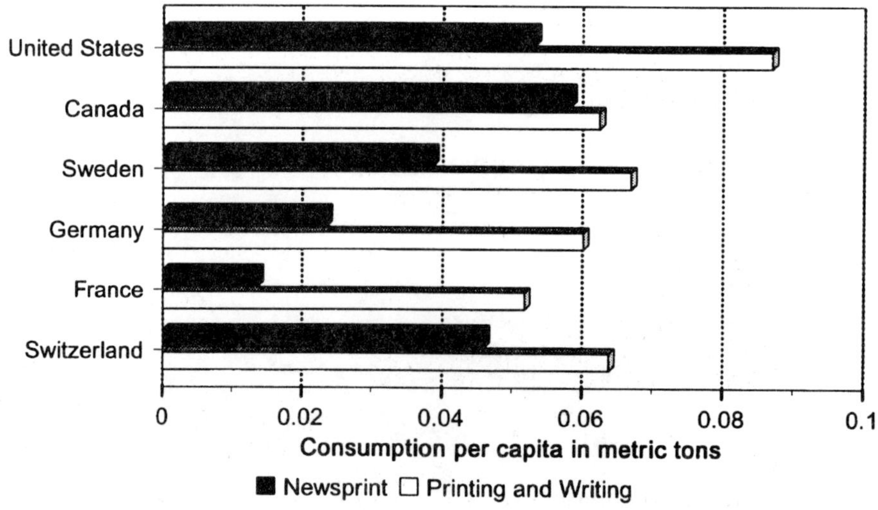

Figure 5. — Annual per capita consumption of paper in selected industrialized countries.

in such places as France. Although the variation in consumption of printing and writing papers is much less, there is obviously considerable room for reducing paper use while maintaining a high living standard. As timber becomes more scarce, consumers of wood products will have an incentive to adopt lifestyles that economize on wood use.

Increased prices and loss of consumer welfare

Increased timber prices will tend to increase the prices of intermediate and final products that contain wood. Increased product prices will provide an incentive for consumers to reduce their consumption of wood products by finding suitable substitutes. To the extent that consumer needs can be well met by these substitutes products, there is no loss of consumer welfare. However, the substitute products generally carry higher prices or poorer performance characteristics. As a consequence, price-induced substitution generally hurts consumers by taking away some of the income that they would otherwise have had for purchasing other products or for saving. Assuming that all of the environmental costs of alternative products are fully recognized in their prices, this income effect is a good measure of the social cost to consumers of increased timber scarcity.

In his study of harvest reductions in the northwest coastal regions of North America, Perez-Garcia (13) estimated this income effect — the loss of consumer welfare due to the higher product prices associated with a reduced timber harvests. The total loss amounts to nearly \$2.5 billion per year, or about \$75 per m^3. Figure 6 shows how this total will be distributed around the world. About 40 percent of the total impact will fall on the regions where the harvest reductions occurred (the United States and Canada). The remainder will hit consumers throughout the world as product prices rise everywhere.

Conclusions: timber famine or not?

Physical depletion of timber, conversion of forest land to other uses and environmental restrictions on timber harvesting in the remaining forested areas will limit future timber supply from many regions that have traditionally produced much of the world's timber. Fast-growing plantations established in the low and middle latitudes will not fully offset the reductions in timber harvest in the traditional supply regions. At the same time, rising population and per capita income insure that the demand for wood products will grow. Estimates of the additional demand range from about 20 to 125 million m^3 per year, with the median lying nearer the low end of the range at about 30 million m^3 per year.

The confluence of rising demand and limited supply will induce a variety of adaptive responses. Timber and wood products prices will rise, especially in the short run. These price increases will induce some short-term increases in timber supply, particularly as areas that were previously too expensive to log become economically attractive, and as rotation ages and growing-stock inventories are reduced. In the longer run, higher timber prices will provoke new plantations and more intensive timber management.

Society can use wood more efficiently in many ways. The forest sector has successfully developed many new products and processes that reduce wood consumption while meeting consumer needs. Accelerating the pace that we create these new knowledge-intensive wood products and processes remains

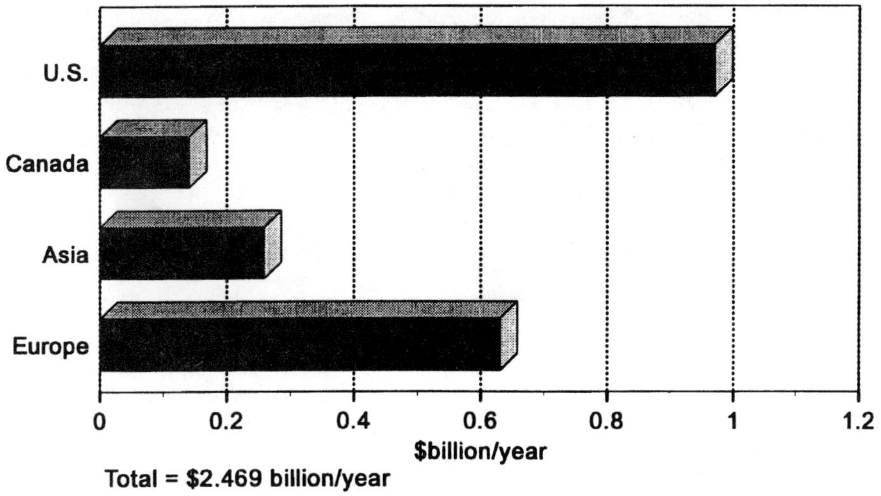

Figure 6. — Loss of consumer welfare due to reduced northwest coastal North American timber harvests (13).

a major challenge for governments, industry, and research institutions.

Other kinds of adjustments carry economic, environmental, and social costs. Substituting such energy- and pollution-intensive materials as concrete, steel, and aluminum for wood will exacerbate local, regional, and global environmental problems.

Despite the wide latitude for adapting to increased timber scarcity, timber demand will still grow more rapidly than supply. Even the most conservative estimates suggest that timber demand will grow by 20 million m^3 per year (16). Such increases will strain the capacity of the world's forests. As a consequence, timber prices are apt to continue their long-term upward real trend, albeit at a rate less than modern-day Gifford Pinchots would have us think.

Manufacturers of wood products will be squeezed between a rising floor of increased raw-material costs and a fixed price ceiling for traditional products set by the cost of available substitute wood and nonwood products. Improved technology in wood products and production processes is the only way to sustain profitability in the face of the these twin pressures. The critical need for improved technology raises important questions concerning appropriate strategies and expenditure levels for R&D, but those are stories for another day.

Literature cited

1. Boyd, C. 1992. Pacific Rim timber supply. Presentation in Santiago, Chile.
2. Cardellichio, P.A., Y.C. Youn, D.M Adams, R.W. Joo, and J.T. Chemelik. A preliminary analysis of timber and timber products production, consumption, trade and prices in the Pacific Rim until 2000. CINTRAFOR Working Paper 22. Univ. of Washington, Seattle, Wash.
3. ________, C.S. Binkley, and V.K. Zausaev. 1990. Sawlog exports from the Soviet Far East. J. Forestry 88:12-36.
4. FAO. 1979. FAO world outlook for timber supply, phase V. FAO, Rome, Italy.
5. ________. 1982. World forest products: demand and supply, 1990 and 2000. Forestry Paper 29. FAO, Rome, Italy.
6. Haynes, R.W., R.J. Alig, and E. Moore. 1992. Alternative simulations of forestry scenarios involving carbon sequestration options. Draft rept. USEPA, Washington, D.C.
7. Hyde, W.F., D.H. Newman, and B.J. Seldon. 1992. The economic benefits of forestry research. Iowa State Univ. Press, Ames, Iowa.
8. Kallio, M., D.P. Dykstra, and C.S. Binkley. 1987. The global forest sector: an analytical perspective. John Wiley, London, England.
9. McKillop, W.L., T.W. Stuart, and P.J. Geissler. 1980. Competition between wood products and substitute structural products: an econometric analysis. Forest Sci. 29:813-822.
10. Moffett, J. 1993. A comparison of product diffusion and distributed-lag models for estimating wood/nonwood substitution in the U.S. window market. CINTRAFOR Working Paper 42. Univ. of Washington, Seattle, Wash.
11. Nelson, J.D. and J.S. Hackett. 1993. An economic analysis of timber harvesting regulations for coastal British Columbia. Univ. of British Columbia Forest Economics and Policy Analysis Unit Working Paper, Vancouver, B.C., Canada.
12. ________ and C. Winsor. 1992. The effects of harvesting guidelines on timber supply and delivered wood costs in the Tangiers drainage. Contract report for the Revelstoke (B.C.) forest District, 24 December.
13. Perez-Garcia, J.M. 1993. Global forestry impacts of reducing softwood supplies from North America. CINTRAFOR Working Paper 43. Univ. of Washington, Seattle, Wash.
14. Pinchot, G. 1907. Conservation of natural resources. The Outlook. 12 October 1907.
15. Prins, R.G. 1993. The economic and environmental impacts of reduced timber harvest and increased softwood lumber prices. Working Paper 29. Forestry Canada, Economic Studies Div., Ottawa, Ont., Canada.
16. Sedjo, R.A. and K.S. Lyons. 1990. The long-term adequacy of world timber supply. Resources for the Future, Washington, D.C.
17. ________ and J. Wisniewski. 1993. Managing carbon via forestry: an economic synthesis. Draft MS.
18. Sidabutar, H.P. and C. Chandrasekharan. 1992. Outlook for the Indonesian forestry sector. *In:* Forest sector analysis for developing countries. R. Haynes, P. Harou, and J. Mikowski, eds. CINTRAFOR Special Paper SP 10. Univ. of Washington, Seattle, Wash. pp. 180-195.
19. SRI International. 1979. The outlook for the world's forest products industries. Palo Alto, Calif.
20. Sutton, W.R.J. 1992. The world's need for wood. Paper presented at Maruia Society meeting "New Zealand and the World's Forests," Wellington, N.Z.
21. Uutela, E. 1984. Income and price elasticities of demand for paper and board. Inter. Working Paper WP-84-102. Inst. for Applied Systems Analysis, Laxenburg, Austria.
22. Vincent, J.R. and C.S. Binkley. 1992. Forest-based industrialization: a dynamic perspective. *In:* Managing the World's Forests, N.P. Sharma, ed. Kendall-Hunt, Dubuque, Iowa.
23. World Bank. 1979. EIS Paper 98. IBRD, London, England.

SESSION II:

Material Substitution — Response to Changing Fiber Supply

The ongoing process of material substitution: some global observations

G.K. Marshall

This paper will present "some global observations" regarding the ongoing process of material substitution or, in other words, "what is happening in the trenches, with a look down the road to what we should expect next" overseas and with a market perspective.

To do this I am going to take you on a visit to both Europe and the Pacific Rim. In both of these areas, we are going to look at a cross section of countries and products.

We will also briefly tour a few buyer's yards in each of these areas to see what sort of material substitution has already occurred in the last 12 months.

Before we start our journey, there is one thing that I must stress. The process of material substitution in the wood products industry has been a constant factor, however, very recently we entered into a different era. In late 1992 and early 1993, the world wood products industry experienced a series of events which transformed us from a regional to a global business. This also shifted the process of substitution to a global basis, making it more complex and sophisticated. Furthermore, just like everything else these days, the process is now occurring much more rapidly.

When we look at Europe, we are clearly confronted by the new market realities. Our old friend Europe, long a traditional customer of North American wood products, has suddenly and dramatically altered its purchasing patterns.

The author is President, Balfour Guthrie Forest Products Inc., Vancouver, B.C., Canada.

We can find many reasons for this — recession, foreign exchange movements, stronger dollar and devalued Scandinavian currencies, plant health problems and restrictions, higher North American prices — but the real, main underlying reason is there is too much European home-grown timber now reaching maturity. Substitution by European wood for North American construction products is not a short-term event, but a long-term trend that will only increase over time.

Our first visit to Europe will take a closer look at this process in the panel products business in the Netherlands. For many years the Dutch market has been used by many industries as a testing ground for new products. The relative size and location of the country, plus the openness of the Dutch people, provide the ideal setting to launch new concepts in automobiles, VCRs, TVs, etc. So this is where

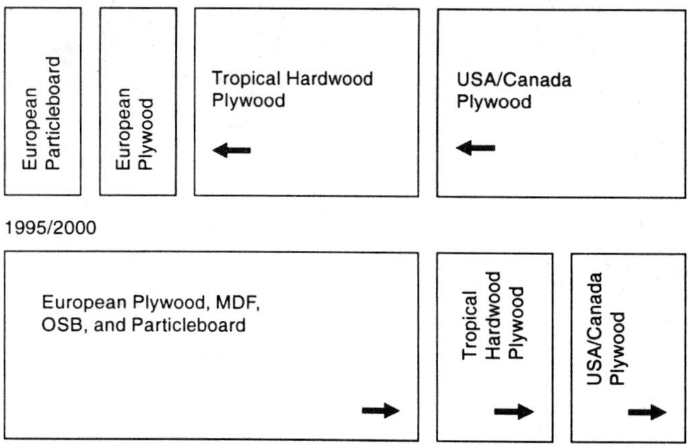

Figure 1. — Netherlands panels and imports.

you can catch trends that will be later seen elsewhere in Europe. This is very true when it comes to panel products, which are virtually all imported.

In the 1970s and 1980s, the gradual movement from Douglas-fir plywood to Canadian softwood plywood and, more recently, southern yellow pine plywood, was the dominant force for a construction panel. Tropical hardwood plywood mainly from Asia (slowly shifting away from Japan, Korea, and the Philippines, to Indonesia and Malaysia) controlled most of the balance of the demand. Small volumes of European production survived in niche corners.

With the 1990s problems of exchange, European recession, environmental pressures, and shortages of supply and price, we are seeing a market quickly shifting its dependence to something few ever considered 5 years ago: a European supply dominance in medium-density fiberboard (MDF), oriented strandboard (OSB), particleboard, and plywood.

North American and tropical hardwood plywood is losing its market share and will soon be in rapid retreat as European mills working with a growing fiber base can more effectively meet the buyer's needs. The niches will, in time, be the only foothold left to the non-European producer.

Figure 2. — U.K. timber frame components.

If we cross the English Channel to the United Kingdom and visit one of the leading European timber frame builders, the cold light of today's market is once again upon us. In 1992, a walk through this manufacturer's facility would have displayed a showcase of North American wood products. High quality Canadian kiln-dried dimension lumber and North American plywood were the two major building products.

A scant 12 months later, "our customer" has prospered through a long recession by improving his product, timber frame structures. He can now build cheaper and better, and deliver on stricter schedules. His factory is busier, bigger, and still growing. Indeed, without ever visiting Japan, they have produced a system very similar to what you see in many places in Tokyo Bay.

Figure 3. — Scandinavian dimension lumber.

Figure 4. — U.K. MDF stairs.

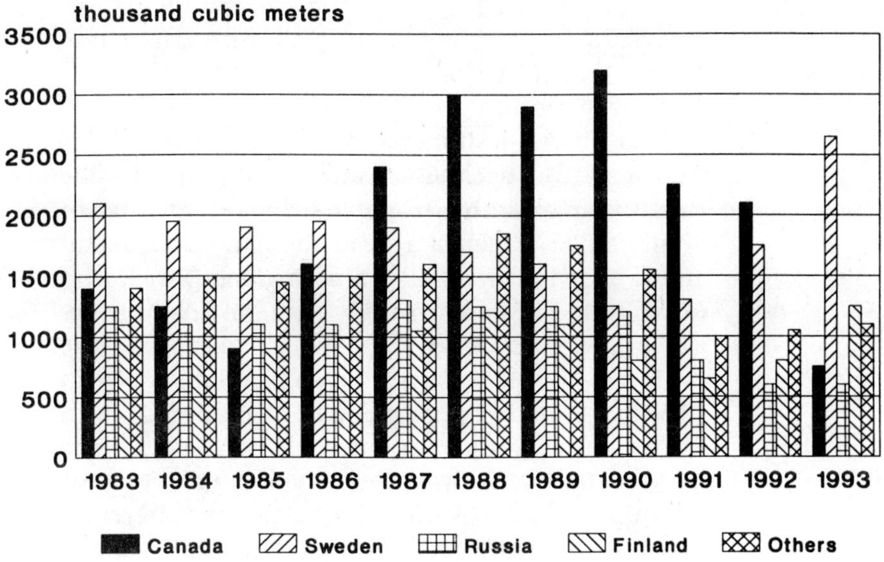

Figure 5. — U.K. imports of softwood lumber. "Others" includes Portugal, Czechoslovakia, Poland, Republic of Ireland, Chile, Norway, United States, Brazil, Austria, Federal Republic of Germany, Benelux, South Africa, and former East Germany.

Figure 6. — German windows.

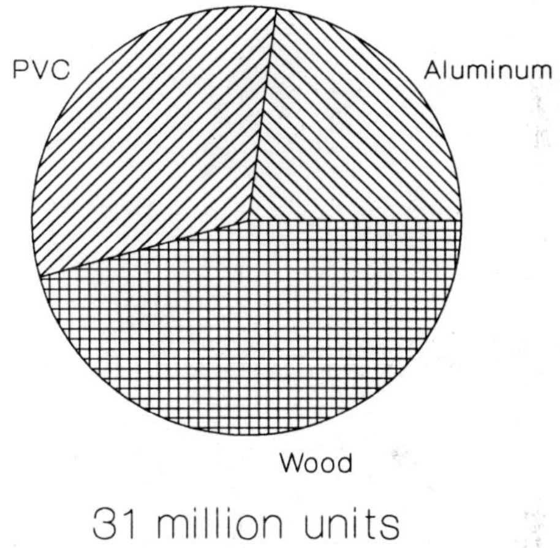

Figure 7. — German window market type and main end uses. Wood: private and public residential housing; aluminum: offices, industrial buildings, public works; PVC: apartment/rental buildings.

The only despairing sight is that North American products have been completely substituted by European products. Swedish and Finnish kiln-dried dimension lumber, plus Scottish OSB and MDF, are now the building products. All are performing well and growing with the needs of "our former customer."

This figure demonstrates how dramatic the shift has been in the United Kingdom. The difference between 1990 and 1993 for Canadian lumber imports will be approximately 1 billion board feet. This production is now either back in the North American market or in the Pacific Rim.

Not wanting to leave Europe on a totally negative note, allow me one more stop: once more cross the English Channel to view the German window market.

Germany manufactures a lot of windows, even in a recession. This year the number will be about 31 million.

Very generally, the market is structured into three product lines: aluminium, PVC, and wood. Aluminium, and a combination of aluminium and wood, which has about 15 percent of the market, is mainly used in offices, industrial buildings, and public works. PVC, which is increasing at the expense of wood, has about a 45 percent share, the main uses being apartment and rental buildings. Wood represents about 34 percent and is principally used in private and public residential housing.

Even though wood is losing its market share,

there are other substitution dynamics occurring that present some interesting opportunities.

Supply problems and rising costs, plus environmental concerns regarding meranti, are shifting demand towards other species. Consumers preferring a lighter colored window favor substitution by spruce, pine, and hemlock.

However, there are two other factors that affect the German wooden window market. The first is the trend to a 2- or 3-piece laminated product. Where before 1988 less than 30 percent of the wooden windows were laminated, in 1993 the market is now close to 50 percent and growing.

The second factor is the increasing production of this laminated product from Eastern Europe. The need for hard currency and the entrepreneur spirit has created a new competitor. Unfortunately, it is a *very* competitive one.

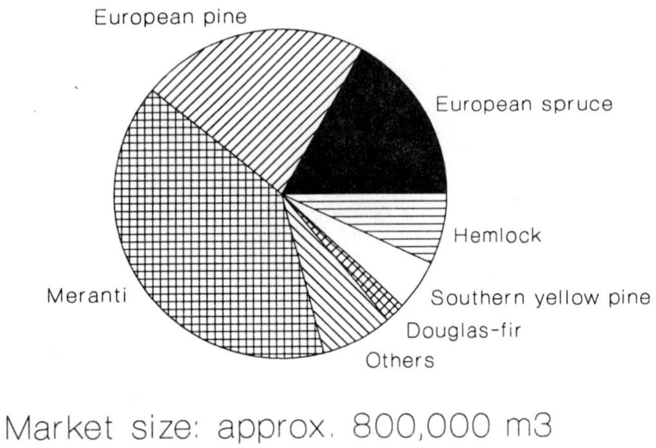

Figure 8. — German window market, percentage of species used. "Others" includes white spruce, radiata pine, and mercussi.

If we move on to the Pacific Rim, we can see quite a different situation developing. The combination of the decline in supply from the Pacific Northwest and the tropic hardwood regions of Southeast Asia against a steady demand from Japan and a growing demand from Taiwan, Korea, and China is creating a new supply/demand relationship.

An interesting country in which to observe the evolution of this new relationship is Taiwan. With a relatively modest population of 21 million, Taiwan has created a strong economic and financial presence in Asia. Taiwan is, however, a very small country, roughly the size of Vancouver Island, and therefore is one of the more densely populated areas of the world.

In the mid-1980s, Taiwan's need for wood was growing. However, domestic logging was causing such serious erosion problems that the government placed restrictions which effectively eliminated this practice.

Since that time, Taiwan has been forced to import all their wood requirements. The regional supply of Southeast Asian tropical hardwoods was the initial main source for Taiwan's needs.

When Taiwan commenced on an ambitious 6-year plan to rebuild the country's transportation infrastructure, there came an immediate need to increase purchasing just as Pacific Rim supply was tightening.

Now, even though the 6-year plan is being extended to 10 years and the expenditures scaled down from US $300 billion to US $200, the requirements from Taiwan are significant and trending upwards.

Figure 9. — Taiwan hardwood log sawmill.

Figure 10. — Taiwan hardwood log sawmill.

The substitution by softwood for hardwood has already started and is expected to continue through the 1990s.

Meanwhile, the sourcing of softwood is also in transition. The North American market share is expected to decline as the plantation forests of South America and New Zealand bring more fiber onto the market.

An interesting side note is the volumes from mainland China. Because of "unfair competition" (a nice phrase for dumping), in May 1993 Taiwan stopped issuing import licenses for China fir, a reminder to all about pricing strategies.

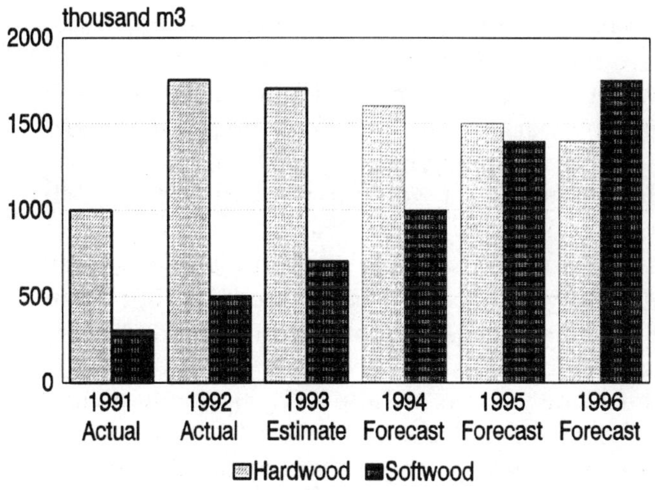

Figure 11. — Taiwan's total wood consumption.

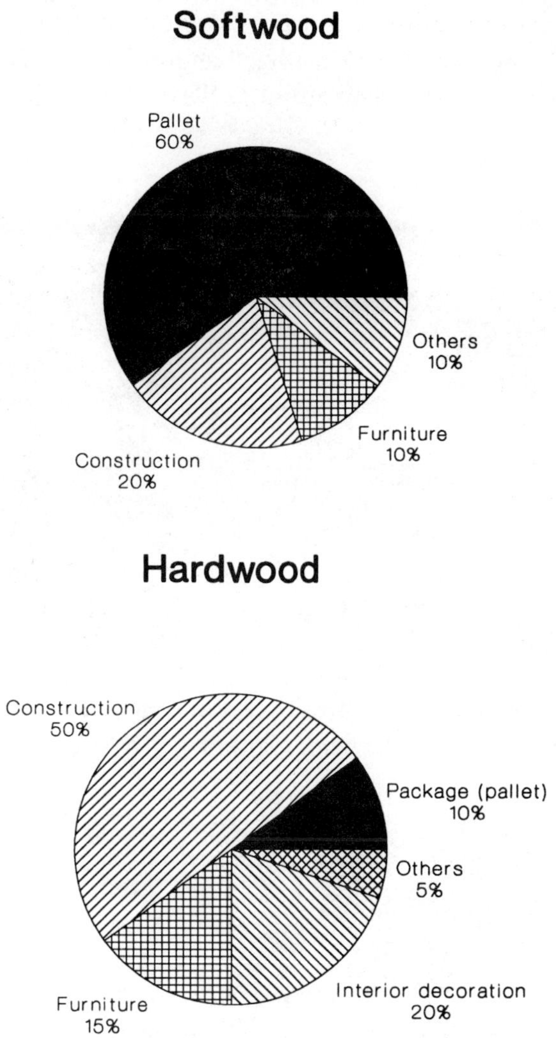

Figure 13. — Taiwan's wood consumption 1993.

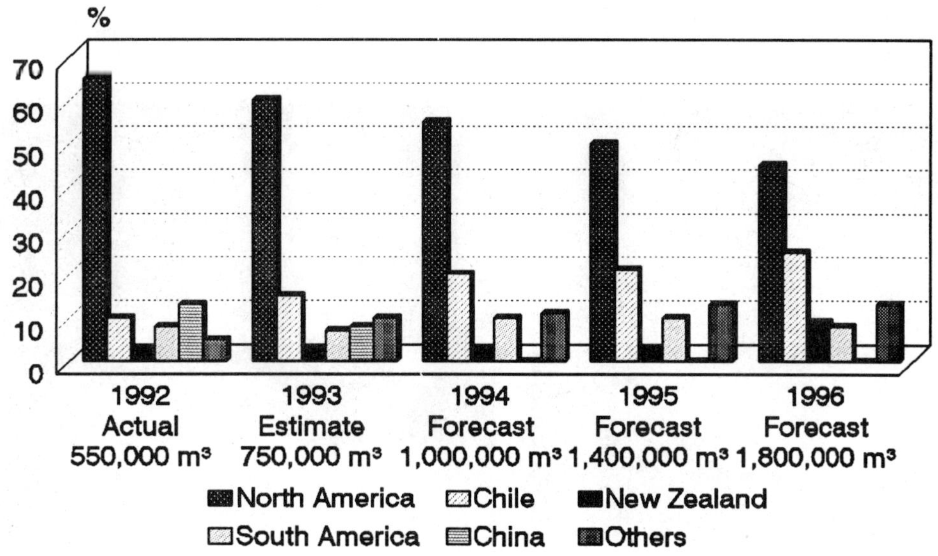

Figure 12. — Softwood sources for Taiwan. "Others" includes Russia.

The end uses in Taiwan for timber fall into four main areas: construction/concrete forms, packaging/pallets, furniture, and interior decorating/taruki.

Figure 14. — Taiwan concrete forms.

Figure 15. — Taiwan pallets.

The hardwood areas that will see the continued substitution by softwood will be in the packaging, interior finishing, and some construction uses.

As you travel through Asia and in particular Japan, you constantly see large usage of moulding. Historically, abundant supply of high grade tropical hardwood and softwood provided for a sizeable natural wood moulding market.

The Japanese, with a strong currency and desire for wood, have been able to maintain this consumption pattern while in most other parts of the world, patterns have changed. The realities of supply of this solid material are now forcing a swifter change to alternative material.

The use of MDF has been growing quite steady in Japan for the last 5 to 7 years. However, when compared to other countries, there is less usage. This is now all changing. Substitution by MDF is now set to grow at an accelerated pace, and the moulding market, plus the thin lauan plywood market, are the primary targets.

To end our trip, I would like to just briefly touch on a few areas in Japan. The amount of product substitution is remarkable. Not only are logs being replaced by lumber and panel products, but there is a high degree of development in laminated products, such as posts and beams.

Softwood veneers are also being used in plywood manufactured in Japan.

Also the trend to more manufactured houses is opening the door to all types of innovation in reconstituted panels, lamination, finger jointing, etc. So what conclusions can be drawn from the observations of the recent events?

The first is that we are in an new era. Many

Figure 16. — Taiwan taruki.

traditional relationships have been permanently altered. However, at the same time, many new windows of opportunity have opened. With all the recent developments, customers are more receptive to new products and concepts.

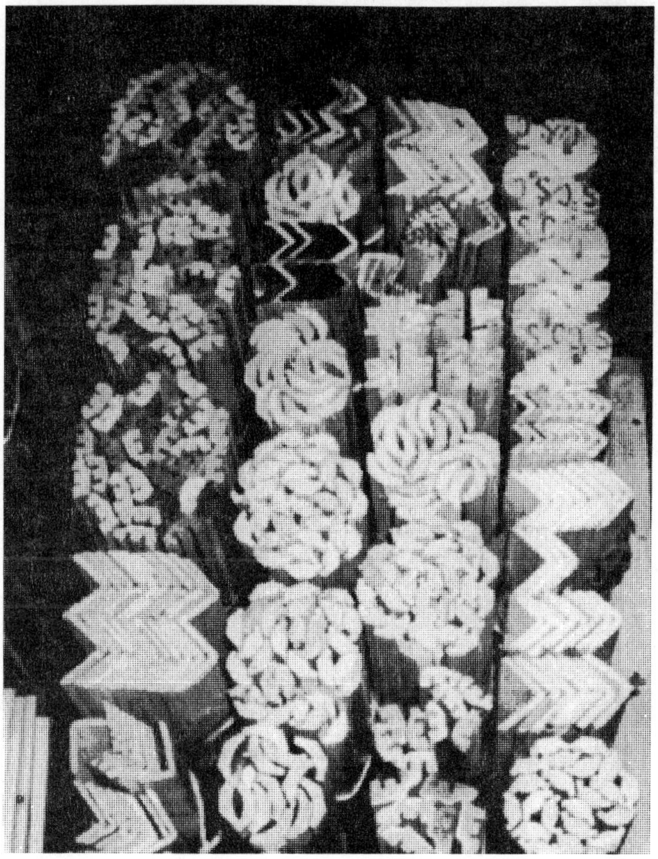

Figure 17. — Japan mouldings.

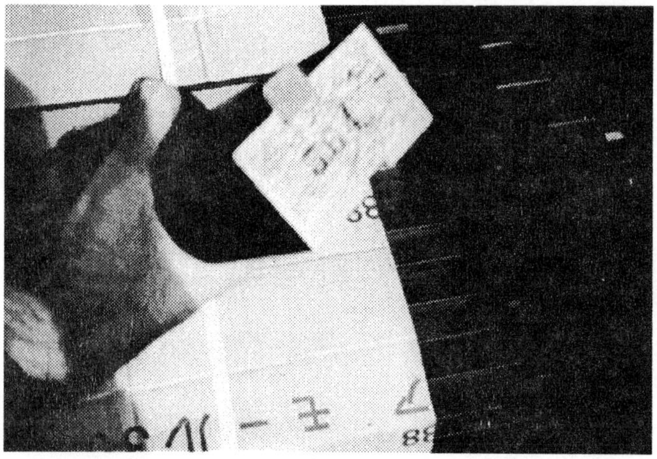

Figure 18. — Japan mouldings.

The next thing is that we now are in a global market. Therefore, our knowledge level has to be dramatically increased. We have to understand more, not only about demand, but also about the global fiber base. Where are significant increases and decreases occurring? What will be the impact? This involves a much more sophisticated understanding of age, quality, strength factors, useage, etc. of all world fiber.

Finally, as the process of substitution continues, there is going to be more nonwood product involvement. However, from my experience and observation, because the culture of our customers is one of *wood*, there is a natural preference to continue with wood-based products.

The task falls on the shoulders of the forest products industry to keep up with the evolving demands

Figure 19. — Japan laminated posts.

Figure 20. — Japan softwood plywood.

of the global market and provide the wood products to fit these needs.

Just before I end my talk, I have one more picture that provides a fun look at global marketing and substitution. Two benches: the same price, £25. One is made from Scandinavian whitewood in England, the other from clear hardwood and metal, made in China. Which one would you buy?

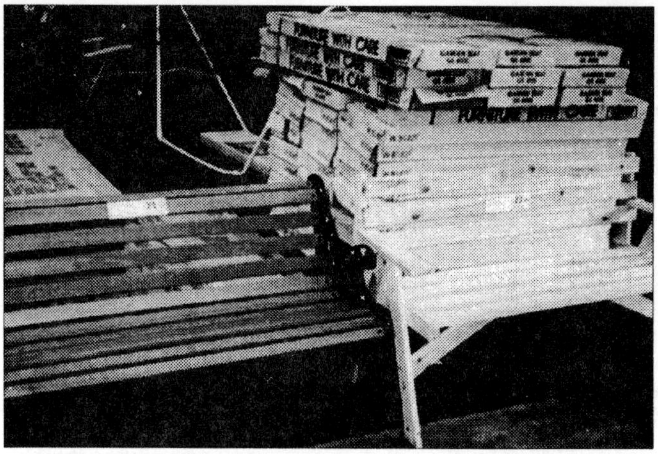

Figure 21. — U.K. DIY garden benches.

Environmental measures as substitution criteria for wood and nonwood building products

J.K. Meil

Abstract

Historically, technical utility and price have been the primary drivers behind substitution of wood and nonwood products. Renewed environmental awareness, pressures from environmental activists, and direct government regulation will increasingly combine to affect building material choices in the future. In the interim, however, the current price system fails to reflect the relative environmental impact associated with alternative product choices and serves to skew these choices. Since all building products offer certain environmental benefits while entailing environmental costs, one result is likely to be the use of a wider mix of materials in a given structure. In other words, building material producers can expect to see more fragmented markets and will have to identify and pursue market niches where they have a competitive edge, with competition defined by environmental as well as traditional price and product utility criteria. Unfortunately, our present environmental knowledge and piecemeal measures used to gauge product impacts don't tell the whole story.

This paper summarizes the work to-date of a Canadian research alliance of public and private sector organizations and individuals who are looking beyond simple environmental measures and impacts to create a better understanding of the underlying environmental trade-offs inherent in decisions to use or not use specific building products. The alliance's goal is to develop a systems model to assess the environmental consequences of using alternative materials for specific building designs in a holistic, cradle-to-grave framework. Their initial focus is on vertical and horizontal structural assemblies using wood, steel, and concrete products in industrial and commercial buildings. The paper presents some preliminary results comparing wood and steel in an exterior infill wall application and then applies environmental externality costs to the resultant absolute environmental measures to provide an approximate measure of the true environmental cost of each assembly. The paper also makes some observations concerning steel recycling.

Introduction

Two years ago, I attended the FPRS "Wood Product Demand and the Environment" conference held in Vancouver. At the completion of that conference it was clear to me, and many others, that we as a global industry were in for some sweeping changes. The drive behind this immense change was and remains the environmental movement. Today's environmental movement is without boundaries as it stretches around the globe and intervenes not only on public lands, but is likely to soon engulf private property as well. What it has changed is the public's perception about wood products and our forestry practices. It has also shifted the wood products industry's preoccupation away from demand-side economics towards supply-side concerns. It has created windfall profits for some and economic hardship for others. Primarily, what the environmental movement has done is create a lot of uncertainty. At present we see this uncertainty manifest itself in higher wood product prices and more volatile price swings. According to *Random Lengths,* the 1993 framing lumber market has been about three times more

The author is a Business Economist, Forintek Canada Corp., Ottawa, Ont., Canada.

volatile than in the previous 10 years with weekly price changes averaging about $15.

But what this environmentally motivated uncertainty has really done is open the door to unprecedented possibilities for substituting alternative materials for wood. We're not talking about baseball bats and hockey sticks or even doors and windows — we're talking about the soul of our industry's product line, basic wood materials used in building construction.

This paper addresses this complex issue of the environment and substitution by describing what environmentally motivated substitution may mean for the building materials industry and, in particular, wood building materials. The paper sets the stage by briefly discussing a Canadian research alliance project which Forintek has spearheaded to address this and other environmental issues related to building materials. The paper concludes with some preliminary results of alliance's work.

Traditional substitution, the price system, and environmental claims

Historically, product performance and price have been the primary drivers behind substitution of wood and nonwood products. Many analysts contend that government regulation motivated by the environmental movement will increasingly affect building material substitution in the future. One result is likely to be the use of a wider mix of materials in any given structure as we begin to distinguish between each product's environmental benefits and cost. In other words, competition between alternative building materials will be redefined to include environmental as well as traditional price and product performance criteria.

Unfortunately, the prevailing price system is incapable of accounting for the environment and may actually send perverse signals about the relative environmental benefits or costs associated with competing products, particularly when they are made of different materials. Although in time we can expect to see more environmental factors embodied in the price system (e.g., carbon or Btu taxes) and therefore in building material costs, the use of market forces to achieve environmental objectives is not likely to occur quickly or on a broad front. The price system and its failures is revisited later in the paper.

The environmental cloud hanging over the wood products industry, coupled with the recent price volatility in wood products and the demise of the nonresidential construction market, has the steel and concrete industries doubling their efforts to capture a share of the residential construction market. Subsequently, each industry has begun to position itself against the other as the obvious "green" choice. The goal of the steel industry is to have a 25 percent share of the residential framing market before the turn of the century. While this is probably overly optimistic, just setting the target gives a pretty clear indication as to the direction they are heading. So what we have here is a lot of environmental information being put out by all three material groups, each with their environmental claim to fame, which only tells a small portion of the whole story. For instance, steel's major environmental claim is that it is North America's most recycled material. Wood chooses to underscore its renewability trait, and concrete counters with the claim of being produced locally from common and abundant natural resources.

And then we have the environmentalists who are rightfully concerned, but rather short-sighted. They have not yet asked themselves that if timber is not to be a product of the forest than what do we replace timber with, and if these alternatives more or less environmentally benign.

Building materials in the context of sustainable development

The growing environmental movement, the inefficiencies built into current price system, and the piecemeal measures presently used to gauge product environmental impacts all provided the impetus for a project we initiated 2 years ago. We call the project "Building Materials in the Context of Sustainable Development."

Much has been written and said about sustainable development, and it is clear that the philosophy has captured the imaginations of government, business, and concerned individuals alike. The concept, however, is neither easy to initially grasp nor does it readily lend itself to 20-second sound bites. But suffice it to say that sustainable development links the common themes of maintaining, sustaining, and preserving environmental processes and of dealing with environmental constraints when meeting present and future needs. Perhaps the number one problem that plagues the concept is that it is not a one-time goal that can be precisely defined and achieved in a certain period of time. Rather, it is more like a moving target — as the global population continues to grow, it puts additional pressures on our planet's resources and biosphere. Because sustainable development is a "living issue," it makes it difficult to measure any progress towards attaining its objective.

The concept's primary value is its broad scope and implied shift away from short-term, economically expedient decisions in favor of something of a more thoughtful and balanced orientation in the long term.

While the alliance's project has a number of secondary objectives, the ultimate goal is to make available a simple model which will enable the building community to assess the relative environmental implications of using various building materials in defined applications — again, within this framework of sustainable development. The goal here is not to mount an all-out offensive in favor of wood, but rather to encourage the wider use of all materials in any particular building design such that together they minimize the building's environmental impact.

Now the sustainable development doctrine demands that we look at the environmental consequences of an area of activity such as building construction on a broad level (Fig. 1). The alliance's approach is both broad and holistic. It starts with the idea that five main activity stages define the life-cycle of a building. These include resource extraction, product manufacturing, on-site construction, building occupancy and maintenance, and demolition and disposal of the building. Transportation services within and between these activity stages are also included. This life-cycle approach is often referred to as a "cradle-to-grave" framework.

What we are concerned with is how various building products interact with and affect the environment during these five stages of activity. To make the concept of sustainable development operational, an operational objective must be posed. It has been defined to be the minimization of physical environmental inputs and outputs required over the life of the building. Minimizing the absolute volumes of these environmental inputs and outputs will go along way in meeting the sustainable development objective, but it is still incomplete. These inputs and outputs must also be considered in the context of various sustainable development criteria such as resource availability and renewability, product durability and recyclability, as well as substance toxicity and pollution abatement levels.

The physical environmental inputs and outputs themselves are relatively straightforward with perhaps the exception of the term "ecological capacity." Ecological capacity denotes the many other effects caused by human interaction with the physical environment during the extraction phase of the life-cycle. For instance, timber harvesting can affect carbon cycling within the forest ecosystem much in the

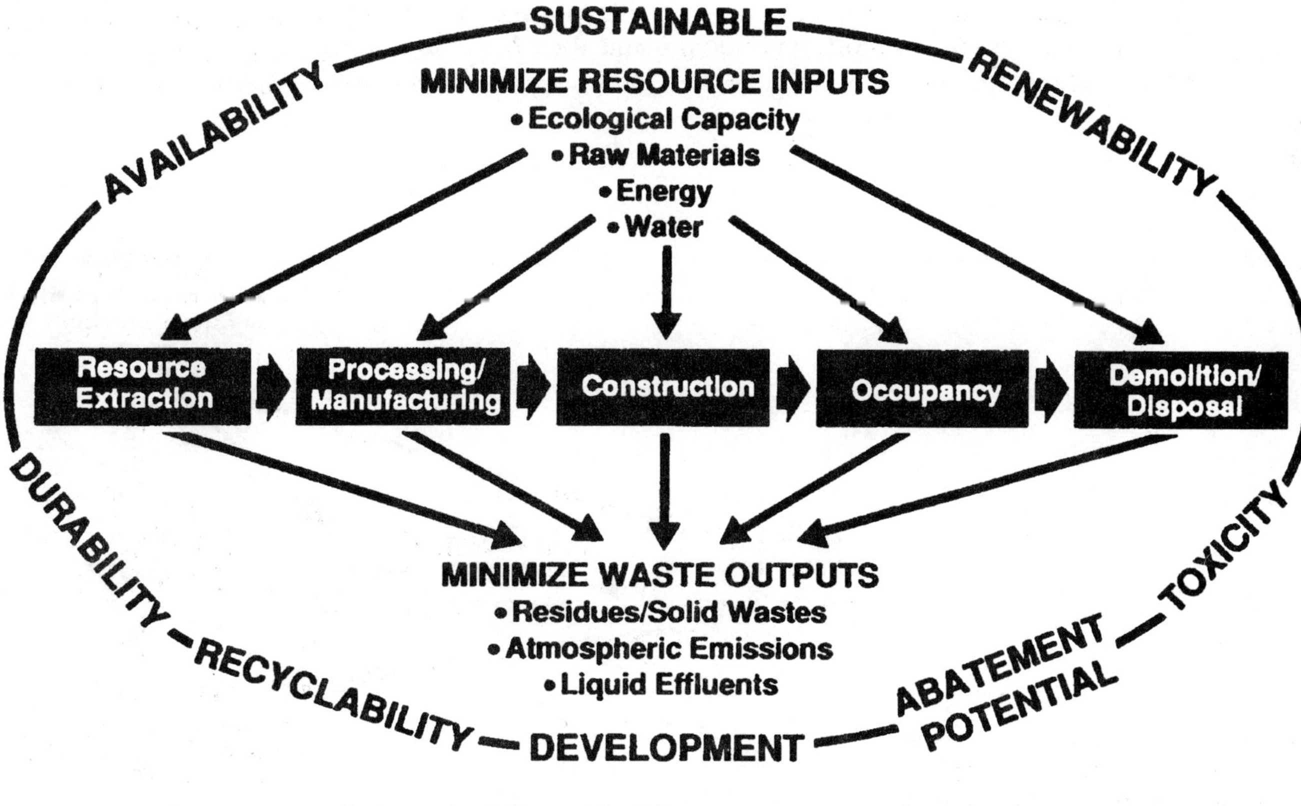

Figure 1. — The sustainability of building construction: objectives and criteria.

same manner as iron ore mining can affect ground water quality. Unfortunately, these effects are not readily quantifiable and available information is sketchy. However, a major task of the alliance's study is to assess some of these key effects which will then be incorporated into the model. So far the ongoing research on ecological capacity suggests that the raw materials for wood, steel, and concrete are physically sustainable; the outstanding question is at what *relative* environmental cost?

Over a period of 6 months the research alliance was formed and includes eight organizations comprised of three university programs, two private consulting firms, a government agency, and another private research firm similar to Forintek to undertake this multifaceted project. Great lengths have been taken to structure the alliance so that the objectivity and integrity of the study is not compromised. Over the last 12 months the research alliance has completed the major analytical task. The initial focus of this work has been on vertical and horizontal structural assemblies using a number of wood, steel, and concrete products in light commercial and residential buildings.

So far the alliance has developed what has been termed "unit factor estimates" (Fig. 2) for a number of wood, steel, and concrete products in terms of their raw material requirements, embodied energy, demand for water, solid wastes produced, and a select number of atmospheric emissions and liquid effluents. These estimates are from the resource extraction stage to completion of a structural assembly on a construction site. An example of a unit factor estimate would be gigajoules of energy used per tonne of steel or kilograms of CO_2 emitted per thousand board meters (MBM) of softwood lumber. In essence, what the alliance has been able to do is three-fold:

1. Provided an updated and standardized benchmark for environmental assessment and comparison of both traditional and new building materials.
2. Incorporated advances in conversion efficiency, pollution abatement, and energy conservation into the environmental measures which extend beyond the prevailing preoccupation with atmospheric emissions and energy use to include raw material requirements, solid wastes, water use, and a number of liquid effluents.
3. Moved the analysis forward by explicitly considering the construction activity stage which has not been previously done in any other similar study.

They have also begun to construct the environmental design computer model (Fig. 3). It contains a menu of typical floor, roof, and wall assemblies for the three materials from which the user specifies the building's geographical location, the actual space to be modeled, and working loads and spans. Given this information, the model proceeds to determine and then break down the required assemblies into their respective materials, applies the unit factor

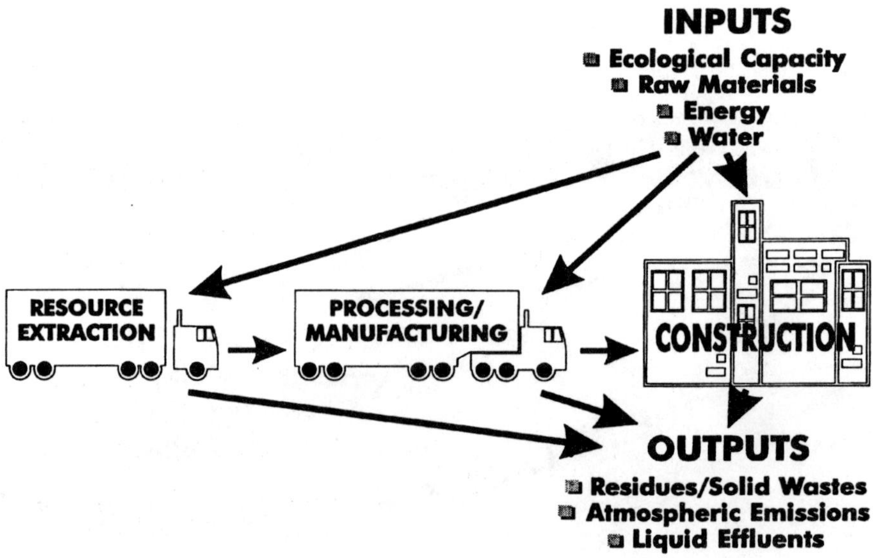

Figure 2. — Unit factor estimates of environmental resource inputs and waste outputs for wood, steel, and concrete products.

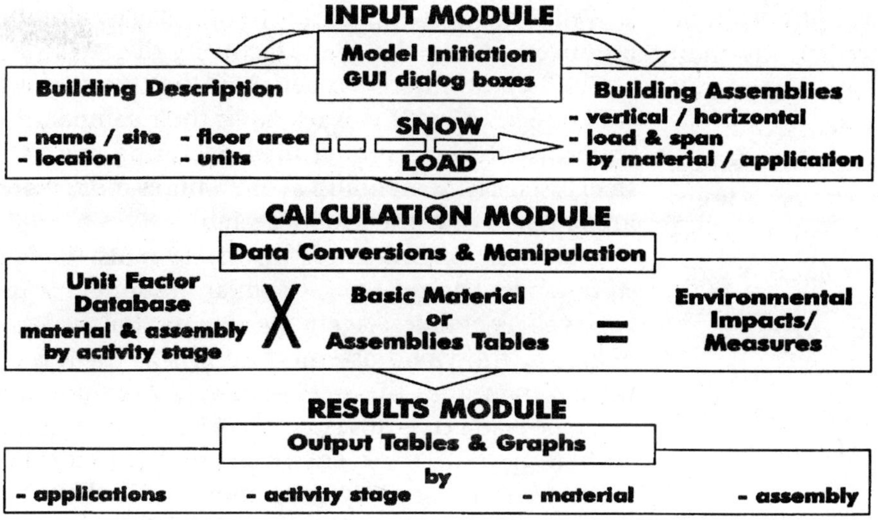

Figure 3. — Simpified flow chart of the system model.

estimates to the materials used, and compiles the total environmental inputs and outputs for that particular design. The user can then specify how the output results should be displayed, either graphically or in a table, by assembly, life-cycle stage, or material. The real value of the model is that it allows the building designer to focus on what aspects of the design are particularly environmentally costly and try alternative assemblies and materials to reduce the buildings environmental load or cost.

The model focuses on comparative building assembly designs rather than on comparing materials directly. This focus is necessary because unit measures such as a pound of wood versus a pound of steel or concrete is for all practical purposes meaningless, as such measures do not consider the functional capability of each of the materials in use.

Preliminary model results

This section presents an actual, but preliminary, assembly comparison using the model and data for two materials in a simple application. The application for comparison is a typical exterior infill wall assembly used in light commercial structures. Usually a building incorporating this type of assembly would be built using steel as the post-and-beam supporting structure. The exterior opening can then be infilled using a number of products. For the purpose of material comparison, 20-gauge nonstructural steel studs produced in an integrated or virgin steel plant was specified as one example and 2 by 4 wood studs as the second alternative. The infill space to be modeled was arbitrarily set at 3 meters high by 30 meters long or approximately 10 feet by 100 feet.

Table 1 summarizes the resource demands for the two alternative infill wall methods. It took over 600 kg of raw materials to produce the softwood lumber and nails used in the wood wall assembly. On a mass basis the wood design is only 75 percent as resource intensive as the steel wall assembly.

Table 1. — Comparative resource requirements for wood and steel wall assemblies.

Raw material required	Wood	Steel
	(kg)	
Roundwood	597	
Limestone	2	77
Iron ore	12	509
Coal	6	252
Zinc		2
TOTAL	617	840

In terms of embodied energy (Fig. 4), the steel wall is three times more energy intensive than the comparable wood wall. The steel wall is more energy intensive in terms of resource extraction and manufacturing, but is slightly less energy intensive than wood during the actual process of construction.

Minimizing energy use is a key sustainable development objective as energy, especially energy from fossil fuels, is perhaps the best environmental impact barometer we have at our fingertips. Moreover, as builders and architects strive to make their buildings more energy efficient, the embodied energy of the building materials used becomes more significant. Table 2 indicates that energy-efficient residential designs relative to conventional designs are about twice as energy efficient and, subsequently, embodied energy expressed as years of operating energy has grown two-fold in importance as well.

All human endeavor results in host atmospheric emissions, some of which are more serious than others. One emission that is especially topical and perhaps even critical is carbon dioxide, because of its role in global warming. Figure 5 shows carbon dioxide emissions for the steel wall are three times that of the wood wall assembly. The relative energy intensiveness of the two products explains a large portion of the carbon dioxide emission difference.

Figure 6 provides further comparisons for some other important emissions. Here again, the wood assembly proves advantageous when comparing carbon monoxide, sulfur dioxide, nitrous oxides, methane, particulates, and volatile organic compound emissions, all of which pose a hazard either to human health or the environment.

Another consideration is solid waste (Fig. 7). The solid waste generated during manufacturing and construction was about 30 kg higher for the wood wall assembly. This outcome is primarily due to the more efficient use of steel on the construction site as well as the growing market for the slag by-product produced during the manufacturing of steel.

The largest difference between the two material wall designs (Fig. 8) was found in their comparative water use during product manufacturing, with the steel assembly demanding over 25 times more water input relative to the wood design.

Water use of course results in effluents. Table 3 summarizes the results for similar effluents for the two wall assemblies. Again we see that for the same effluents, the wood assembly results in far less effluents than does the steel assembly. One problem yet to be resolved is how to categorize these effluents with respect to their environmental toxicity. For instance, is 1 mg of cyanide more toxic than 100 or even 1000 g of suspended solids? By next year the alliance hopes to have an answer so this toxicity dimension can be added to the output of the model.

That's the extent of the alliance's preliminary quantitative results. A lot of work remains to complete the total picture. While caution should be exercised, these preliminary results suggest that there

Table 2. — Comparing embodied energy with operating energy.

House type and location	Heating energy	Embodied energy	Embodied energy expressed as years of heating energy
	(Gj/yr.)	(Gj)	(yr.)
Conventional, Vancouver	107	1,000	9.4
Energy-efficient, Vancouver	60	1,075	17.9
Conventional, Toronto	143	1,000	7.0
Energy-efficient, Toronto	82	1,075	13.1

Source: Cole, R. 1993.

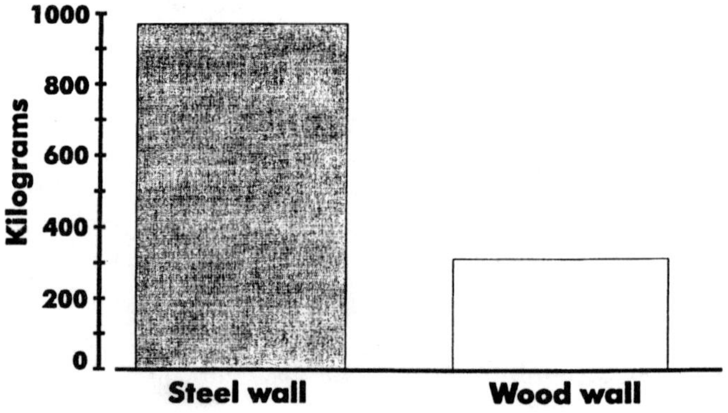

Figure 5. — Comparative carbon dioxide emissions.

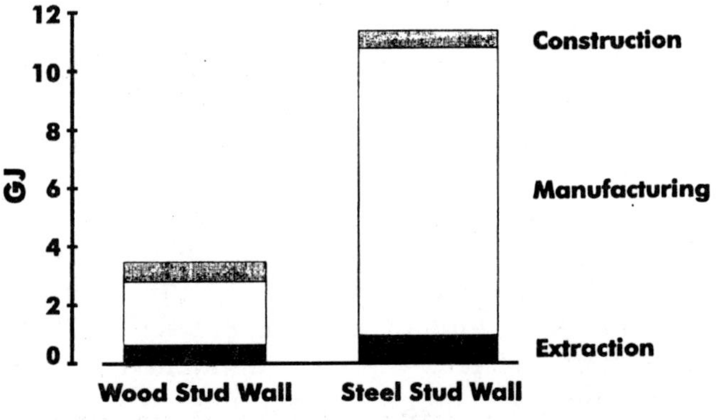

Figure 4. — Comparative energy use.

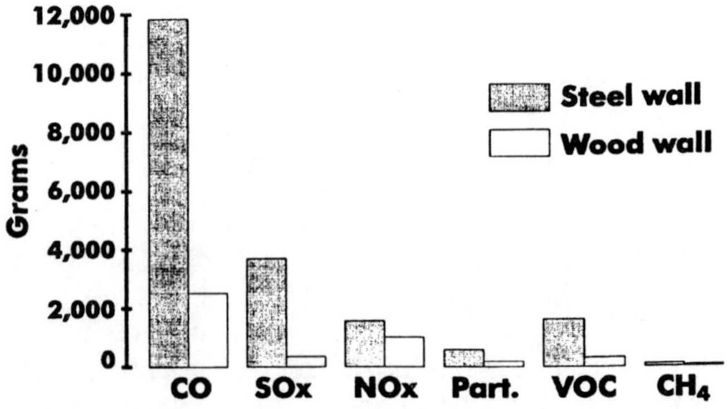

Figure 6. — Other air emissions.

appears to be some absolute environmental advantages of using wood instead of steel from a life-cycle perspective, at least in this particular instance.

How can these results be interpreted in terms of our present price system and the issue of substitution?

Environmental externality costs

To answer the question we have to consider the concept of environmental externality costs. These are the additional costs to society in terms of human health and environmental damage that are not already incorporated into the product's present price.

Essentially, determining environmental externality costs provides a method for estimating the full cost of environmental impacts to society. To ignore these external environmental costs in planning, pricing, and product selection values them and society at zero. Perhaps more importantly, by putting a dollar value on environmental damage provides building designers and policy-makers with a measure they inherently understand and which should encourage all concerned to invest in, and use, less polluting resources and materials. Politically, these externalities amount to pollution fees or taxes similar to the Btu tax President Clinton entertained.

The Pace University Center for Environmental Legal Studies in New York completed a lengthy study of externality costs for electricity generation and arrived at the following dollar values for four important air pollutants as well as three nonfossil fuel electricity generation methods (Table 4). Their work on the external cost of effluents proved inconclusive due to the wide range of values reported in the literature. They indicated, however, that effluent pollution is less damaging than air pollution and so we have incorporated an estimate of effluent discharge cost which is pegged at one third the value of the least costly air pollutant (CO_2). This is a conservative estimate in light of the ranges in values reported in the Pace study.

With the total quantity of pollutants known, plus the quantity of non-fossil fuel electricity used, it is possible to estimate and compare the environmental externality cost of the two alternative infill wall assemblies. Table 5 summarizes the direct material costs as well as the externality costs of the two alternative wall assemblies constructed in Toronto.

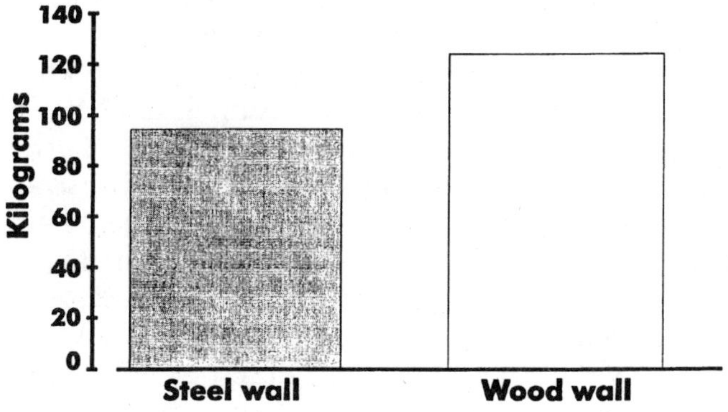

Figure 7. — Comparative solid wastes.

Table 3. — Comparative manufacturing effluents.

Effluents	Wood wall	Steel wall
Suspended solids (g)	12,180	495,640
Nonferrous metals (mg)	62	2,532
Cyanide (mg)	99	4,051
Phenols (mg)	17,715	725,994
Ammonia and ammonium (mg)	1,310	53,665
Halogenated organics (mg)	507	20,758
Oil and grease (mg)	1,421	58,222
Sulfides (mg)	13	507
Iron (mg)	507	20,758

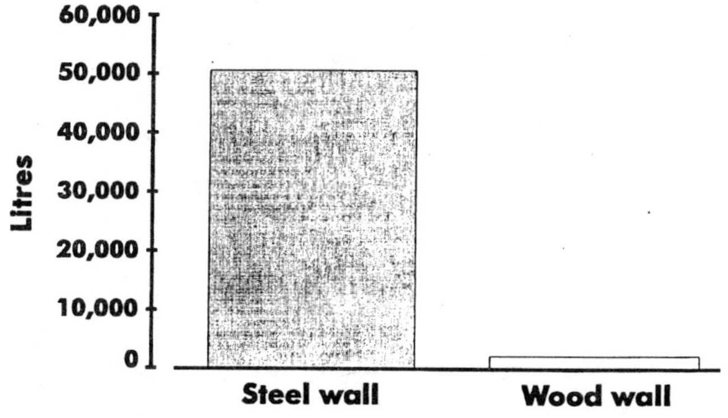

Figure 8. — Water demand.

Table 4. — Environmental externality cost estimates.

Pollutant	Cost
	(C$/kg)
Carbon dioxide	0.15
Sulfur dioxide	1.80
Nitrogen dioxide	4.47
Particulates	2.62
Nuclear	8.08
Hydro	0.33
Renewable	1.11
Effluents[a]	**0.05**

[a] Estimated by Forintek from Pace study.
Source: Pace University, 1990.

Table 5. — Direct material and indirect externality costs (wood and steel stud infill walls, 3 by 30 m).

	Wood wall	Steel wall
	(C$)	
Cost of material	272.80[a]	240.00[b]
Externality costs (inclusive of fasteners)		
Electricity (nonfossil fuel)	1.46	4.67
CO_2	47.00	145.65
SO_2	0.66	6.65
NO_x	4.52	7.04
Particulates	0.49	1.55
Effluents	0.61	24.80
Subtotal	54.74	190.57
TOTAL	327.54	430.57

[a] Inclusive of studs, top/bottom plate, blocking, and waste @ $400/MBM Toronto.

[b] Inclusive of studs, top/bottom plate, blocking, and waste @ $0.24/lin. ft. Toronto.

Table 6. — Embodied energy comparison of structural wood and recycled steel stud walls.

Structural wood wall

- add top plate to nonstructural example, total embodied energy increases to 4,020 Mj

Structural steel wall

- change stud/track gauge from 20 to 14, increases weight by 100%
- assume 50/50 split between recycled and virgin material (virgin material input 100% of nonstructural example with an embodied energy of 11,400 Mj)
- add in the energy needed to process recycled scrap content and reheat virgin material (4,500 Mj)

Difference

- the structural steel wall is about 4 times more energy intensive than its wood counterpart

As indicated, the wood wall is not presently cost competitive with the steel alternative at a cost C$400/MBM of softwood lumber. Generally, at prices above C$375/MBM for lumber wood has been found to noncompetitive with steel studs in nonstructural applications. However, from an environmental cost accounting perspective, the wood wall assembly is only a quarter of the cost of the steel wall or is some four times more environmentally benign. Much of the gain for the wood wall assembly can be ascribed to its lower carbon dioxide emission level.

Taken together, the direct material and environmental costs sum to a significant cost advantage for wood and given the price sensitivity of builders, any builder confronted with these true costs would likely choose the wood stud wall over the steel alternative. Indeed, if this was the present day situation, substitution would not be a major issue.

Recycling

Lastly, a few words on recycling. Let's put this issue in the context of a structural wall assembly by extrapolating from the nonstructural infill example. The wood products industry is perhaps more concerned about the threat of steel substituting in structural framing more so than in nonstructural applications. To make the nonstructural wood wall example load-bearing requires that a top plate be added, which adds 16 percent more energy input to our nonstructural example for a revised embodied energy total of about 4,000 Mj for the structural application (Table 6). To transform the nonload-bearing steel wall assembly into a structural wall assembly is a little more complicated. First the stud gauge will have to change from 20 gauge to 14 gauge, which doubles the weight of steel incorporated into the wall. And while the recycled content of steel studs may approach upwards of 60 percent recycled material, for simplicity it is assumed that the split between recycled and virgin steel content is 50/50. So in essence, the new structural steel wall assembly incorporates as much virgin steel input as the previous nonstructural example and has a similar embodied energy of about 11,400 Mj. The mini-mill energy used to process the recycled scrap content as well as reheat the semifinished virgin material must also be added, which is estimated at another 4,500 Mj for a combined total of close to 16,000 Mj. So, relative to the structural wood wall assembly the structural steel wall is some four times as energy intensive and correspondingly is at least that much more environmentally damaging, despite its recycled steel content.

Conclusion

It should be kept in mind that these results are still preliminary as the model has only been up and running for a short while. We think, however, that this type of information is going to be crucial to a more constructive dialogue with the public, the building community and policy-makers alike — especially if we as an industry are going to eventually sway present public perception away from its narrow and sometimes ill-informed focus on our forestry practices to one favoring a more objective, holistic, and balanced orientation that includes the life-cycle cost of the products we choose to use.

Economic dynamics of material substitution: overview and application to the market for wood siding

Lloyd C. Irland

Abstract

Forces such as consumer preferences, builder preferences, relative installed costs, life-cycle costs, and comparative technical features all affect choices between competing construction materials. Materials choices are made on the basis of installed costs and not on materials costs alone. The effect of changes in relative materials prices is muted by the role of labor and overhead in installed costs and by the many forces making for conservatism in the building industry. The diversity of competitive situations in the many market segments is another reason why materials substitution responds slowly to changes in relative prices. The forces affecting materials substitution operate within individual market segments or submarkets. The operation of these forces is illustrated through a brief discussion of the market for wood siding products.

Introduction

The 1990s will be a time of exceptional dynamism in world wood products markets. Tighter and less reliable log supplies, often accompanied by declines in piece size and quality, are the themes of the decade. These factors are affecting wood users at all levels of the marketing chain from the logging contractors to the renters of buildings built with wood siding.

Manufacturers, architects, engineers, equipment producers, builders, mill designers, building code writers, and researchers and technologists will participate in myriad shifts in product designs, replacements with new materials, improvements in performance, and commercialization of pioneering technologies. To help these people understand the likely direction of technological change at a detailed level, economic analyses have been developed that can help us predict how end-users will change raw materials usage in response to changing price ratios, long-run technological changes, or other factors (2,8,10,11,14). Unfortunately, in the past there was often little variance in these ratios and, in the 1990s, the variances will exceed past experience. Also, the range of options is now much wider, and the social criteria and constraints being applied to raw materials specification are more complex (5,12). For these reasons, statistical relationships based on past experience may not help us much in the future unless re-estimated and reformulated to account for these new realities.

In this short paper, I will outline a broad approach for analyzing the economic dynamics of materials substitution in construction, emphasizing how these considerations apply to wood. These general considerations are then applied to a mini-case study of the market for wood siding.

Economic dynamics: overview

Market segments

The siding market is divided into a variety of segments. It is within these segments that the economic and technical dynamics of substitution actually operate. The way in which these forces affect product choices can often be different in each segment. Market segmentation variables might include:

- Commercial and governmental construction
- Residential, single or multi-unit

The author is President, The Irland Group, Winthrop, Maine. The author would like to express his thanks to Dr. Con Schallau, who delivered this talk at the conference.

- Residential, rental versus owner-occupied
- Cost of building
- Regional styles
- Urban or rural areas
- Rental versus owner-occupied
- Repair/remodeling versus new construction

Market segments will be comprised of the various cross-classifications generated by what amounts to a multi-dimensional matrix organizing all of the factors together. Each segment that is defined by a combination of these factors is likely to respond in different ways to changes in relative prices, relative performance, or to regulatory factors.

A general model of materials substitution in construction

The factors that affect choices between competing raw materials are many and complex. Table 1 summarizes a general model of materials substitution, which attempts to organize those factors in a useful way. In some instances, considerations presented focus on a single level of the market; in others, they come into play at several points in the marketing chain. These factors interact in a complex fashion. Working out this model empirically (e.g., by simulating the way the factors affect decisions on a single project by regulators, architects, specifiers, builders, owners, and renters) will show that what is actually happening is more like a spider's web than a linear process.

If econometric modeling of these factors is to be attempted, close attention must be paid to the problems of aggregating across many different market segments and to how different factors change in importance over time. The relevant comparisons can be viewed as pairwise in nature, but usually this works only within a given market segment. For example, the low-end residential new construction market segment (e.g., "FHA raised ranch") is extremely price sensitive for siding materials, while the high-end market is not. Also, life-cycle cost factors may receive more attention in certain government projects and in high-end residential segments than in others.

Significantly, engineering and technical traits, while critically important at the design level, are not often critical in themselves in tipping materials choices in one direction or the other. Rather, their significance is in the way in which they affect initial installed costs (and lifetime costs where applicable).

Many unpriced factors can influence materials choices. Stability of supply and predictability of pricing have tipped starkly against lumber in recent months, granting a strong advantage to competitors. Onsite handling factors that enhance predictability and flexibility for the builder may be critical, especially in builder-financed speculative construction (which is in a cyclical trough at present). Siding products that are prefinished, generate less on-site waste, are easily cut to size, are manufactured with high precision, and are easily stored on-site make builders' lives easier.

As in the case of children's toys, in construction the specifier is not normally the end-user. For this

Table 1. — Factors in a general model of materials substitution in construction.

Consideration	Comparative factor between competing materials
Product traits and performance factors	Weight per EFU[a]
	Durability
	Finish retention and colorfastness
	Fastening traits
	Uniformity
	Heat/fire resistance
Builder factors	Consistency of supply
	Consistency of pricing
	Packaging and technical support
	Consistency of product quality
	Onsite handling, weight, waste, difficulty of cutting, etc.
End user factors	Specific end user benefits
	Color/texture preferences
	Architectural styles/designs
	Maintenance demands over time
Economic factors	Materials cost, delivered ("own price")
	Nonprice costs/benefits to builder
	Installed Cost
	Life-cycle Cost
	As perceived by building owner
	As perceived by end user/tenant
Regulatory and other	Fire codes
	Building codes
	Historic preservation
	Health and safety
	"Green" building regulations
Materials policy questions	Energy intensity
	Import dependence
	Recyclability/reusability
	Environmental and other externalities of materials cycle

[a] EFU = equivalent functional unit, or the amount of product that meets the same performance requirement, as square feet of siding area or load supporting capacity for a given span. Comparisons per cubic foot, lineal foot, or per ton do not compare equivalent functional units.

reason, a variety of considerations of little relevance to the final user come into play. Just as the child is not usually concerned about the toy's cost, so it is that in construction, materials prices may enter into decisions in strange ways. For siding, we have the unusual case in which the vinyl product, properly installed, has a decisive advantage over most competitors in both initial cost and in lifetime cost. For the homeowner, avoiding disagreeable expenditures of time scraping and painting is a major, if unpriced, advantage.

More often than not, however, lower lifetime costs have to be paid through higher initial costs. In our society, where many income properties are owned by syndicates of absentees or even speculators, and where homeowners may move every 5 years, lifetime costs have only a muted impact on materials choices, if any at all.

At the same time, considerations of broader social significance such as recyclability, energy intensity, and some safety and health considerations come to bear upon decisions through various regulations, if at all. Product traits related to safety and inflammability are enforced by local codes or by insurance companies, through rate differentials. Ensuring that wood products receive fair treatment based on objective facts keeps technical representatives of manufacturers and trade associations fully employed.

In sum, the factors affecting materials substitution are complex. They interact in ways that differ over time and by market segment. To illustrate this in a more concrete manner, let us look quickly at one end use application — the market for siding.

Application: the wood siding market

In the siding market, wood competes in four principal forms: shingles, lumber in the form of clapboards or barnboards, plywood products, and fiber-based hardboard clapboarding (Table 2). Leading direct competitors are aluminum, vinyl, and brick and cement-bonded products. Many of the siding products attempt to provide an appearance duplicating that of clapboards, illustrating the importance of retaining the texture and appearance of the original wood clapboard. Sizable volumes of value-added wood products are used in the new residential siding market (Table 3).

New residential construction

The American Plywood Association estimates that total wall area for new residential siding is about 2 billion ft.^{2} per year. In this market, wood is the leader in square footage but it has been losing share slowly for the past 5 years (Figs. 1 and 2). Aluminum has also been declining. Brick continued its long decline and then rebounded a bit, and vinyl took a share away from all of the others.

Wood's share in 1983 was highest in the Midwest, followed by the Northeast and the West (Fig. 3). These differences emerge from differences in income levels, architectural styles, climates, and other factors. By 1992, however, regional differences had narrowed and the shares had fallen dramatically. Data are not available at this level of detail on how siding market shares have fared in other new construction markets.

Repair and remodeling market

Repair and remodeling includes a variety of activities, from replacing or restoring siding to adding rooms or wings to existing buildings. The particular demands of siding replacement often dictate the choice of materials, in view of cost limits and the

Table 2. — Wood siding types and principal competitors.

Products	Expected lifetime	Comments
	(yr.)	
Lumber	10 to 100	Traditional in many designs
Shingles	30 to 50	Strong tie to regions and styles
Veneer/fiber-based	- -[a]	Imitates lumber appearance but improves on finish and engineering properties
Aluminum	20 to 50	Especially suited to remodeling uses; share declining
Vinyl	50	Low initial and life cycle cost; imitate wood look; good for initial and remodeling
Cement-bonded	- -	Durable; brittle
Brick	Lifetime	Durable; low insulation value, costly
Stucco	- -	Strong regional style influence

[a] Unable to locate estimates for this project, but leading manufacturers offer 25-year warranties.

Table 3. — Wood products used in siding applications: new construction.

Softwood plywood, OSB, waferboard	1 billion ft.^{2} (3/8-in.)
Nonstructural panels	424 million ft.^{2} (3/8-in.)
Softwood lumber	511 million board feet
Shingles	300,000 squares

Sources: (7), p. 16, 18; (1), p. 28; trade sources, p. 28. The estimate for plywood is by Adair; McKeever & Anderson show 870 million ft.^{2} (3/8-in. basis) for 1988, plus 230 million ft.^{2} of OSB/waferboard, which essentially matches Adair's estimate.

expense of removing existing siding. These considerations usually favor aluminum or vinyl (6).

The repair and remodeling market is about $100 billion per year, of which about one-third is for materials. The nation contains about 105 million housing units. In recent years, about 5.5 percent, or some 3.2 million households added or replaced siding each year (3).

Cost comparisons

Installed costs vary widely for the different siding products. On the basis of installed costs alone, we would expect that wood's market share might slowly erode in most market segments. The apparent strong advantage of vinyl in lifetime cost is likely to reinforce this trend.

Life-cycle costs

Differences in life-cycle costs do not always affect choices of construction materials or building systems. The markets are too diverse, complex, and strongly driven by initial costs. Periods of building ownership may be brief in many instances, and building owners do not always bear all relevant life-cycle operating costs or structural costs. In the siding market, several interesting patterns emerge. First, brick, which has the highest initial costs, has very small annual maintenance costs. Vinyl has the unusual feature of having both low initial costs and low life-cycle costs.

At what price do they switch?

When lumber prices act has they have been doing recently, the question is often asked, "what's the price where they will all switch to steel or vinyl?" There is no "point estimate" that will answer this question. The impact of direct materials cost is muted by the proportion of other costs in installed costs. For wood and immediate substitutes, materials are about one-third of installed costs. In addition to the muting effect of other costs, the specific effect will depend on the application being considered and which immediate substitute is closest in installed cost, considering other relevant performance factors as well. The siding market is a somewhat more uniform end-use category than many others. But

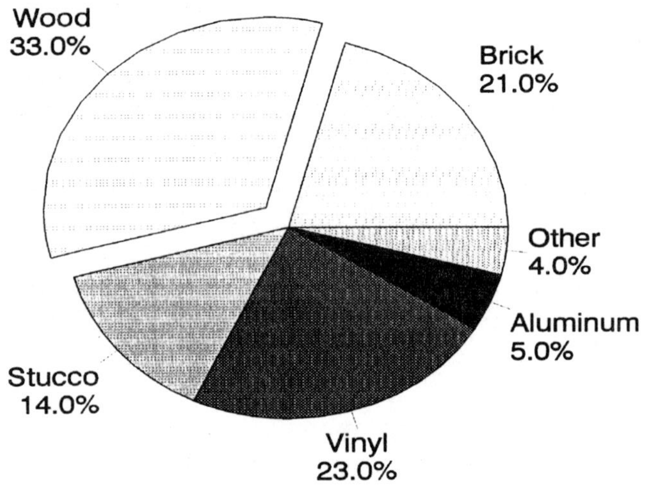

Figure 1. — Siding market shares 1992, by number of new homes.

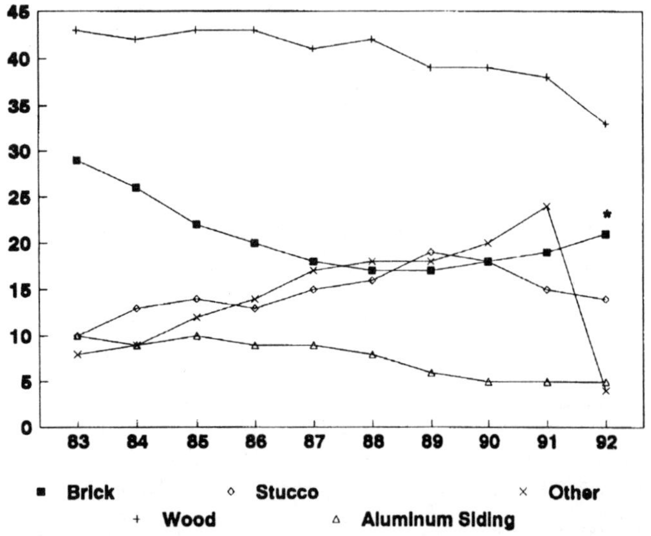

Figure 2. — Siding market shares by type in the United States, by number of new homes.

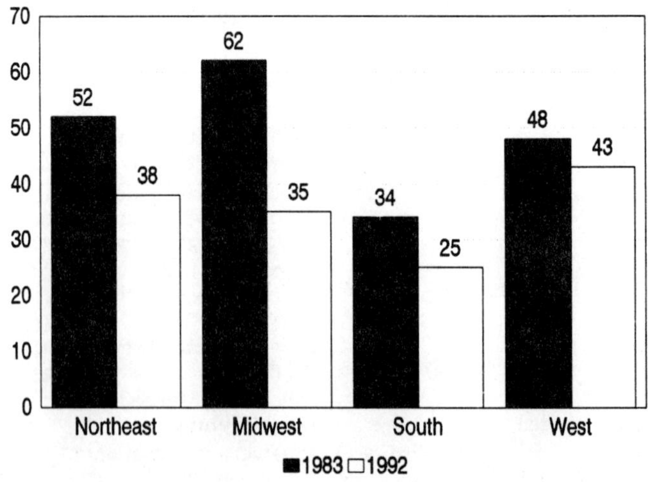

Figure 3. — Wood market share by region, 1983 and 1992 (by number of new homes).

it is not uncommon for a single house to combine a variety of siding types. This factor also slows down the impact of material cost changes. Since many siding competitors imitate the appearance of wood (often in great detail), the role of shifting architectural styles is also muted.

The price range at which substitution is triggered will vary by market segment. Some segments are highly price sensitive and others are not. It is this variation among segments which accounts for the generally modest pace of substitution in construction materials. A builder in my area observes that if wood siding prices go up from recent levels, it will eliminate the use of wood entirely at the low end, where its share is currently small. But in the large, high end houses, there will be no effect at all. Wood retains its image there as the most desirable material, even for homebuyers that could afford brick and could afford to invest to gain lower life-cycle costs.

Then there is the overall conservatism of the construction industry's marketing chain, customers, and regulatory system. Even in the face of large cost differentials, changes in materials use occur slowly (10).

In sum, even within what seems to be a straightforward end-use application, there is a good deal of complexity involved in determining how wood prices and other performance factors will affect substitution.

Conclusions

The process of materials substitution is complex. Materials choices are affected by a complex web of considerations. Technical traits of the products may be of secondary importance in determining these choices. Relative prices of competing prices may have only a limited influence in any brief period of time. This is because of the many forces that make for inertia in the typical bill of materials for buildings.

Wood products prices have reached extraordinary levels in recent years, and have attained an unprecedented degree of volatility. These conditions seem likely to persist during the 1990s. High price volatility is becoming a major competitive disadvantage for wood. Architects and builders are motivated to seek alternatives to an unusual degree, compared to any time in the recent past. From the standpoint of market forecasting, past statistical results on market responsiveness to changing price ratios may no longer be relevant, since future prices are likely to be outside of their past range.

The example of wood siding and its immediate competitors illustrates how the economic dynamics of materials choices differs from one market segment to another. In particular, from segment to segment, the relevant competing product may be different. Based on its advantages in installed cost and in maintenance costs, vinyl has taken a share from several of its competitors, and seems likely to continue doing so. While data are not available, it seems clear that wood's share in the remodeling applications is already low. How swiftly these changes in share will proceed is not easily predicted.

We are entering a period of unprecedented change in timber supply, regional sources of production, world trade, and technology for producing building products from wood. Foreseeing how these changes will develop, and forecasting the most promising business opportunities they will create, must be done with differences between market segments in mind.

Table 4. — Construction cost per square foot for siding items (installed cost).

Siding material	Cost
	($/ft.^{2})
Hardboard, 7/16-inch primed	1.66
Vinyl siding PVC	1.87
Plywood, T-111 fir 5/8-inch factory stained	1.92
Rough sawn cedarboards 3/8-inch, natural	2.10
White pine 1 by 8, rough 1 by 8, natural stained	2.20
White cedar shingles (extras), 16 inches with 7-1/2-inch exposure	2.21
Aluminum siding, painted (.019-in.)	2.43
Red cedar shingles (no. 1), 16 inches with 7-1/2-inch exposure	2.69
Cedar boards, A grade 1/2- by 6-inch	3.52
Stucco, w. mesh 1-inch thick	4.33
Redwood, clear, VG 1/2- by 4-inch	4.43
Commonbrick	8.45

Source: (10), pp. 111, 175, 178-9, 229.

Table 5. — Example: a vinyl siding manufacturer's life-cycle cost comparison.

Material	Life-cycle cost index (generic vinyl = 100)	Typical warranty
Generic vinyl siding	100.0	50 years or lifetime
Wolverine restoration vinyl siding	107.8	Lifetime
Stucco, 2 coats	116.1	None
Hardboard	122.4	25 years substrate
Aluminum	123.9	50 years or lifetime
Cedar boards Grade A	163.8	None

Source: Wolverine Technologies, Livonia Michigan, Siding cost comparison. Comparisons are sensitive to assumed lifetimes and maintenance regimes used. Installation costs based on RS Means data. See also (3).

Literature cited

1. Adair, C. 1993. End use marketing profiles for structural panels. American Plywood Assoc., Tacoma, Wash. 99 pp.
2. Alexander, S., and B. Greber. 1991. Environmental ramifications of various materials used in construction and manufacture in the USDA Forest Service. Gen. Tech. Rept. PNW-GTR-277. Pacific Northwest Res. Sta. 21 pp.
3. Aluwalia, G. 1993. Is remodeling activity shifting? Housing Economics (Nat. Assoc. of Home Builders), October 1993. pp. 15-16.
4. ________ and A. Shackford. 1993. Life expectancy of housing components. Housing Economics (Nat. Assoc. of Home Builders), August 1993. pp. 5-9.
5. Irland, L.C. 1993. Wood producers face green marketing era. Wood Tech. (March/April) pp. 34-36.
6. Leger, G. 1993. Succeed with wood siding: do's and don'ts of wood and hardboard siding. Remodeling News (May) p. N6 ff.
7. McKeever, D.B. and R.G. Anderson. 1992. Timber products used to build U.S. single-family houses in 1988. Forest Prod. J. 42(4):11-18.
8. McKillop, W., T.W. Stuart, and P.J. Geisler. 1980. Competition between wood products and substitute structural products. Forest Sci. 26(1):134-148.
9. Means, R.S. 1993. Means building construction cost data, 52d annual edition. R. S. Means Co., Kingston, Mass. 552 pp.
10. Moffett, J. 1993. Comparison of product diffusion and distributed log models for estimating wood, nonwood substitution in the U.S. window market. CINTRAFOR Working Paper 42. Univ. of Wash. 29 pp.
11. Spelter, H. 1985. A product diffusion approach to modeling softwood lumber demand. Forest Sci. 31(3):685-700.
12. U.S. Congress, Office of Technology Assessment. 1992. Green products by design: choices for a cleaner environment. OTA-E-541. Govt. Printing Office, Washington, D.C. 117 pp.
13. U.S. Dept. of Commerce. Characteristics of new housing. Current Construction Rept. C25. Annual, various years.
14. Williams, R.H. 1992. Trends in the consumption of basic materials in the U.S. and elsewhere. *In:* Wood products demand and the environment. Forest Prod. Soc., Madison, Wis. pp. 26-34.

Steel: not a new idea

Richard Haws

Introduction

I would like to give you a little history of steel framing in residential construction and also talk briefly about developments in Japan. Japan has a very strong residential steel construction market. About one-third of the homes built in Japan are built using steel. I will also talk about the AISI Residential Advisory Group and the projects we are doing.

Steel is not a new construction material, even in the residential market. In 1933, steel-framed homes were exhibited in the "A Century of Progress" display at the Chicago World's Fair. In the early 1950s, the White House remodeled using steel, and Lustron homes of Columbus, Ohio built a plant to premanufacture steel housing. In the late 1960s, US Steel, Bethlehem Steel, and several other steel companies developed systems for residential construction. This effort had the most significant impact.

The steel product developed in the 1960s for the residential market was not a heavy hot-rolled section using for bridges or buildings, but was instead a lightweight steel Cee section. This Cee stud and accessory members were designed so they could be used in a manner similar to wood studs. I think it is important that I mention at this time that these framing members are made by the customers of the steel mills. AISI member companies (the steel mills) only make the steel coils, not the finished product. The framing manufacturers purchase steel coils and make the steel products. I believe this is different than what is common in the wood industry. Georgia Pacific and others typically own land, the sawmills, and distribution centers.

Why didn't steel framing catch on? Carpenters had no idea how to install steel studs. Fastener techniques were slow and cumbersome. Nuts and bolts required drilling and welding required special tools and skills. There was little educational literature available. And, finally, the distribution and availability of the product was essentially confined to subsidiaries of the steel producers.

Steel companies were alone in promoting steel in the residential market. The steel producers knew little about residential construction. And, because they were in the business of selling steel, the more steel in a home the better. If a corner detail needed three studs, six studs must be twice as good! In addition, wood was plentiful and less expensive. The builder had no reason to change.

The positive side

Something did come of all this effort. The cold-formed Cee section became widely accepted by commercial construction. In the 1970s, engineers, architects, and construction workers, including skilled trades, became accustomed to working with steel studs.

Many of the early applications used both wood and steel. In this instance, steel joists were used over load-bearing wood stud walls. Notice that the joists are spaced 24 inches on center while the wall studs are 16 inches. In most cases, steel framing was used on spacings equal to that used for wood framing, either 16 inches or 24 inches on center.

In some cases, steel stud walls were used with wood joists. Even wood 2 by 4s were used to stiffen steel joist walls.

Steel joists and studs typically come with holes prepunched. But these holes are not always where they are needed. So, tools to do such common tasks as drilling holes for plumbing were developed. A standard drill with a carbide tip blade is used. This operation is similar to that for wood joists.

Some steel systems were developed which did not have the primary load carrying member at 16 or 24 inches on center. These systems had columns

The author is Program Manager, Cold-Frame Steel Construction, American Iron and Steel Inst., Washington, D.C.

spaced 6 or 8 feet apart. Horizontal members were spaced at 24 or 16 inches on center to support the exterior and interior sheathing. These systems are not direct substitutions for wood framing.

Japan's use of steel

In 1990, it was brought to our attention that several foreign countries were using steel in their residential markets. In December 1990, representatives of AISI went to Japan to see for themselves. They met with the Koazi Club in Tokyo. The Koazi Club is an association of companies which promote and encourage the use of steel. The part of the Koazi Club that promotes the use of steel in residential construction involved every possible stakeholder: from steel producers to distributors, builders, manufacturers, code officials, insurance, tools, and transportation, to name just a few.

The driving forces to move Japan away from timber was a framing material were:

1. Nearly all framing was imported, primarily from the United States and Canada.
2. Lot sizes in Tokyo are small (about 20 by 25 feet), very expensive, and new home buyers are demanding larger homes. The solution was to build upward (three- and four-story homes are common).
3. Many residential streets are too narrow to allow any type of construction vehicle. Every piece of framing must be carried and lifted into place.

In Japan, many homes are premanufactured. The design of the premanufactured homes was such that only the exterior walls were load bearing. Wall panels were constructed in roughly 2-foot, 4-foot, and 6-foot modules, with a variety of inserted doors, windows, and sheathing material. The potential home buyer would sit down with a salesperson and literally design the house on a CAD station. Interior walls could be easily moved because they are not load bearing. This flexibility allowed the home buyer to custom design their own home using a limited set of standard modules. Computer drawings were automatically generated, a bill of material sent to the factory, and the house was then panelized at the factory. Within several weeks, the materials were shipped to the construction. Since 12 million people live in the metropolitan area of Tokyo, shipping is not a significant factor.

Advisory group formed

Based on what was seen in Japan, it was decided to discuss the formation of a similar group to explore the North American residential market. On February 9, 1991, 21 people attended a meeting to explore the possibilities. Twelve of the attendees were not steel producers. The group unanimously decided to form the "Steel in Residential Construction Advisory Group." It was also decided that roofing and siding would be included with framing.

At the next meeting, four subcommittees were formed: Market Research, Education and Promotion, Technological Research, and Codes, Standards, and Specifications. It was initially decided not to form subcommittees based on products. However, earlier this year it was decided that residential steel roofing merited its own subcommittee.

As people heard about our group, we continued to receive requests to join. Like the Koazi Club, we wanted to involve all stakeholders, so the Advisory Group has maintained an open membership policy. This listing shows the diverse membership of the group. Today, there are 300 members of the Advisory Group.

In May 1991, AISI decided to participate in the National Association of Home Builders' (NAHB) Resource Conservation House. This home was built by the NAHB Research Center is a research park outside of Washington, D.C. The purpose of the home was to showcase building material alternatives which were recycled, recyclable, and durable. NAHB chose steel because steel is 66 percent recycled and 100 percent recyclable. The roof was made of recycled computer casing, the siding was recycled wood fiber.

The home used steel framing for the walls, steel floor joists, and steel trusses. All doors and appliances were also steel. The framing members were 4 inches deep and were spaced 24 inches on center. All of the framing material and the engineering were donated by our new Advisory Group. In addition, the Institute of the Ironworking Industry donated the labor to erect the steel frame.

The Advisory Group was only 3 months old and this was their first project. The home has been featured on "CBS This Morning," the cover of *Popular Science* magazine, and *Better Homes and Gardens* magazine, as well as hundreds of other magazines and newspapers.

NAHB builder survey

Interest in steel framing began to grow since the Resource Conservation House opened in December 1992. A NAHB builder survey on the West Coast in January 1993 found that 6 percent of the respondents indicated an interest in steel framing as an alternative to lumber. The same question asked this past summer resulted in 45 percent saying they were interested in using steel.

Metal Home Digest

A new magazine was first published in July 1993

which features residential steel construction. *Metal Home Digest,* with its first issue, distributed 30,000 copies.

Educational how-to manual

To satisfy the inquiries generated by this publicity, the Advisory Group has focused on education and has developed educational brochures and videos. A videotape titled, "Build It With Steel" has been completed and the *Residential Framing Manual* was completed in August. The manual includes an introduction to steel framing, a directory of steel framing and roofing manufacturers, a listing of steel home builders, fastener guidelines, construction specifications, and standard framing details. The manual includes 70 details showing all common applications of steel framing. These details are three-dimensional and are available on computer disk as well. We want to make sure the builders and code officials use the information in their decision-making process.

An 800 number has been established to handle the inquiries. We now average 900 inquiries per month. We have no budget for the placement of advertising. Today, we have been very fortunate in relying on newspapers, trade journals, radio, and TV to feature steel.

Now, I would like to tell you what some of the projects are which we will be doing in the next several months.

Destroyed homes

In August of 1992, Hurricane Andrew swept through southern Florida and destroyed nearly 100,000 homes. This was the largest natural disaster in American history. Hundreds of thousands of people were left homeless, many of whom are still without homes even now.

Masterserve, a not-for-profit construction company located in Fort Lauderdale, Fla. asked AISI to help build 200 homes in Florida City, a suburb of Homestead. AISI member companies have pledged to provide nearly one thousand tons of steel, enough steel for all the framing, roofing, and siding. Members of the Advisory Group have offered to roll-form the framing, siding, and roofing products at no charge.

The project is named Andrew Estates South and consists of 200 homes which will be built over a 2-year period. The homes will be built with volunteer labor from 48 states. Each volunteer typically contributes 1 to 2 weeks. The project was developed by and will be managed by Masterserve Construction. Coral Ridge Ministries, which is carried by over 250 television stations, has offered support for the project. Coral Ridge will ask for volunteers for this project in coming broadcasts.

The homes will be valued at $55,000 and the mortgage will be $30,000 to $25,000. Each buyer will be required to donate 700 hours of community service. The buyer must have lost a home during Hurricane Andrew. The homes will be 1,000 square feet minimum, single story with no basement. Builders from across the country will be welcome to visit the site to learn firsthand about building with steel.

Since many of the donators are located from Florida, the American Trucking Association, Gifts in Kind, and the Christian Broadcasting Network have offered to donate the transportation of material to the construction site.

The New American Home

Another project is The New American Home 1994. This is a demonstration home in Las Vegas, Nev., intended to demonstrate the latest in home technology. The home is constructed of steel framing and also Conform, a system of styrofoam blocks filled with concrete. The home has 5,200 square feet of living space.

The steel portion of the home consisted of both lightweight galvanized steel framing and heavier, hot-rolled structural steels. The home is being built by the National Council of the Housing Industry, a division of NAHB, and was available for tours during the NAHB Convention in Las Vegas in January 1994. The home will be featured in *Builder Magazine, Popular Science, Ladies Home Journal,* and many others.

Demo house

AISI will also build a demonstration house in the parking lot of the Las Vegas Convention Center during the Builders' Show. This house will be much simpler than the New American Home, being single story with a two-car garage. The home will be open for tours and seminars during the show.

There is a need for a distribution network which knows the residential builder and one with which the builders are familiar. This represents an opportunity for existing suppliers to the home builders to expand their market. In fact, several major distributors of wood framing also offer steel framing. Georgia Pacific and Boise Cascade are on the Advisory Group mailing list and receive all reports and associated information.

SESSION III:

Technology and Processes — The Means for Change

Advances in wood protection and drying

B. Gfeller

Abstract

Human activities are increasingly influenced by ecological concerns. One of the main challenges of the modern entrepreneur is to combine economic and ecological aims. Wood preservation and kiln-drying are two sectors of the forest products industry where advances are needed. In the field of chemical preservation of wood, the chemical industry and some enterprises in the wood industry are developing new chrome- and arsenic-free substances. Due to the fact that the conventional, and controversial, substances and processes have excellent properties as far as efficiency is concerned, the development of new products and production methods is extremely difficult. Yet some research has been reported, and some results have been presented. In the field of kiln-drying, economic and ecological aspects must also be successfully combined. These aspects are not opposites in this case as they were in wood preservation. In fact, modern kiln-drying is less costly under some conditions and the amount of energy consumed is smaller. The vacuum dryer is discussed. New procedures and installations with high capacities are new developments. Vacuum dryers can be used for many wood species with a variety of dimensions. The amount of energy consumed is significantly smaller than with conventional dryers. This is especially the case with the continuous vacuum dryer working with steam. It is important for the wood industry to make better use of its own wood energy. In Europe, most primary wood industries are at present almost self sufficient in energy supply. With heat engines and the possibility of supplying the power grid with a surplus of electrical current, the use of residual wood for the production of heat and electricity becomes more profitable. And, as a consequence, the environmental compatibility of wood-based industries improves.

The author is Professor, Swiss Inst. of Wood Technology, Biel, Switzerland.

Introduction

As an organic and hygroscopic material, wood is attacked by microorganisms and insects, and moisture pickup and loss can cause considerable variation in dimensions. The organic nature of wood and its hygroscopicity have a significant impact on its performance in structures. The goal of wood preservation is to protect wood against the attack of insects and microorganisms. In many applications wood must also be dried and exposure to moisture controlled to avoid damage.

Modern architecture works with forms that submit building materials to extreme environments and stresses. Due to the fact that architects are not willing to accept strong restrictions in relation to material properties, the architectural design process demands very careful choices of material design and use. The choice of an appropriate wood species is very important. The wood has to be protected against weather during transport and assembly to avoid unnecessary moisture absorption. Only when all these measures are unable to prevent the attack of microorganism or insects should chemical treatment of wood be considered.

Advances in chemical wood preservation

Methods for the treatment of wood against the attack of microorganisms and insects have been known and practiced for centuries. Concerns about the impact of these treatments on the environment, however, are relatively recent. The consequences of an increasing public consciousness regarding environmental impacts of chemicals are extremely important. For example, a petition recently was presented in the Belgian Parliament to levy an "eco-tax" on products affecting the environment. If this petition is accepted, the price of CCA-treated wood in Belgium would increase from 500 to 700 DM/m^3 up to about 3,000 DM/m^3 (5). According to new legislation in Switzerland, wood treated with chromium must be burned in special incineration installations at a cost of about 30 to 60 Fr/m^3. Chemically treated wood is receiving closer scrutiny by wood users. The Swiss Railway Company reduced orders for wood railroad ties treated with creosote from 100,000 ties per year to 30,000 due to the controversy over the possible impact of creosote on the environment. In the context of compatibility with the environment, Leisser (14) developed the life-cycle schematic shown in Figure 1. It is interesting to note that the life-cycle circuit for untreated wood is closed while the circuit for preservative-treated wood is open and contains many question marks.

In addition to continued developments in wood preserving processes, a great deal of emphasis throughout the global research community is being placed upon studies related to answering questions of compatibility of preservatives with the environment. A few articles that are representative of this type of research are included in the Literature Cited (6,8,12,15). In this presentation we will be concerned only with water-soluble, salt-based preservatives since they are being used on an increasing basis and comprise the bulk of new wood preservatives coming into the market. Use of organic compounds in solvents, such as creosote, will decrease as pressures to impose VOC taxes increase. The result will be that preservative-treated wood ties will be replaced on an increasing basis by concrete or steel.

In Europe, preservative salts based on arsenic trioxide and chromium are often used, however some countries have prohibited the use of arsenic, and salts of chromium are being viewed negatively due to their high heavy metal content. For this reason the chemical industry is working hard to develop preservative salts that do not contain these elements. These new products must have a broad spectrum effect against microorganism and insects, high leach resistance, and good toxicological and ecotoxicological properties. In addition, they should be able to be applied in conventional installations and they should not be corrosive to connectors in use. Preservative retentions to achieve the same level of protection that is afforded by arsenic- and chromium-based salts should be the same or less, and leach resistance is extremely important. Leisser (14) presents a summary of the leaching and loss of gases of different wood preservative salts (Table 1).

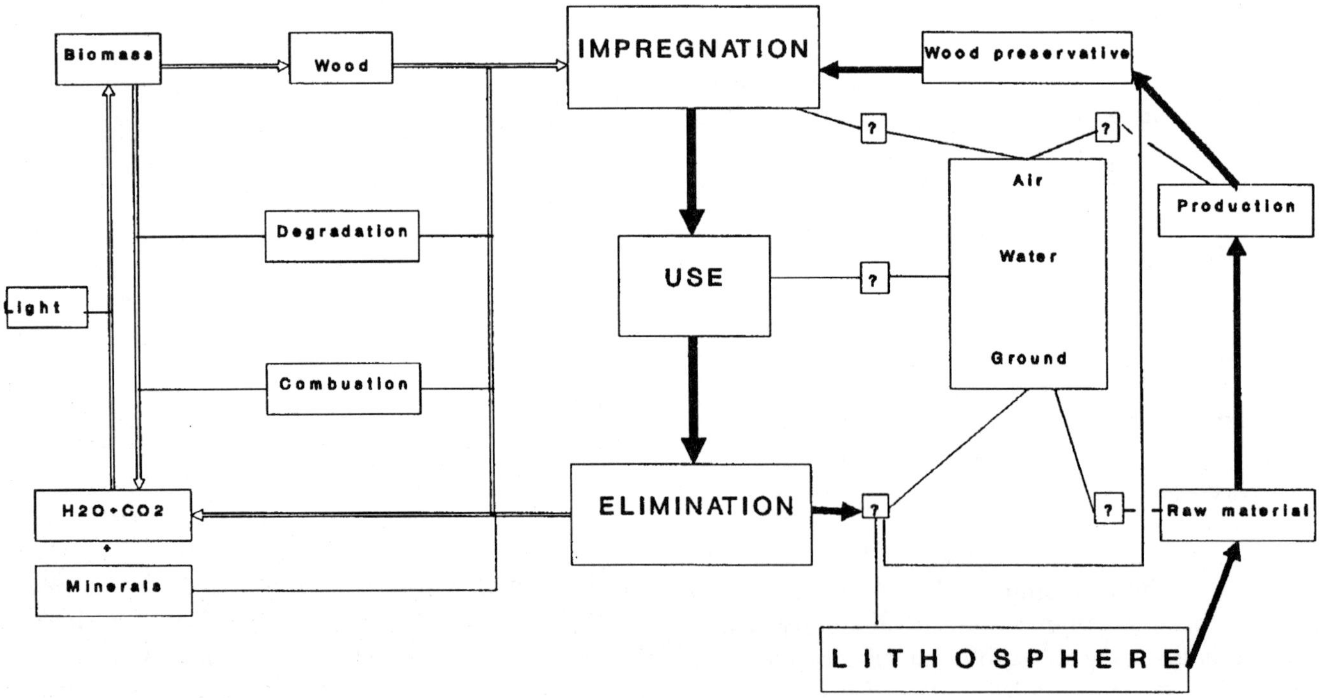

Figure 1. — Simplified schematic of the untreated wood life-cycle circuit (left) and preservative life-cycle circuit (right) (14).

Since the salts in Table 1 were tested under different conditions, no direct comparisons between the preservatives can be made. However, Leisser reports a case in which the leaching of arsenic in wood used in a fodder bin amounted to 75 percent after 2 years in use.

Initial results of tests on a promising new preservative, copper polymeric-betain, based on boron and copper compounds, indicates that it may be effective in the same or even smaller quantities than CCA or CCB salts (Fig. 2). An intensive discussion about the penetration depth of some of the new products under consideration and their compatibility with normal production processes is now taking place. Due to the rapid fixation of some of these salt formulations, the diffusion of the salt and the penetration depth may be reduced (14). The question of leach resistance also has not been definitively clarified up to now and that is why authorities in certain countries do not currently accept these new products. Some users assert that several of the new substances could be corrosive to timber connectors.

The challenges facing the chemical industry and the wood industry are great. Pressures from environmentalists will increase and the chemical industry and the wood industry will have to present a definitive solution for the problems of environmental compatibility of their products.

Advances in the drying of wood

The properties of materials are to an increasing extent being considered from a life-cycle approach, including the energy necessary for production, use, and reuse of the material and possible pollution of air and soil. Wood has very good properties in this respect. Approximately 80 percent of the energy used in the production of most wood products consists of process heat. A large part of this proportion is consumed during kiln-drying.

Wood drying is becoming increasingly important in Europe due to the fact that the demand for kiln-dried wood in residential building is increasing. Air drying and conventional kiln drying as it is practiced in much of Europe does not allow economical drying of the large dimensions of softwood and hardwood timber that are commonly used in European home construction. As a result, dry kiln manufacturers have placed a great deal of emphasis in recent years on the development of vacuum dryers. At the Ligna Fair '93 in Hanover, Germany, most dryer manufacturers exhibited vacuum drying installations. Because of the significance of these developments in the European market, the remainder of this presentation will concentrate on vacuum drying technologies.

Vacuum drying technology

In 1990, a talk at the University of Rosenheim was entitled "Unconventional drying of structural wood" (1). Three years later this "unconventional" method which involves vacuum drying is considered standard practice. The primary basis for the application of a vacuum during the drying process lies in the fact that the boiling point of water (and therefore the evaporation point) is reduced in a vacuum, and the use of a vacuum increases the rate of moisture diffusion by amplifying the gradient. Reducing the partial air pressure also has the effect

Table 1. — Leaching and gas loss data for various water-soluble, salt-based wood preservatives (14).

Wood preservative	Leaching	Gas loss
	(%)	
CF salt	F 25 Cr 6-10	F 30-35
CFB salt	F 30-50 B 60 Cr 10	F 35
CCB salt	B 90 Cu 7 Cr 7	
CCF salt	F 60 Cu 5 Cr 5	F 30-35
CCA salt	Cu 5 As 20 Cr 5	
C-HDO	Cu 15-20 B 60-90	

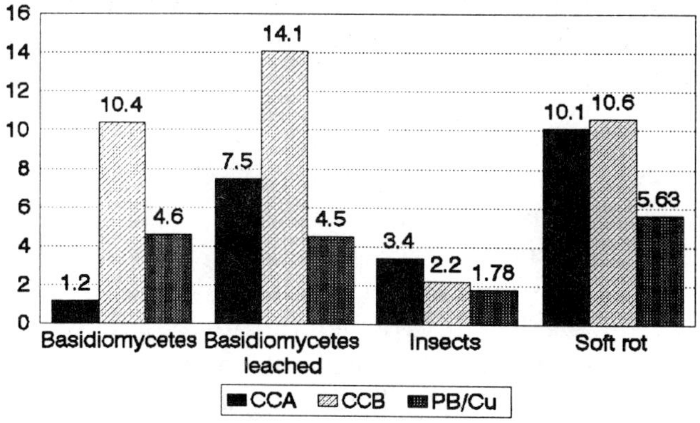

Figure 2. — Threshold limit values of some salts for pressure impregnation against insects and fungi (7).

Table 2. — Qualitative comparison of three vacuum drying technologies.

Technology parameters	Continuous heat plates	Continuous convention	Discontinuous
Heating	Contact with heating plates	Convection in remaining air or steam	Convection in normal atmosphere
Drying of the air	Condensators	Condensators	Condensators
Piling time	Long	Short	Short
Need for electrical energy for fans	Small	Large	Medium
Total need of heat	Small	Small	Large
Dryer capacity	Small	Large	Medium

Table 3. — Main components of vacuum dryers (18).

Compound	Function	Requirements, remarks
Chamber	Vacuum-resistant drying box	Material must be resistant to corrosion or coated
Dehumidification system	Drying of the air	Plates must be resistant to corrosion
Humidification	Humidification of drying air	Fast production of steam
Heating system (plates, heating elements)	Heating of wood and air	Heating the plates (aluminum) or the heating elements
Fans	Circulation of heating agent	High air velocity
Vacuum pumps	Evacuation of the chamber, retention of air pressure	Adapted to air and condensing vapor, resistant to corrosion
Switchboard and processor	Control of the drying parameters	Hand-controlled, semiautomatic, automatic, or computer-controlled processor
Sensors	Control of the drying conditions	Temperature, pressure, velocity of air, moisture control of wood

of decreasing discoloration of the wood caused by oxidation and blue-stain (9). Table 2 lists some qualitative comparisons of three types of vacuum drying processes (from (10) and (17)). The comparisons in Table 2 provide only some general indications, and specific quantitative results are not yet available. However, it is obvious that the continuous vacuum process with convective heating is the technology of the future due to the fact that large capacities are possible. In the following discussion, only this technology will be considered.

The steps in the vacuum drying process with convective heating are (1):

1. Evacuation empties the boiler of almost all the air.
2. Heat-up phase: heating of wood to the desired drying temperature.
3. Drying: pumping steam out of the boiler and reducing the equilibrium moisture content (MC). This results in differences of absolute pressure between the inside of the timber and the chamber atmosphere.
4. Conditioning as the final step, to equalize MC differences.

The drying conditions (pressure and temperature) are adapted to the desired final MC and to the species and dimension of the timber being dried. Normally the air pressure in the chamber lies between 50 and 200 mbar. This corresponds to a dew-point temperature of 40° to 65°C.

Description of dryer components

The main components of the vacuum dryer with convective heating are shown in Table 3 and a schematic of the dryer is shown in Figure 3 (18).

Drying chambers are generally cylindrical. As a consequence, holding capacity depends on the dimensions of the pile, the diameter and length of the chamber, and the dimensions of the apparatus in the chamber. The desired capacity of the installation will depend on the species and dimensions of material to be dried. Presently it can be stated that the capacity of chambers for vacuum drying with convection heating is generally only 15 to 20 percent of the capacity of chambers for convection drying, however, vacuum dryers with a capacity of 200 m^3 are coming on the market. This capacity, when considered in relation to material throughput, corresponds to a conventional convection drying chamber of about 1200 to 1500 m^3 capacity. Up to this point, vacuum drying chambers have been completely assembled in the manufacturer's workshop, and then delivered in one piece. The reason for this is that quality welding can be accomplished more easily in the workshop than on site. Larger chambers must be welded on site or fabricated from an expensive concrete-steel construction, which has the drawback of problems related to corrosion. In Europe, rectangular chambers made from aluminum are now being provided which provide a high ratio of chamber volume to holding capacity.

Energy consumption

On the basis of many investigations on the energy consumption of vacuum dryers, it can be stated that the specific heat consumption is smaller than for conventional convection dryers. The specific power consumption, however, is considered to be higher. Results of energy analyses by different researchers are displayed in Table 4. These results lead to the following conclusions:

- The differences in results of analyses between researchers are quite large, due to the fact that the drying conditions were very different.
- In a direct comparison, the heat consumption is lower than in conventional convection dryers but power consumption is higher.
- Drying times are shorter (80%) than in conventional convection dryers.

It must be considered that the initial MC of the wood (u_a) and the final MC (u_e) are very important. The lower the initial MC is, the higher the specific heat consumption will be, due to the fact that the proportion of bound water versus free water that must be evaporated increases.

Drying quality

There is not much information about drying quality in literature. However, in general it can be stated that vacuum drying processes can be applied to a wide range of species and dimensions with the result of decreased drying times without loss in quality. Species that are prone to discoloration (especially hardwoods) should be dried at low temperatures (2°C), especially when the MC is above the fiber saturation point. According to Brunner (11), large

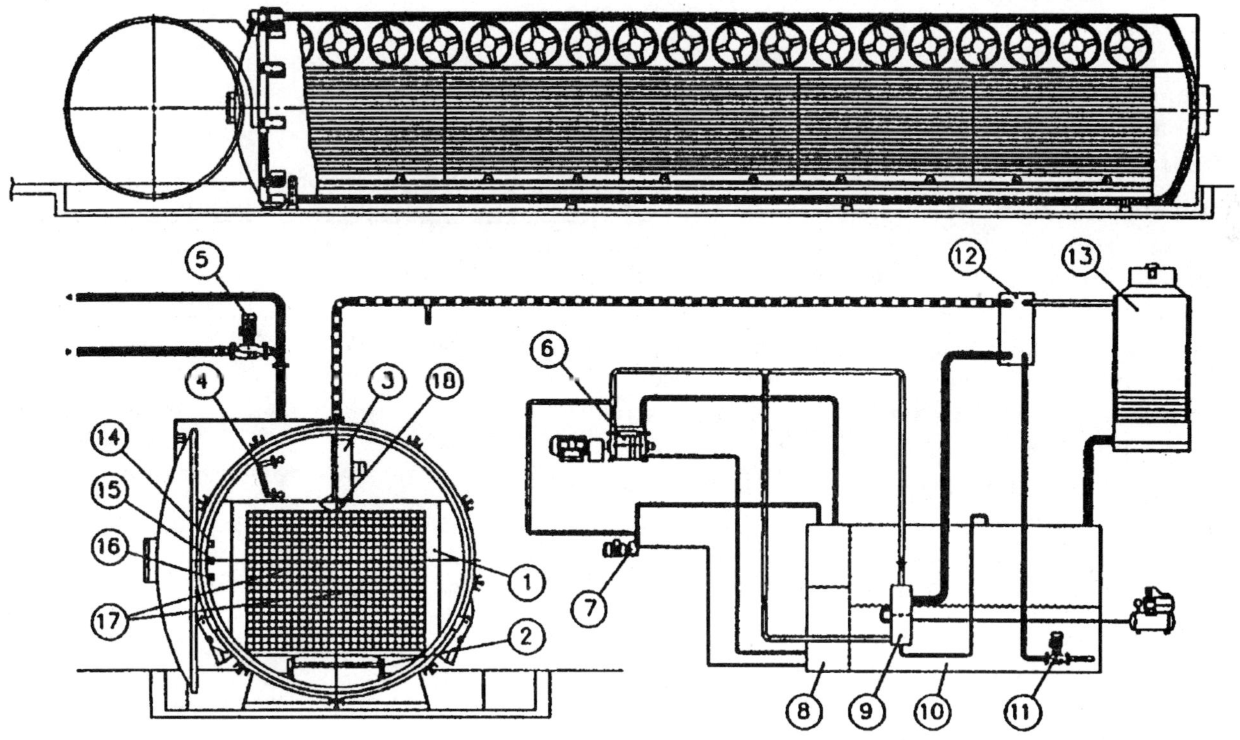

Figure 3. — Schematic drawing of a vacuum dryer. 1) vacuum chamber; 2) stacking carriage; 3) fans; 4) heating elements; 5) pump; 6) vacuum pump for evacuation of air; 7) vacuum pump for drying; 8) cooler; 9) condensate measurer; 10) steam trap; 11) cooler pump; 12) cooling plate; 13) cooling tower; 14-16) sensors for measurement of pressure; temperature; and chamber climate; 17) sensor for MC; and 18) air bag. (From ref. 18).

Table 4. — Results of energy analyses by different researchers for vacuum and conventional kiln-drying processes: u_a = initial MC, u_e = final MC.

Description of material type and conditions of drying	Drying process	Heat consumption	Power consumption	Reference
		(kJ/kg water)	(kWh/kg water)	
Oak, 50 mm u_a 70-80, u_e 8 Capacity 1300 (m^3/a) t_{Tr} (h)	Vacuum (16) Conventional (90)	Vacuum (504) Conventional (840)	Vacuum (0.24) Conventional (0.20)	(10)
Fir, 23-78 mm Different (11 tests)	Vacuum Conventional	Vacuum (2400) Conventional (3600)	Vacuum (0.51) Conventional (0.07-0.52)	(19)
Hardwood u_a 50%, u_e 8%	Vacuum	800	0.38	(16)
Oak u_a 35%, u_e 8%	Vacuum t_{Tr} 195h	2300	0.52	(2)
Fir u_a 80%, u_e 16%	Vacuum t_{Tr} 120h	2460	0.17	(2)
Beech u_a 70%, u_e 8%	Vacuum t_{Tr} 144h	2770	0.20	(2)
Oak, 27 mm u_a 21%, u_e 9.2% Volume 30 m^3	Vacuum discontinuous	2.74 (kWh/kg)		(3)
Oak, 27 mm u_a 20.8%, u_e 8.6% Volume 65 m^3	Vacuum discontinuous (tandem)	2.07 (kWh/kg)		(3)
Oak, 54 mm u_a 26.9%, u_e 15.8% Volume 67 m^3	Vacuum discontinuous (tandem)	2.16 (kWh/kg)		(3)

variations in the initial MC can not be readily equalized by applying vacuum drying technology, especially when the final target MC is not very low.

Drying costs

Drying costs for various species, dimensions, and initial and final MCs as reported by a number of researchers are shown in Table 5. Since the conditions were not similar in each case, comparisons between the results can not be made. However, various users of vacuum drying processes have reported that costs are not substantially higher than in conventional dryers. When considering costs, it is very important to take into account that vacuum drying technology makes it possible to dry large dimension structural lumber successfully, which is not possible in convection chambers.

Conclusions

The wood industry is interested by kiln drying processes applicable to many species and different dimensions. Continuous vacuum drying technology has been intensively developed in the past few years, and is presently ready for industrial application. This technology allows shorter drying times and makes relatively rapid drying of large dimensions possible.

Table 6. — Costs for kiln drying as reported by different researchers.

Description of material type and conditions of drying	Costs	Drying process	Reference
Lumber, 150 mm u_a 60%, u_e 20%	DM 25.–/m^3	Vacuum	(1)
Hardwood u_a 50%, u_e 8%	DM 37.–/m^3	Vacuum	(17)
Oak, 27 mm u_a 30%, u_e 10%	DM 87,60/m^3	Vacuum	(3)
Oak, 54 mm u_a 40%, u_e 14%	DM 142,02/m^3	Vacuum	(3)
Beech, 50 mm	Fr 131,72/m^3	Vacuum	(13)
Oak, 50 mm u_a 75%, u_e 8%	DM 126,35/m^3	Vacuum	(10)
Oak, 50 mm u_a 75%, u_e 8%	DM 167,20/m^3	Conventional	(10)
Structural lumber	DM 10.–/1 cm Dicke	Vacuum	(4)

Literature Cited

1. Anonymous. 1990. Beschleunigte Schnittholztrocknung im Unterdruckverfahren. Holz-Zentralblatt (114):1723-1730.
2. ________. Firmenprospekt der Firma IWT, Danmark.
3. ________. 1993. Die Vakuumtrocknung von Laubholz. Holz-Zentralblatt (28):434-435.
4. ________. 1992. Informationstag zum Thema Bauholztrocknung. Holz-Zentralbaltt (34):543.
5. ________. 1993. Öko-Steuern auf behandeltes Holz? Holz-Zentralblatt (73):1178.
6. ________. 1993. Gesundheit- und Umweltaspekte beim Holzschutz. Holz-Zentralblatt (3/4):33.
7. Barth, V., and H. Härtner. 1993. A new type of biocide suitable for use in different fields of wood preservation. IRG Doc. No. IRG/WP 93-0014. 24th Ann. Meeting Inter. Res. Group on Wood Pres.
8. Bona, T. 1991. Chemischer Holzschutz auf Speilplätzen und Freizeitanlagen. Holz-Zentralblatt (101):1548.
9. Brunner, R. 1991. Die Schnittholztrocknung im Vakuum-eine echte Alternative bietet sich an, Teil 1 Grundlagen der Vakuumtrocknung, Stand der bisherigen Verfahrenstechnik. Holz-Zentralblatt (54):909-918.
10. ________. 1991. Die Schnittholztrocknung im Vakuum-eine echte Alternative bietet sich an, Teil 2 Technische Beschreibung des neuen High-Vac-Trockners. Holz-Zentralblatt (55/56):936-937.
11. ________. 1993. Holztrocknung im Vakuum-oder bleibt alles konventionell? Teil 2 Die Streuung der Holzfeuchtigkeit bei Trocknungsende verlangt besondere Aufmerksamkeit. Holz-Zentralblatt (60/61):1018.
12. Deppe, H.J. 1992. Die Umweltschutzgesetzgebung in ihren Auswirkungen auf das Recycling von Rest-und Altholz. Holz-Zentralblatt (73/74):1201,1202,1206, 1208.
13. Gautschi. H.P. 1990. Trocknungsseminar Schweizerische Ingenieur-und Technikerschule für die Holzwirtschaft.
14. Leisser, B. 1992. Holzschutzmittel im Einsatz. Bauverlag GmbH Wiesbaden und Berlin.
15. Marutzy, R. 1991. Kesseldruckimprägnierte Hölzer für Spielplatzgeräte: Auswaschungs-und Entsorgungsprobleme. Holz-Zentralblatt (116):1806.
16. Niemz, P. 1993. Physik des Holzes und der Holzwerkstoffe. DRW-Verlag Stuttgart.
17. Opel, A. 1991. Vakuumtrocknung als Lösung für das Problem der Trocknung von Bauholz mit grossen Querschnitten. Holz-Zentralblatt (109):1692,1696.
18. Ressel, J.B. 1993. Zum aktuellen Stand der Vakuumtrocknung von Schnittholz, Teil 1 Grundsätzliches zur Verfahrenstechnik und zum Bau der Trocknungsanlagen. Holz-Zentralblatt (78):1274,1276,1278.
19. ________. 1993. Zum aktuellen Stand der Vakuumtrocknung von Schnittholz, Teil 2 Beschreibung der auf der Messe vorgetslletn Anlagen. Holz-Zentralblatt (79):1313, 1314,1316.
20. Trübswetter, T. and M. Weber. 1992. Die Heissdampftrocknung im Vakuum-Untersuchungen zum Energieverbrauch. Holz-Zentralblatt (34):541-542.

Radio frequency/vacuum drying of thick lumber

Robert Zwick and Stavros Avramidis

Abstract

The Council of Forest Industries (COFI) radio frequency/vacuum (RF/V) kiln project is a $2.9 million investigation into the commercial feasibility of RF/V drying for softwoods. The project participants are a collection of stakeholders who would benefit from the technology, and includes the British Columbia lumber industry (the end user); B.C. Hydro (local electric utility); the Canadian Electrical Association (promotion of electro-technologies); Salton Fabrication Ltd. (local supplier of conventional kilns); and the federal government (promotion of energy conservation technology, industrial development). A previous COFI/MacMillan Bloedel Research investigation showed that drying thick sections of softwood lumber could be viable in a RF/V kiln. A description is given on how the present COFI project is structured, as well as the activities involved. Advantages of using RF technology versus other heating methods are reviewed. Some of the results of the present investigation are shown, including the kiln drying of timbers. The project will focus on a competitive analysis of RF/V drying costs as compared to conventional drying methods.

Background

In early 1989, an investigation was initiated by MacMillan Bloedel Research (MBR) and the Council of Forest Industries (COFI) to examine the effects of radio frequency/vacuum (RF/V) drying on west coast softwoods. Much of the reasoning which initiated the work was that there were no adequate methods to dry high-grade western hemlock and amabilis fir, the species group normally referred to as Hem-Fir. This problem has been compounded with serious pressures in overseas markets for the lumber to be either kiln-dried or pasteurized.

Hem-Fir lumber was trucked to New York and dried in a commercial hardwood RF/V kiln at a red oak sawmill. The softwood lumber grade and dimensions were what the Japanese market typically terms as "baby squares," which is at least a Standard and Better grade, 4-1/2 by 4-1/4 inches and in lengths of 10 to 13 feet. Subsequent to the first two kiln-drying runs of baby squares, one additional run of Hem-Fir was done in New York, as well as one run of Douglas-fir, and one run of western red cedar. The results of the investigation were reported in Avramidis and Zwick (1).

What the investigation demonstrated was that the RF/V drying of lumber has drying times which are independent of thickness and at the same time the wood exhibits unusually high drying quality. The fundamental mechanism of moisture movement was shown to be longitudinal, and out the ends of the lumber. This is because the longitudinal water vapor flow coefficient is in the order of 104 to 105 times greater than either in the radial or tangential direction, as discussed by Siau (3). The study also showed that drying times for hemlock could be achieved for any thickness in less than 2 days, and the energy efficiency of the process appeared to be much higher than previously thought. The study therefore concluded that RF/V kiln-drying could be commercially feasible for thick softwood lumber and that further investigations were warranted.

Initiation of the COFI RF/V kiln project

Based on only five RF/V kiln-drying runs in eastern United States, a project proposal to build a RF/V kiln in western Canada was put together (4). This

---

The authors are, respectively, Manager, Lumber Technical Services, Council of Forest Industries, Vancouver, B.C., Canada; and Associate Professor, Dept. of Wood Sci., Faculty of Forestry, Univ. of British Columbia, Vancouver, B.C., Canada.

proposal suggested a project structure where the lumber industry (through COFI) would be the lead organization to make arrangements for financing. After the proposal was put together, the solicitation of funds were undertaken in 1990.

During this time the lumber companies were experiencing a massive recession, even though they were convinced that the RF/V kiln project was a worthwhile investigation. There was a need to lessen the risk capital for the research project. The approach was to find other companies, organizations, or funding agencies who would have a stake in seeing the successful initiation of the project. Ultimately, a funding scheme was successfully put together, with an overall budget of $2.92 million (Fig. 1). The lumber companies who are participating in the project contribute on a percentage of their coastal lumber production. Their availability of time in the RF/V kiln is also based on this percentage (Fig. 2).

The project proposed that two RF/V kilns be built. One would be a laboratory RF/V kiln used to investigate the fundamental mechanism of drying with this technology and determine how fast softwood lumber could be dried without wood damage and degrade. The second kiln would be a full-scale commercial prototype, one that could be economically viable and practical in operation for a west coast sawmill. A full-scale kiln would allow for a representative investigation into the economics of its operation, and a proper market study of the RF/V dried product could be done in the marketplace. Following is a list of objectives that were set out for the project, including the commercial aspects of the research investigation:

The objective of this project is to design, construct, and operate a commerical RF/V kiln for drying softwood lumber.

Applied Research

1. RF generator circuit design that can self-tune to the load dielectric characteristics.
2. Automate the control of the RF/V kilns to dry west coast softwood lumber.

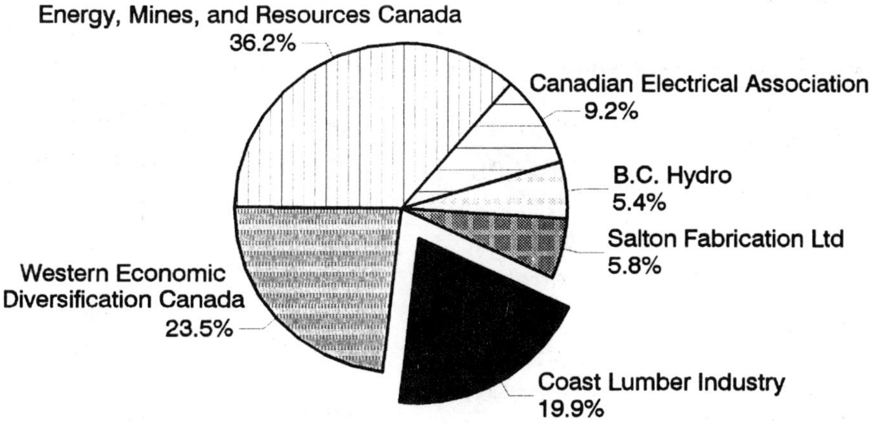

Figure 1. — Funding partners of the COFI RF/V kiln project (total project costs = $2.91 million).

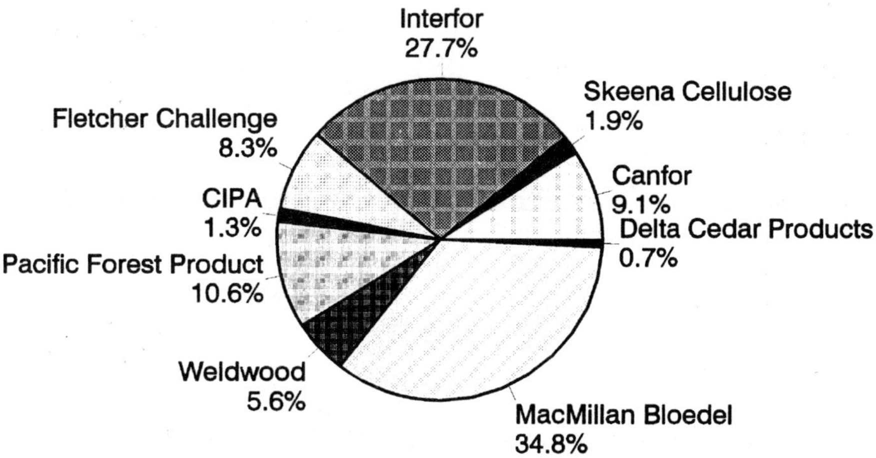

Figure 2. — Percentage of available RF/V kiln time for lumber industry participants, based on coastal wood production.

Development

3. Construction of a full-scale 350 kW unit with a 24-thousand board feet (MBF) chamber.
4. Design of a material/package handling system for this size of unit.
5. Detailed analysis of RF/V dried lumber quality, structural properties, degrade, and operating costs.
6. Development of RF/V kiln schedules.
7. Evaluate energy efficiency of the RF/V drying process.
8. Monitor and evaluate environmental effects (including RF radiation and kiln effluent).
9. Determine the RF/V schedules necessary to pasteurize a lumber package from pinewood nematodes.

Commercial

10. Market investigation of end-user acceptance to RF/V dried lumber.
11. Develop a manufacturing base in British Columbia to design, build, and service RF/V kilns.

With the successful procurement of funds and the acceptance of the objectives proposed for the project, a management structure was put together and the activities for the project were planned out. The outline in Figure 3 shows how parallel activities were put together to fast-track the project schedule. A critical path for the project was put together, and it prioritized the activities necessary for its most efficient execution. Throughout the project, the biggest priority was the procurement, construction, and commissioning of the RF generator for the commercial prototype RF/V kiln.

Responsibilities were divided among the distinct activities within the project: RF Generator, Site Activities, Vacuum Chamber/Material Handling, Main Kiln Power Supply, Laboratory Kiln (including its experimental program), and Main Kiln Process Control/Experimental Program. People involved with the project are shown in Table 1. They consist of professionals with a variety of backgrounds: industry, government, academics, and consultants.

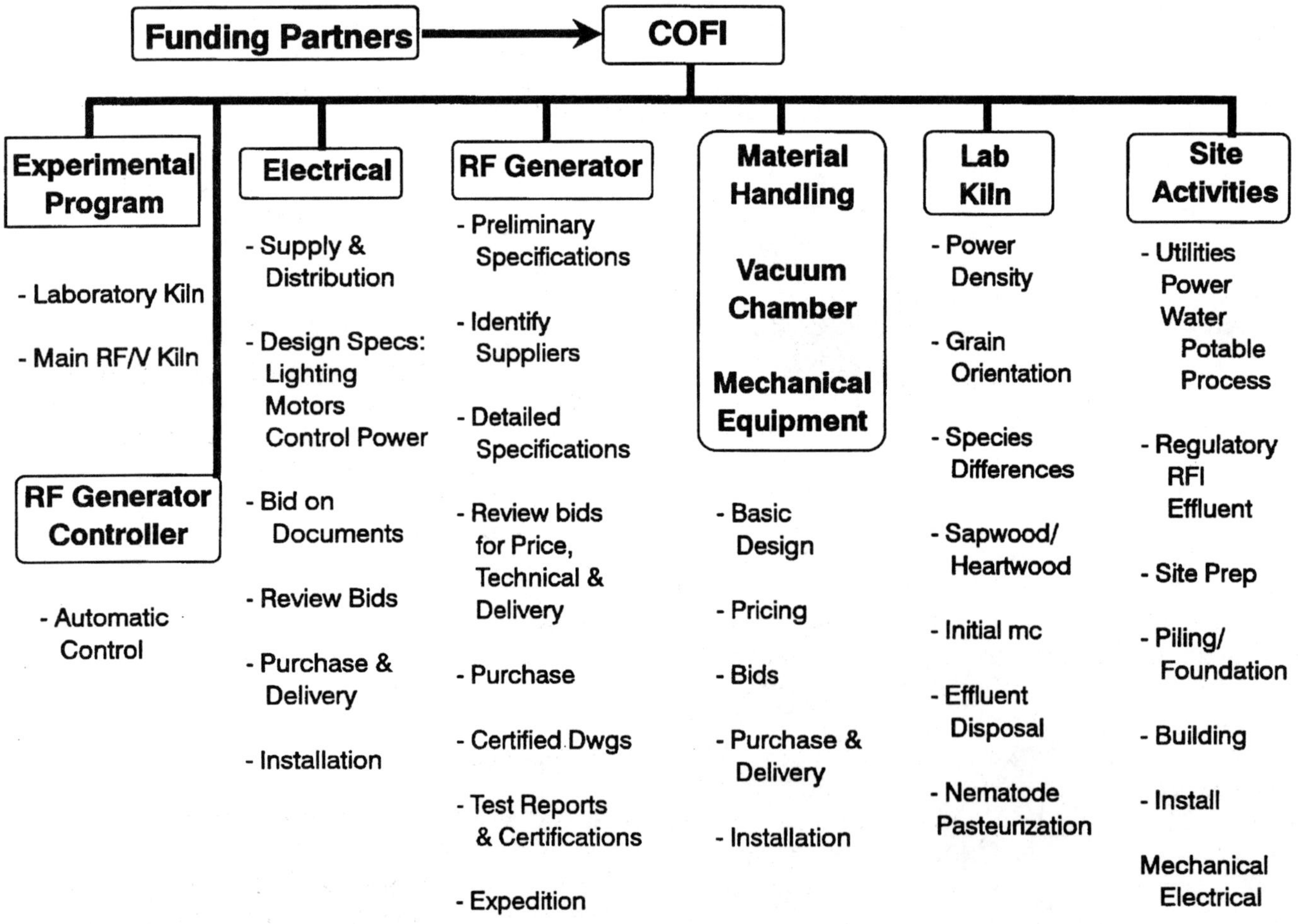

Figure 3. — Parallel activities were drawn up to fast-track the RF/V kiln project.

Laboratory RF/V kiln

The laboratory RF/V kiln, located at the Department of Wood Science, University of British Columbia, Vancouver, B.C. was built to develop dry kiln schedules for the main RF/V kiln. Having the ability to RF/V dry lumber in controlled laboratory conditions allows for the improved investigation into the fundamental drying physics. Because of time and cost restraints, the laboratory kiln was not built to be a scale model of the main RF/V kiln, nor was it built to evaluate energy efficiency of the drying process.

The kiln was constructed of a 9-foot-long, 30-inch mild steel pipe. It has a shell-wall thickness of slightly less than 1/2 inch, which can easily handle the 35 torr absolute pressure inside the chamber. The cylinder has conventional slip-on flanges for a 30-inch-diameter pipe on each end. Hinged cover plates were hung on each end, allowing for easy access into the chamber. The interior wall of the pipe was coated with an epoxy paint in order to resist corrosion.

In addition to the two doors on each end, flanged openings had to be made into the cylinder wall for the RF vacuum feedthroughs, condensate drain, steam vent into the condenser, and vacuum feedthroughs for the fiberoptic temperature probes and the teflon tube pressure probes. In relative terms, a chamber under vacuum is not operating at a very large pressure. There is no need for the chamber to be certified as a pressure vessel, and all the piping, flanges, etc. are off-the-shelf components.

The RF generator is a used Model 400 Mann-Russell unit that was built circa 1960. The machine was dusted off, new solid state rectifiers replaced the old vacuum tubes, and the unit was operational within 6 weeks. It operates at a frequency of 13.56 MHz, a standard frequency allocated for industrial use. It is capable of a 5 kV maximum electrode plate voltage at 2 amperes. The RF power output is 10 kW into a total load of 58 BF (0.14 m^3) which translates to a maximum power density of over 70 kW/m^3. This is a power density in the order of ten times what is presently thought to be necessary for drying softwood lumber.

The vacuum pump is a conventional SIHI Model 60 liquid ring pump with a 3-hp motor. It came preassembled with the condenser, water separator, etc. on a common frame. The unit was fitted with an air ejector that allows the pump to operate at pressures as low as 15 torr, which is considerably lower than what commercial RF/V kilns are expected to operate at. The condensate (which comes out

Table 1. — COFI RF/V kiln project teams.

Funding Partners (Monitoring Committee)
Carl Rathburn (Chairman) — Weldwood
Ken McKeen & Bryan Hannay — Council of Forest Industries
Rob Zwick (Project Manager) — Council of Forest Industries
Don Skinner & Norm Benoit — Energy, Mines, and Resources
Steve Oda, Rejean Girard, & Jacob Roiz — Canadian Electrical Association
Mike Carter — Salton Fabrication Ltd.
Eric Shum — Western Economic Diversification Canada
Brian Donnelly — B.C. Hydro
Sergio Barbon — MacMillan Bloedel
Jan Aune — MacMillan Bloedel Research
Suezone Chow & Jake Whitehead — Canfor
Hew Kidston — Interfor
Electrical
Bruce Neilson — Powertech
Rob Zwick — Council of Forest Industries
John English — H.A. Simons
Radio-Frequency Generator
Rob Zwick — Council of Forest Industries
Bruce Neilson — Powertech
Mark Churchland — MacMillan Bloedel Research
Martin Pearson — MacMillan Bloedel Research
Roger Poirer — TRIUMF
Roy McDonald — Salton Fabrication Ltd.
Brian Donnelly — B.C. Hydro
Material Handling/Vacuum Chamber/Mechanical
Roy McDonald — Salton Fabrication Ltd.
Rob Zwick — Council of Forest Industries
Al Moul — H.A. Simons
Bruce Neilson — Powertech
Lab Kiln
Stavros Avramidis — University of British Columbia
Bruce Neilson — Powertech
Rob Zwick — Council of Forest Industries
Jan Aune — MacMillan Bloedel Research
Frank Liu — University of British Columbia
Graham Mackay — Forintek
Site Activities
Rob Zwick — Council of Forest Industries
Al Moul — H.A. Simons
Jake Whitehead — Canfor
Bruce Neilson — Powertech
Experimental Program and Process Control
Rob Zwick — Council of Forest Industries
Stavros Avramidis — University of British Columbia
Graham Mackay — Forintek
Geoff Middleton — Council of Forest Industries
Bruce Neilson — Powertech

in the form of condensed steam, and as condensation from the kiln walls) is collected in one receiving tank in order to monitor moisture content (MC) removal rates via a pressure differential transducer. In addition, it allows for the later evaluation of the kiln effluent for environmental purposes.

The aluminum electrode supports the equivalent of a 10- by 10-inch by 7-foot timber dried in the kiln. This, in fact, has already been done for hemlock, western red cedar, and Douglas-fir, all with spectacular results. Fiberoptic probes are used to measure temperature since it is very difficult to run thermocouple wires inside the RF field. Internal wood temperature and pressure (which is measured with teflon tubes drilled into the lumber) are monitored throughout the drying schedule. All necessary data is collected on a PC acquisition system. The laboratory RF/V kiln is operated manually.

Laboratory RF/V kiln experimental program

The purpose of the laboratory RF/V kiln was to address three objectives:

1. Development of RF/V kiln schedules for the commercial species and sizes of interest in the main RF/V kiln.
2. Monitor and evaluate environmental effects of the kiln effluent and come up with strategies on how to ensure the main RF/V kiln water discharge will meet all B.C. Ministry of Environment waste regulations.
3. Determine the RF/V kiln schedules necessary to pasteurize a lumber package from pinewood nematodes.

The purpose of laboratory RF/V kiln was not to evaluate energy efficiency nor the economics of RF/V drying. The unit also was not built to develop an automated control system nor to be a scaled-down model that could simulate RF voltage fields on the main RF/V kiln. The reasons for not investigating these areas with the laboratory kiln was primarily the cost and tight project schedule.

Results from the laboratory RF/V kiln

The experimental component with the laboratory RF/V kiln at the University of British Columbia was initiated in September 1992. A year later, the following runs have been completed and the experimental data analyzed:

- 1 run with 10- by 10-inch western red cedar
- 15 runs with 4- by 4-inch western red cedar
- 1 run with 10- by 10-inch western hemlock
- 11 runs with 4- by 8-inch western hemlock
- 16 runs with 4- by 4-inch western hemlock
- 2 runs with 10- by 10-inch Douglas-fir
- 3 runs with 2- by 4-inch southern yellow pine for the pinewood nematode pasteurization study.

All runs were carried out with either constant or variable electrode plate voltage which directly affects the heat generated within the lumber load. The ambient pressure within the kiln was maintained between 35 and 70 torr.

From the above experiments, the following conclusions can be drawn:

1. Western red cedar 4 by 4 inches square in cross-section can be dried with a constant electrode voltage (0.8 kV) in about 24 hours with very little degrade and very flat horizontal MC distribution.
2. Western hemlock 4 by 4 inches square in cross section can be dried with variable electrode plate voltage in about 32 hours with no checking, lumber shape distortion, or discoloration (Fig. 4). Flat horizontal MC distribution was also observed.
3. In both species, the total lumber degrade was a fraction of what is commonly observed in conventional drying. The total shrinkage was about 40 to 50 percent of that in convective drying.
4. Pinewood nematode pasteurization of wood can be achieved in less than an hour with no moisture loss from the lumber load.
5. The cross-sectional dimension of the lumber has no effect on the total drying time, contrary to what is observed in conventional drying. Drying times were greatly affected in large timbers (4 by 8 in., 10 by 10 in.) by the sapwood/heartwood orientation within the electromagnetic field.

Experimentation is continuing with schedule development for Douglas-fir, effluent evaluation and

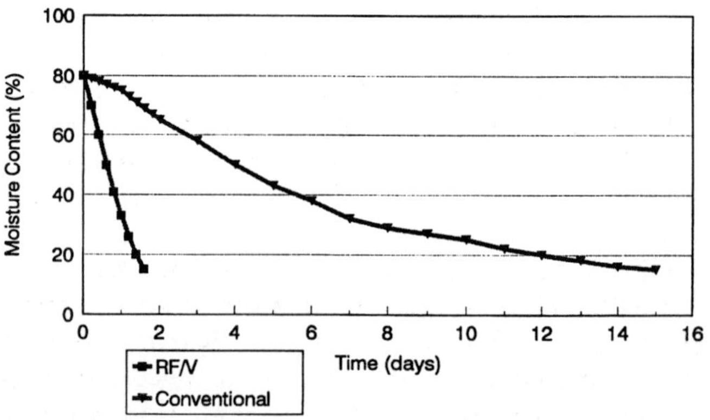

Figure 4. — Drying curves for RF/V versus conventional dry kilns.

neutralization, and mathematical modeling of the RF/V drying process.

Commercial prototype RF/V kiln

The main RF/V kiln was designed with the intention of commercial operation at a west coast softwood sawmill. Previous RF/V kiln designs were primarily for specialty hardwood sawmills. They have, in relative terms, a low capacity and are very cumbersome to load and unload. Therefore the softwood kiln had to be larger, and with a totally redesigned material handling system.

The COFI design resulted in having vertical electrodes rather than horizontal electrodes used in previous kiln designs. This allows for the rapid loading of the kiln carts with existing sawmill forklift equipment. The RF/V kiln can also have doorways on both ends of the kilns which allow for back-and-forth shuttling of the carts to improve loading and unloading. With very rapid drying times, it is essential for the RF/V kiln to be utilized to its maximum.

The prototype kiln dimensions are 45 feet long by 13 feet wide by 11 feet high. It was designed to withstand a total vacuum pressure of –14.7 psi gauge. The kiln chamber is capable of holding approximately 24 MBF or 56.6 m^3 of wood, making it the largest RF/V kiln in the world. The packages do not require any stickering, use conventional dunnage, and are loaded just as they would be onto a truck. The kiln chamber and material handling system was supplied by Salton Fabrication Ltd., a local supplier of conventional kilns. If the technology proves to be commercially feasible, Salton will be well positioned as a supplier of RF/V kilns for the lumber industry.

The RF generator is a triode oscillator with a circuit designed to operate at a frequency which "floats" between 8 MHz and 5 MHz, depending upon the MC of the wood. The RF voltage varies between 7 kV when the wood is wet, to approximately 40 kV with a dry load. The generator is capable of providing a 350 kW RF output and operates with a conversion efficiency of approximately 65 percent. This means the AC power input is approximately 540 kW.

The control system on the RF generator is designed to produce a requested RF power output at any settable value. The RF/V kiln was originally sized to supply a power density of approximately 6.4 kW m^3. Experience developed from the laboratory program has shown that this power density could be somewhat larger, which directly improves drying times.

The vacuum system is a conventional helical screw pump, and is a novel application for this kiln of pump. The steam is brought through a heat exchanger and the water is condensed out. The air stream then is sent through a cyclone or knockout tank, in order to remove any remaining slugs of water. All water is collected into a holding tank and then discharged into the storm sewer. The main advantage of this type of vacuum system is that no water is required (such as what is found in the liquid ring vacuum pump). Another advantage is its energy efficiency.

Water that is condensed from the kiln chamber walls is collected at one end of the kiln and drained into a collection tank. It is also discharged directly into the storm sewer. The amount of water drained from this tank depends on the ambient air temperature, but typically varies between 20 to 50 percent of the kiln discharge. A further advantage of this closed system is the ability to keep track of the total amount of water removed.

Some results from the commercial prototype RF/V kiln

The objectives for the experimental program with the main RF/V kiln is focused primarily on investigating the commercial feasibility of the drying process. Therefore, the kiln has instrumentation to determine its electric energy consumption and its water consumption. In addition, experimental data is developed on each drying for initial MC, final MC, shrinkage, and lumber degrade.

A further objective of the project is to develop an automated control system to operate the kiln. This involves research into how the MC of the kiln charge can be determined with the operating characteristics of the RF generator. This procedure has been successfully developed in the laboratory kiln, and now is being applied to the main RF/V kiln

Drying times in the RF/V kiln are typically in the range of 20 to 40 hours for softwood lumber, regardless of its thickness. It is however, highly and directly dependent upon initial MC. Therefore species such as Douglas-fir (low initial MC) will take less than 24 hours, whereas sapwood hemlock will take much longer. Drying times in RF/V kilns are directly related to the power density which is the amount of RF power going into a given volume of wood. The maximum power density that can be used depends on the species of wood being dried and its longitudinal vapor flow coefficient.

As mentioned previously, total average energy efficiency in RF/V drying is approximately 65 percent, which is very good. Conventional kilns drying 2-inch lumber typically have an energy efficiency of about 50 percent and this is drastically reduced if drying thicknesses are increased. Electricity costs are about $15 to $30/MBF (electricity rates in British Columbia are only $0.028/kWh), depending upon the initial MC of the wood.

Lumber quality is very good with the RF/V drying process. Typically for Hem-Fir, the lumber degrade has been in the range of 5 to 10 percent of the wood product value being dried. Approximately 10 to 15 percent of the pieces usually exhibit some grade change. These degrade values, however, do not include high MC or "wet" pieces, which can also be interpreted as being "off grade." In drying Hem-Fir, wet pieces have been a problem. The major causes are thought to be the large variability with the initial MC, and the low permeability of some hemlock pieces. The low permeability problem has been attributed to possibly being wood which has been infected by bacteria (2).

Remaining activities on the COFI RF/V kiln-drying project

With the successful proof of concept on this project, the remaining activity is to thoroughly investigate the commercial feasibility of the RF/V drying process. In addition, software is being developed in order to completely automate the kiln operation. In this way, future RF/V kilns will have the software already available for operation, backed by the research experience from the project.

Lumber dried in the RF/V kiln has been shipped to overseas customers in Japan and Germany. Customer feedback will provide some idea what price the RF/V dried wood product would fetch in the marketplace. This, too, will be part of the commercial feasibility study into softwood RF/V drying.

Another issue is the need for a local source of RF equipment, service, and expertise to supply the anticipated demand. The infrastructure for a local RF manufacturing firm has been put into place, should there become a high enough demand for RF/V kilns. It became painfully apparent during the project that there exists a lack of a competitive supply for high-power RF generators, including an unacceptably slow delivery for the equipment and the poor availability of service technicians.

Conclusions

There appears to be a place for RF/V kilns in the commercial drying of softwood lumber. The COFI project is focusing on commercial softwoods from British Columbia which are large-sectioned, high grade timbers and typically are for nonstructural end-uses. Results from this research investigation will allow lumber companies who are funding the project to make a rational decision on the extent they should invest into RF/V dry kilns at their present operations.

The poor availability of a competitive supply for RF equipment will ultimately translate to an excellent opportunity for some RF equipment manufacturers who are willing to rise to the challenge of servicing the customers' needs. Commercial investigations will likely result in the lumber industry investing a large amount of capital for RF/V kilns. Should this be the case, RF/V drying technology might allow the western Canadian lumber industry to remain competitive for the years ahead.

The unique funding and management infrastructure of the COFI RF/V kiln project has been a positive experience for the western Canadian lumber industry. This was a high-risk, capital-intensive research project with a very aggressive project schedule. It was therefore necessary to bring in various stakeholder organizations that had an interest in the commercial development of RF/V drying technology. It allowed for the alleviation of risk amongst the lumber companies, as well as the reduction of capital necessary to fund the project.

The lumber industry would like to extend their gratitude to Salton Fabrication Ltd., B.C. Hydro, the Canadian Electrical Assoc., Western Economic Diversification Canada, and Energy, Mines, and Resources Canada for their participation in the RF/V kiln project.

Literature cited

1. Avramidis, S., and R.L. Zwick. 1992. Exploratory radio frequency/vacuum drying of three B.C. softwoods. Forest Prod. J. 42(7/8):17-24.
2. Kozlik, C.J. 1969. Problems of drying western hemlock heartwood to a uniform final moisture content. 21st Annual Meeting of the Western Dry Kiln Club, West Coast Dry Kiln Assoc.
3. Siau, J.F. 1984. Transport Processes in Wood. Springer-Verlag, Berlin, Germany. 245 pp.
4. Zwick, R.L., A. Redokop, B. Neilson, L. Oliveira, and S. Avramidis. 1991. Development of a commercial radio frequency/vacuum kiln for drying softwood lumber. Lumber Tech. Serv. Rept., Council of Forest Industries.

Future directions in the technology and processing of biocomposites

David V. Plackett

Abstract

Biocomposites are high-performance products derived from wood or other naturally occurring lignocellulosic materials in fiber, particle, or veneer form combined with adhesives. Examples include medium density fiberboard, oriented strandboard, and veneer-based products such as plywood and laminated veneer lumber. This paper describes opportunities to add value to these products as well as to a new generation of biocomposites including molded fiberboard, products incorporating recycled materials, and engineered wood products. Technologies based on chemical modification or in-line chemical treatment to improve product properties are considered to have significant future potential. Collaboration between organizations engaged in biocomposites research and industrial partners will be essential if these future developments in biocomposites are to be realized.

Introduction

Biocomposites are defined as products derived from wood or other naturally occurring lignocellulosic materials and consisting of fibers, particles, flakes, strands, or veneer bonded together by synthetic resins. In the broadest sense, biocomposites could also include paper and a wide variety of products derived from wood pulp. This paper focuses mainly on decorative or structural composite products including fiberboard, particleboard, oriented strandboard (OSB), and plywood. In addition, reference is made to a new generation of engineered wood products that are now becoming available. Technologies on the horizon offer much promise for new high-performance biocomposites in the 21st century; some applications are described. This paper also addresses products manufactured from nonwood materials since these products play a significant role in countries that are short on forest resources.

Background

In North America, the abundant supply of timber has historically meant that there has been little reason to consider alternative fiber sources for composite products. The most significant wood composite in North America has in the past been softwood plywood; however, this is now changing. In particular, Canadian waferboard/OSB production has grown rapidly in recent years and now represents more than 50 percent by volume of Canadian wood panel production. A recent analysis predicts that OSB consumption in the United States in 1993 will exceed 1992 consumption by 15 percent and the OSB share of the American structural panel market is expected to climb from 30 percent in 1992 to 33 percent in 1993 (12). Particleboard and medium density fiberboard (MDF) products have also shown growth in recent consumption trends in the United States (13).

Japan, one of the world's major wood-importing countries, has consumed approximately 14 to 15 million m^3 of plywood annually in recent years and most of this has been of southeast Asian origin. Indonesian plywood currently represents more than 30 percent of the plywood consumed in Japan. The decreasing availability of tropical plywood logs has been a major trade issue and Japanese companies have been turning to other countries (e.g., New Zealand) to access materials for composites such as

The author is Manager, Composites and Treated Wood Products., Forintek Canada Corp., Vancouver, B.C., Canada.

plywood. Japanese links to New Zealand include ownership of New Zealand-based processing companies and rights to cut radiata pine in certain plantation areas. Examples include Sumitomo Forestry, the owner of the Nelson Pine Industries MDF mill in the South Island of New Zealand, and Juken Nissho Ltd., a company that owns a laminated lumber plant and the Triboard plant in the North Island. Triboard, a three-layer structural product consisting of MDF surface layers with a strandboard core, is manufactured exclusively at the Juken Nissho plant in New Zealand (10). A recent paper indicated that wood panel manufacturers in New Zealand could use a significant portion of the radiata pine resource that is expected to come on stream in the next 10 to 15 years (3). MDF, plywood, and veneer from New Zealand could capture growing markets in the Asia Pacific region although prospects may be better for the latter two. New Zealand MDF retains a reputation as a high-quality product and the industry has a history of innovative developments that continues to this day. Notable firsts in the New Zealand MDF industry include the first plant in the southern hemisphere, the first plant to use a single refiner, and the first instance of adhesive injection into the fiber stream immediately after the refiner.

Where wood is not so widely available, such as in the People's Republic of China, various nonwood materials are now used to manufacture biocomposite panel products. Approximately 200 processing plants are in operation in China using a variety of raw materials to produce composite panel products, including hemp stalks, rice and wheat straw, bagasse, cotton stalks, rice hulls, and bamboo.

Future technologies

Current world demand for biocomposite products from wood or other lignocellulosics is based upon the use of such products in a wide variety of decorative and structural applications. In addition, engineered wood products such as laminated veneer lumber (LVL), Parallam®, and Timberstrand®, meet structural product needs in a world where supplies of tight-grain, high-strength wood raw materials are decreasing. In order to examine future possible processes and products, it is useful to consider the current position and development needs for various types of composite products.

Medium density fiberboard

MDF is a relative newcomer to the wood composites scene, since the first plant was built only 25 years ago (2). The product combines good machining properties with a smooth surface and good edge-holding characteristics. A light-colored board is considered desirable in some markets and this is achievable depending upon the wood species used to provide the raw material. Expansion in MDF production in the Asia/Oceania region has been particularly notable with roughly 100 percent growth in capacity in the last 5 years. The European industry has also grown rapidly with an estimated 15 percent increase in production from 1991 to 1992.

Growth in application areas for MDF will depend upon continuing product improvements. For example, in New Zealand, MDF products with average densities as low as 480 kg/m^3 are now manufactured with properties nearly the same as boards with standard densities approaching 725 kg/m^3 (21). Such low-density products offer the high-performance attributes of conventional MDF combined with excellent handling characteristics. In contrast, as shown by a recent development in Australia, high-density panels (>1,000 kg/m^3) with surface properties approaching those of high-pressure laminates may also offer much potential for the future. The typical MDF density profile in which a high-density surface overlays a lower density core is an important characteristic, and new products might potentially be opened up if higher surface densities could be achieved through physical or chemical means.

The quality of MDF is linked to the quality of the incoming raw material. In other words, the maintenance of standards for chip size and cleanliness, and limits to the quantity of sawdust, shavings, and other residues incorporated in MDF, will contribute to the quality of the final product. As a result, a number of mills worldwide are moving to adopt appropriate ISO standards defining the quality of the incoming raw material and the procedures for individual operations within the mill. Modifications to the refining process to allow more residue utilization could be of benefit if processing costs can be reduced while overall product quality can be maintained.

Perhaps the biggest single opportunity in the further development of fiber-based composites such as MDF lies in methods to improve dimensional stability and water resistance of the wood substrate. A successful development of this type could expand uses for MDF in joinery and interior construction applications. The difficulty with this issue lies in ensuring that an improved MDF product will perform satisfactorily in exterior applications in the long term. A successful product will need not only good dimensional stability but also appropriate insect and fungal resistance. Although wet-process fiberboard (e.g., Masonite) has been used as sheathing in the

United States for a number of years, the use of lower density MDF as a long-term exterior product has yet to occur. Medite of Europe has introduced a 10-year performance guarantee for its isocyanate-bonded Medite Exterior boards which are widely used as exterior paneling and doors, shop fronts, signs, and garden furniture. The company states that the use and application of the board must be in accordance with instructions specified with the guarantee.

Technologies do exist to produce an exterior-grade MDF through various processing options and these have been summarized (11). For example, Shen has developed a process in which fibers are steam-treated at high pressures and temperatures to generate a binderless board held together by wood sugars liberated during the process (20). This process is now being commercialized using agricultural waste materials in China. The resulting product is claimed to have good dimensional stability; however, the product is dark in color and this will be a disadvantage limiting the development of such products in other markets.

Chemical modification of fiber to covalently bond reactive chemicals and to provide enhanced product properties has considerable potential. This approach has been widely studied and acetylation in particular appears to be an attractive, relatively low-cost option (14,16,17). Sheen recently reported the development of a pilot plant operation in the United Kingdom, and work in collaboration with the Biocomposites Center at the University of Wales at Bangor has identified the optimum process conditions and resin type to give a true exterior-grade board with excellent machining properties and good appearance (18). The first industrial application of acetylation has taken place in Japan and pilot plant studies to produce exterior-grade commercial siding, doors, and windows, interior automobile parts, and furniture are now in progress in the United Kingdom and Sweden.

The uptake of acetylation technology for wood products will likely depend upon a number of factors, both technical and nontechnical. For example, a less capital-intensive process could be an advantage if acetylated products are to be produced on a large scale in the future. The idea of modifying the fiber to intermediate weight gains instead of to the maximum achievable weight gain (e.g., about 20% for softwoods) should receive more attention as it could provide some of the desired wood property improvements at lower cost. Acetylation is also potentially applicable to a variety of non-wood fibers; for example, improved strength, decay resistance, and water repellency has been demonstrated with jute (1). Rowell and Keany demonstrated that fiberboards made from acetylated bagasse fiber showed much slower swelling rates on exposure to high humidity when compared with untreated control boards (15). In the same study, internal bond strength was found to be higher in acetylated boards while modulus of rupture and modulus of elasticity were slightly lower.

Rowell has commented that the breakthrough in chemical modification and its application to composites will occur when the process can be applied during the hot-pressing stage of manufacturing. The use of gas-injection during mat processing can be foreseen as a possible development evolving from steam pressing, allowing chemical modification of the raw material during the pressing stage (5). Clearly, any such development would need to address environmental and safety issues associated with use of wood modifying agents in-line.

Oriented strandboard

As indicated earlier, OSB now has a very large portion of the North American structural panel market. The product is very cost-competitive with plywood for applications such as sheathing and roofing and has the major advantage that low-cost roundwood can be used as the raw material. OSB is made from either softwoods or low-density hardwoods. Aspen in particular is widely used because of its availability in the north-central United States and in parts of Canada.

In Canada, an alliance between Forintek Canada Corp. and the Alberta Research Council has been established with support from the Structural Board Association, the industry body representing OSB manufacturers. The purpose of the alliance is to coordinate research and effectively utilize available resources of equipment, personnel, and funds to maximize research efforts on engineered wood and composites in Canada. The initial focus of the alliance concerns OSB-related projects and issues associated with the resource, material handling properties, and their influence on product properties. The alliance is also addressing one of the main disadvantages of OSB when compared with plywood — its tendency to swell more when exposed to water. In construction this is mainly a concern on floor sheathing where rain-exposed swollen edges can create rework that is not required with plywood. A process developed at the eastern laboratory of Forintek in Ottawa offers some promise to overcome such problems (6). Acetylation can also potentially be applied to strand-based products, and studies of adhesive performance in these types of composites have been completed by Youngquist et al. (24).

The Structural Board Association has also identified two key activities requiring continued research: 1) improvements in fire resistance; and 2) the development of a full package of design information for OSB as a performance-rated product. Work on both these issues is ongoing within the present Forintek/Alberta Research Council alliance.

Veneer-based products

Plywood has been frequently categorized as a mature wood product in the later stages of its product life-cycle. However, as outlined by Lyngcoln, a dedication to quality and responsiveness to market demands has allowed plywood to continue to develop as a product in a number of countries (9). In addition, in Australia, Finland, and elsewhere, the development of LVL has dovetailed well into existing structural plywood operations.

Developments in the past have focused on cost reductions in veneer production and drying through equipment improvements at the fabrication end of the plant. Scope still remains for further cost reductions in veneer production. For example, green veneer recovery can potentially be improved by installing microprocessor-controlled, billet centering devices and installing formalized recovery control programs. Dryer costs can be reduced by gaining an understanding, through practical experimentation, of optimum drier temperature profiles relevant to specific wood species and driers. Improvement in green veneer quality and yield is also possible through pitch control technology linking the knife angle to peeler log position and shape (23). Such technology has considerable potential to add further value to existing plywood operations. Lyngcoln has also described improvements in plywood panel thickness control through the use of easily installed microprocessor-controlled panel thickness control systems after the hot press (9). These systems could allow plywood sanders to be used specifically to clean plywood surfaces rather than as calibrated thickness controllers.

Value-added veneer-based products can potentially be made by densification of surface veneers through physical compression or chemical impregnation techniques. Chemically densified surface veneers on plywood could open up an alternative to Kraft paper-overlaid boards for concrete formwork applications — advantages could perhaps include lower cost and more re-use of boards before disposal. Densification of spruce-pine-fir surface veneers through resin treatment has been shown to raise the modulus of elasticity above 2 million psi, a requirement for some LVL markets (22). The optimization of such resin treatments is presently being studied in a joint project between Forintek and the University of British Columbia Faculty of Forestry.

Steam-injection pressing for composite products has now been commercialized. The advantages of steam-injection pressing for production of particleboard with reduced thickness swelling have already been outlined (7,19). This approach has now been applied to the production of compressed LVL with steam pressing. Work by Kawai et al. has shown that it is possible to manufacture thick compressed LVL with short pressing cycles (8). Both modulus of rupture and modulus of elasticity of compressed LVL increase with increasing product density and these mechanical properties are higher than those of hardwood lumber with the same density range. The technology is potentially applicable to either phenolic- or isocyanate-bonded LVL. Steam-injection technology also opens up possibilities for chemical processing by introducing wood modifying agents through gas-injection processes.

New generation products

As well as the opportunities for value-added enhancements to existing fiber-, strand-, and veneer-based products described in the previous sections, there is also a range of new processes and products that are either under development or could be developed in the next decade. This section of the paper describes a few of these new developments.

Molded fiber products. — Technology is now available for combining wood fibers or fiber bundles with plastics or synthetic fibers to make a variety of new composite products. Non-woven web technology, for example, can provide processing flexibility and allows combinations of wood and synthetic fibers to be formed into panels or various molded shapes.

A molded wood fiber product is currently being manufactured commercially in Vancouver, B.C by Canadian Forest Products Ltd. The product, known as Woodmat, is produced from mechanically refined wood fibers and textile fibers in a mat lay-up. The textile fibers are interwoven amongst the wood fibers to provide mechanical integrity to the mat prior to molding. The product is suitable for automotive door panels and a variety of other interior uses. A similar product based on a wood fiber/textile fiber mixture without adhesive is also being commercialized under the name Take-Home-Turf as a carrier for grass or flower seeds in garden areas (i.e., synthetic turf or flower beds). Further developments in molded fiber products can be expected as technology to improve the "chemical compatibility" of wood and nonwood fibers improves.

Utilization of recycled materials in biocomposites. — A major research program aimed at development of high-value composite products derived from recycled materials is now in progress at the USDA Forest Service, Forest Products Laboratory (FPL) in Madison, Wis. The program includes use of wastepaper, recycled wood or other biomass, and recycled plastics as well as other materials to form either dry- or wet-formed fiber composites. Spaceboard, a development of the FPL, is a molded pulp structural product manufactured by press-drying recycled fiber against compressive rubber molds. The result is a product thin enough for strong, lightweight corrugated containers or thick enough for wall, floor, or furniture applications. The program at the FPL is coordinated with other U.S. government agencies including the Department of Energy and the Environmental Protection Agency.

Opportunities exist for the potential use of treated wood waste materials in the manufacture of new composite wood products; however, the major challenge of handling treated wood waste safely will have to be addressed before such opportunities can be exploited.

Engineered wood products. — Indications are that markets for engineered wood products (e.g., LVL, Parallam, Timberstrand) in North America will grow rapidly in the 1990s. These products can utilize low grade veneers or residues to generate very high value structural materials with uniform, predictable properties. The engineered wood products fill the void created by restrictions on old growth logging and the resulting lack of larger dimension wood products as a result of the transition to smaller logs. In addition to LVL, Parallam, and Timberstrand, engineered wood products also include such materials as lumber/OSB I-beams, LVL/plywood I-beams, and box I-beams. A recent study predicted a doubling of sales of engineered products between 1990 and 2000 with major growth in all categories (4). Another engineered product is an Australian product, utilizing crushed wood scrim in combination with an exterior adhesive in a continuous pressing process to form a structural beam. One Scrimber manufacturing plant has been built in Australia but is not in commercial operation at the time of this writing. Further value-added developments in engineered wood products in the future could include products with enhanced dimensional stability and durability for exterior use.

Summary

Biocomposites will play an important role in filling product needs in a future world of diminishing natural resources and vastly expanded human populations. New technologies have continually changed the face of the wood composites industry in North America and elsewhere during the 20th Century and will change the industry at an accelerated rate in the future.

Key technologies will produce molded products, recycled products, engineered products, and products with enhanced properties including dimensional stability and fire resistance. Ongoing improvements in adhesive technology will also be required if high-performance, environmentally friendly biocomposites are to be commercially successful in the future. Specialty products developed using these key technologies and advanced adhesive technology will form the basis for new commodity products in the next century.

References

1. Andersson, M. and A.-M. Tillman. 1989. Acetylation of jute: Effects on strength, rot resistance, and hydrophobicity. J. Applied Polymer Sci. 37:3437-3447.
2. Clarke, P.A. 1991. Panel products — past, present and future — developments. J. Inst. of Wood Sci. 12(4):233-241.
3. Coakley, P. 1992. The Pacific Rim wood composites market — A New Zealand perspective. *In:* Proc. Pacific Rim Bio-Based Composites Symp., Rotorua, N.Z. pp. 1-10.
4. Comstock, G. 1992. History and evolution of bio-based composites in North America. *In:* Proc. Pacific Rim Bio-Based Composites Symp., Rotorua, N.Z. pp. 11-19.
5. Ho, S.K., C.R. Anderson, and M. Croucher. 1992. Ammonia treatment of strandboard during pressing. *In:* Proc. Pacific Rim Bio-Based Composites Symp., Rotorua, N.Z. pp. 73-82.
6. Hsu, W.E. 1986. Method of making dimensionally stable composite board and composite board produced by such method. Canadian Patent 1215510.
7. ________. 1989. Apparatus for steam pressing compressible mat material. U.S. Patent 4850849.
8. Kawai, S., Q. Wang, M. Sasaki, and M. Tanahashi. 1992. Production of compressed laminated veneer lumber by steam pressing. *In:* Proc. Pacific Rim Bio-Based Composites Symp., Rotorua, N.Z. pp. 121-128.
9. Lyngcoln, K. 1992. LVL and plywood — An Australian overview. *In:* Proc. Pacific Rim Bio-Based Composites Symp., Rotorua, N.Z. pp. 100-112.
10. Plackett, D.V., J.M. McLaughlan, and R.J. Burton. 1991. Fibre-based composites in New Zealand: Past developments and future opportunities. New Zealand J. of Forest Sci. 21(2):246-255.

11. ________. 1992. Dimensionally and ultraviolet-stable composites from lignocellulosics. *In:* Emerging Technologies for Material and Chemicals from Biomass. R.M. Rowell, T.P. Schultz, and R. Narayan, eds. ACS Symp. Series 476. pp. 88-97.
12. Resource Information Systems, Inc. 1993. Wood Products Review — Softwood lumber, panels and board. pp. 107.
13. ________. 1993. Particleboard and MDF Commentary, Wood Products Service. Volume XVIII, No. 8.
14. Rowell, R.M. 1983. Chemical modification of wood. Review. Commonwealth Forestry Bureau. Oxford, England. 6(12):363-382.
15. ________ and F.M. Keany. 1991. Fiberboards made from acetylated bagasse fiber. Wood and Fiber Sci. 23(1):15-22.
16. ________. 1992. Opportunities for lignocellulosic materials and composites. *In:* Emerging Technologies for Materials and Chemicals from Biomass. R.M. Rowell, T.P. Schultz, and R. Narayan, eds. ACS Symp. Series 476. pp. 12-27.
17. ________. 1992. Opportunities for value-added bio-based composites. *In:* Proc. Pacific Rim Bio-Based Composites Symp., Rotorua, N.Z. pp. 244-252.
18. Sheen, A.D. 1992. The preparation of acetylated wood fibre on a commercial scale. *In:* Proc. Symp. on Chemical Modification of Lignocellulosics, Rotorua, N.Z. pp. 1-8.
19. Shen, K.C. 1973. Steam-press process for curing phenolic-bonded particleboard. Forest Prod. J. 23(3):21-29.
20. ________. 1990. Binderless composite panel products. *In:* Proc. Composites Wood Products Symp. R.J. Burton and G.L. Tarlton, eds. Rotorua, N.Z. pp. 105-107.
21. Sturgeon, M. 1992. The future of panel products: medium-density fibreboard. *In:* Proc. Pacific Rim Bio-Based Composites Symp., Rotorua, N.Z. pp. 42-50.
22. Troughton, G.E., D.C. Walser, A.W. Andersen, and L.C. Palka. 1993. Development of high-stiffness laminated veneer lumber made from B.C. softwoods. Forest Prod. J. (in press).
23. Walser, D.C. and R. Clarke. 1987. Veneer lathes having veneer thickness sensor and thickness control. U.S. Patent 4791970.
24. Youngquist, J.A. and R.M. Rowell. 1990. Adhesive bonding of acetylated aspen flakes. Part 3. Inter. J. of Adhesion and Adhesives 10:273-276.

SESSION IV:

New Resources and Products — Best Utilization of Existing Fiber

New engineered wood products for the Japanese market

Frank Michiels

Abstract

The objective of this presentation is to report on the progress of engineered wood products in the Japanese market. An increase in the acceptance and use of such products is occurring. Examples of such products include mechanically rated lam stock and veneer glulam, parallam, laminated veneer lumber, I-joists, laminated decking, and oriented strandboard. These products are supplied by both foreign and domestic manufacturers. The overall prospect for the increasing use of engineered wood products in Japan in the near future is good.

Introduction

The tradition of wood use in building construction in Japan extends back nearly 1000 years to systems originally adopted from China. The use of traditional wood systems, based on post and beam construction which combines esthetics with structural performance in each member, has declined in Japan since World War II, dropping from over 75 percent market share in 1980 to under 40 percent share in 1990. Several factors have led to this decline, including the poor performance of wood structures during the Great Kanto Earthquake and the fire-bombing of World War II, a decline in the supply and rise in the price of quality wood materials, and a decline in the availability of knowledgeable craftsmen.

Introduction of engineered wood products into the Japanese market

Attempts to introduce engineered wood products into the Japanese construction market were begun in the mid-1980s at a time when the use of traditional post and beam wood construction had fallen to an all-time low of 38 percent market share. Engineered wood products faced many entry level problems including:

- Building codes geared to concrete and steel and based on fire zones.
- Import tariffs that were skewed against finished products.
- A lack of appropriate grading standards.
- Expensive and sometimes arbitrary accreditation procedures.
- A lack of recognition of the technical advantages of engineered wood products by academia, government, and industry.
- A lack of expertise among engineers, architects, and contractors.
- Traditional concepts within the wood distribution and construction systems that were inappropriate for engineered wood products.

Even with all of the problems listed above, the recent building code revisions in Japan which allow the use of wood in three-story and larger nonresidential structures have been very important factors in stimulating the market, and the main strength of engineered wood products is that they offer an opportunity to restore credibility to wood construction.

Primary uses of engineered wood products in Japan today

Laminated posts

The laminated post industry began in Japan in late 1970s and has grown to an annual production level of approximately 100,000 m^3 according to the Japan Laminators Association. Currently one laminated post production line dedicated to the Japanese market exists in the United States, and several other U.S. manufacturers produce for the Japanese market on a small scale. Present technical standards for laminated posts require 5-ply layup, however this may be revised to allow 3-ply. The market for laminated posts in Japan appears very strong at this time.

The author is President, Michiels International, Inc., Seattle, Wash.

Glulam beams

Glulam beams have shown very rapid market growth in Japan since 1985, and continued growth is expected in the near future. In 1992, the consumption of glulam beams in Japan included domestic production of 34,434 m^3 (Japan Laminators Assoc.) and imports totaling 23,313 m^3, of which 92.8 percent was supplied by U.S. producers (Japan Plywood Inspection Corp.). The first 6 months of 1993 showed strong growth in imports of glulam beams, totaling 23,726 m^3, with 90.8 percent being supplied by U.S. producers (Japan Customs Bureau). Domestic producers of glulam beams have been growing in number and in production capacity. This growth has been assisted by the development of ancillary industries which has led to the availability of industrial production machinery and engineered connector systems. In addition, the supply of knowledgeable engineers has increased.

Oriented strandboard (OSB)

Problems within the traditional supply sector for panel products in Japan have created a strong opportunity for OSB. In 1992 the consumption of North American OSB in Japan was 23,313 m^3 (Japan Customs Bureau) and in the first half of 1993 North American OSB consumption had already reached 21,548 m^3 (Japan Customs Bureau and American Plywood Assoc.). OSB is used in both in residential and industrial/commercial construction.

LVL and I-joists

One dedicated structural LVL production line has been in existence in Japan since 1992. The primary application for LVL in Japan, however, is in nonstructural core applications. Only one I-joist manufacturer has gained approval to market broadly in Japan.

Other engineered wood products with market potential

Other engineered wood products with market potential in Japan include machine stress-rated lam stock, in which the largest demand is for metric sizes, mechanically rated veneer, Parallam, and structurally rated laminated decking.

Issues affecting further growth of engineered wood products in Japan

A primary issue affecting the potential growth of engineered wood products markets in Japan is the current revision of the Japan Agricultural Standards. The main conflicts that exist revolve around species recognition and differing concepts related to performance standards versus prescriptive standards. The result may be a compromise consisting of mixing visual standards with mechanical standards, however, such a compromise may result in an even stricter standard.

Other issues that may affect future markets include further revisions of national and local building codes, more transparent application procedures for accreditation or "Code 38" approvals, and information dissemination to improve technical recognition of engineered wood products at the university, industry, and government levels.

Product development: the key to global competitiveness

Robert J. Tichy

Abstract

As the world becomes a true global marketplace, the ramifications of these changes are felt in all aspects of business today. Products from around the world are beginning to appear in previously inaccessible markets. With this trend, we must ask ourselves if we are prepared to meet the global market challenge.

This paper will discuss the relationship between business success and product development. For the purposes of this paper, a product will be considered either a tangible object, service, or process. This paper will attempt to show that product development is *not* an option, and that it must simply be conducted efficiently. From product concept to market introduction, the commercialization process must be fast and to the point. This paper will also show that for any company, a sustainable competitive advantage can only be achieved through a process of continuous improvement (i.e., aggressive and successful implementation of technologies). The process of continuous improvement produces an evolving backlog of products awaiting implementation. These are the tools and equipment required to become and remain globally competitive.

Introduction

As communication and transportation technologies improve, the world becomes a true global marketplace. The ramifications of these advances are felt in all aspects of business today. Products from around the world are beginning to appear in previously inaccessible markets. For example, even the once isolated North American housing industry is seeing domestic housing exports and foreign imports. Today more than 30 percent of all building construction (including housing) in the United States is either managed or financed by Japanese interests.

This paper discusses the relationship between business success and product development. It also discusses some barriers and business imperatives that impact that success. For the purposes of this discussion it is assumed that product development is equivalent to what is commonly called research and development (R&D). These terms will be used interchangeably.

The author is President, Technology Management and Implementation, Inc., Federal Way, Washington.

Product development — Why bother?

The first question that should be asked is: Why bother with product development? It is typically an expensive and time-consuming activity. Past experience in the forest products industry has shown that from the initial idea to market implementation the process takes from 10 to 20 years. The astute businessman will ask: Can my money be spent more productively? The answer is "most definitely," if the process takes two decades. However, the answer is "most definitely *not*," if the process is conducted more efficiently! What I hope to prove is that product development activities are *not* an option, they must simply reach commercialization faster. Ways of improving the efficiency of product development activities will not be discussed in this paper. However, the basic need for product development will.

The underlying premise is that continuous improvement (in Total Quality Management terms) of your product, process, or service is the only competitive advantage a company can ever expect to have for the long term. For example, it is difficult to produce a patent that cannot be avoided or a trade secret that will not leak out. With the exception of the round toothpick, the examples are few. This is also true for sales and marketing strategies. The only true competitive advantage is to outrun your competition. This is done by continuously, aggressively, and successfully implementing product change.

Change can be troublesome and sometimes terrifying. I have heard many business leaders suggest "We are all in favor in progress — just as long as we don't have to change anything." However, change is the only thing that we can always count on. It is the only thing we know for sure. Unfortunately, change is what we face daily, therefore it must be managed successfully.

In the business world change is motivated by two factors — fear and greed. More nebulous terms such as competitive threat and profit objective could have been used to describe these motivators, but fear and greed cut right to the basic human response. These are powerful motivators. They drive most business decisions. So the question becomes "How does continuous improvement, change management, fear, and greed relate to product development?" The result of product development efforts are the technologies awaiting implementation. These are the tools and equipment required to outrun your competition, and make no mistake, it is a race.

To emphasize this premise it is useful to look back in time to the Industrial Revolution. It is agreed that the Industrial Revolution was the true beginning of the free enterprise system (i.e., capitalism). What was the catalyst for this dramatic change in commerce? Was it a brilliant marketing plan, a unique sales strategy, or a fantastic advertising campaign? No! It was the implementation of technology! We went from making iron in cup-size ladles to massive iron and steel mills with 50,000-gallon ladles. We went from spinning wheel yarn production to enormous woolen mills. And all within a reasonably short period of time. The point here is that the implementation of technology created an economic structure that feeds on the technology created by R&D efforts.

Let's look briefly at two familiar products to see where the implementation of change has paid handsome dividends. First, the auto industry — it isn't difficult to see who has performed better over the past decade. Toyota, for example, has the ability to implement design changes on the production line on a daily, if not hourly, basis. Second, the watch industry was revolutionized by digital technology. The digital watch technology was a Swiss development. Seiko saw the Swiss demonstrate their new technology at a trade show and the rest is history. Seiko now has approximately 60 percent of the worldwide watch market. The Swiss watch industry simply decided that it would be too much trouble to change what they have perfected over so many years ("if it ain't busted, don't fix it").

Business success

So how does this impact business success? As long as we conduct our business within a free enterprise system, the aggressive implementation of technology is *not* an option. The word "implementation" is used regularly to emphasize the fact that the product is actually commercialized. It is useless to spend any money on R&D if there is no commitment to implement the result. This means that the R&D function must be an integral part of the business planning function so that R&D activities are directed toward fulfilling the company business strategy. The role R&D plays in setting business strategy, however, is the subject for an entirely separate discussion.

Barriers

What are the barriers to achieving or maintaining success in virtually any industry? There are many. The following are just a few, pertinent to the utilization of forest products.

Barrier 1: Ourselves

As an industry, there has been no real commitment to long-term growth as measured by our level of R&D spending. Globally, national R&D expenditures have been slowly increasing for over two decades. Typically, the United States spends approximately 1.8 percent of its Gross Domestic Product (GDP) on nondefense R&D, while both West Germany and Japan spend in excess of 2.5 percent of their GDP. This is nearly 40 percent more than the U.S. total. Traditionally, the U.S. forest products industry has been below the national average and trending negatively. For example, R&D expenditures of less than 0.1 percent of sales are common in the forest products industry, while our counterparts in Japan have been known to spend in excess of 2 percent of sales on R&D. In Japan, 25 out of 30 of the largest construction firms in Japan have building construction R&D organizations. A report by the Civil Engineering Research Foundation describes that in the United States not one construction company had an "identifiable R&D budget." Shimizu Corp., in 1992, spent nearly twice the entire amount the U.S. construction industry spent on R&D. Remember, more than 30 percent of the building construction in the United States was either managed or financed by Japanese interests. The question is: who is preparing for the race?

We often lament the fact that we are a diverse and fragmented industry. Surveys show that our domestic and foreign customers have noticed. These surveys also suggest that we as an industry show little sustainable (long-term) commitment to our customers' needs.

Barrier 2: Our competitors

Beyond the competition between forest products manufacturers comes competition from the non-wood materials such as steel, concrete, and masonry. As nonresidential and commercial construction slows, these competing materials are looking for new markets, just as we are. Where do you suppose they are looking? There are numerous examples of these competing materials making inroads into our traditional market through the implementation of new technologies:

- steel framing concerns are achieving success in marketing their products and systems in the residential housing industry
- the California energy codes accept masonry block systems as thermally efficient
- the Butler buildings organization has purchased timber frame building companies and are staffing up to expand their systems approach into wood building markets
- the Japanese and Swedes are establishing home manufacturing facilities in the United States

All of the above are relatively recent developments. The question is: What changes have occurred, or what technologies have been implemented to allow these intrusions? A closer look at the steel buildings industry may provide at least part of the answer. This industry is a good example of explicitly attending to the customer's needs. Steel buildings have historically provided excellent performance, complete building packages, consistent material supply (along with stable pricing), and readily available engineering support. In virtually any city in the United States, an engineer can be found that can design with steel, concrete, or masonry. This is not true for wood-based structures.

Foreign housing competitors are also not idle. Many joint relationships between Japanese housing manufacturers and U.S. materials suppliers have been formed. Recently, the Swedish Home System has built a housing plant in Eugene, Oreg. The race to build housing in the United States and abroad continues to draw more and more contestants.

Barrier 3: Our regulatory system

Building codes are typically written by nonwood interests. Even so, with the exception of some fire and environmental issues, the building codes themselves do not represent a significant barrier. The problem comes in the application and interpretation of the codes.

For example, many states have a difficult time inspecting closed-panel systems, thus limiting their use. International standards exist (ISO 9000 Series) outlining the principles for in-plant inspections. We must implement the concept of in-plant quality control procedures approved by the building code departments to certify code compliance. This is no different than the quality assurance procedures applied to most building materials.

Frequently, reciprocity between states becomes a bureaucratic nightmare, and often the interpretation of the code means very different things in different locals.

None of these barriers are insurmountable. However, it must be recognized that the process of building structures is changing and the process of regulating them must change too.

Barrier 4: Our customers' perceptions

It is true that wood rots, burns, and swells. Wood, put in that context, can obviously be portrayed in an unfavorable light. However, no construction material is immune to negative connotations. Each material carries a list of both positive and negative attributes. It is the responsibility of the particular material interest to work with their customers to find solutions to their needs. Dwelling on a competing materials' weakness to sell your own does not provide long-term security. For wood and wood-based products, a major survey of architects, engineers, building owners, contractors, and home owners revealed the following list of perceptions that need significant attention. For example, wood is perceived as:

- non-structural material
- flammable
- low durability
- high maintenance
- excessive variability
- inconsistent supply
- poor engineering support

Whether those of us in the forest products industry believe this list or not makes little difference. The bottom line is that our customers do. It is our job to change those perceptions.

Business imperatives

To successfully compete in virtually any business, there are some basic characteristics a company must possess. Several are listed below.

- Recognizing and accepting that the marketplace has changed, and we must compete in a global economy, is first. As previously mentioned, no industry or country is immune to foreign competition.

- A business without a clearly defined and communicated company strategy is a rudderless ship. Without a vision of what you wish to become, how will you know when you get there?
- Appropriate performance measures must be clearly defined. "What gets measured, gets produced." If production volume is how employees are measured — volume is what will result.
- Because of the rapid market changes observed today, a company must acquire the ability to respond to these changes without hesitation. This, in itself, requires change.
- A commitment to the long-term implementation of change is vital. In the spirit of continuous improvement, we must attend to your customers' needs explicitly. Remember, the customer defines quality. This manifests itself in many ways. For example, we must: 1) provide incentives (e.g., price, performance, construction speed, energy efficiencies, warranties); 2) leverage aesthetics and environmental friendliness; 3) ensure consistent product supply and quality; 4) provide technical/engineering support; and 5) educate customers, including materials suppliers, specifiers, code officials, and design educators.
- Establish multi-disciplined problem solving teams. A team should include all relevant business functions and maybe some seemingly irrelevant functions. Often some of the better ideas come from people who are not familiar with the problem at hand. These teams could include labor union representation or building code officials. No boundaries are assumed.
- Aggressively implement technologies. This could involve improved or entirely new classes of materials and systems. More efficient combinations of materials in a given structure are an option (i.e., steel, concrete, and wood in a single building). New design procedures, improved accounting procedures, or project management software are only a few examples.
- Finally, some of the most effective ingredients for success are already in place. We, as a nation, are creative and resourceful, under tremendous pressure to improve business performance, and angry. This is a powerful combination to motivate successful change.

Conclusions

- Change is inevitable. The only thing we know for sure is that we know nothing for sure.
- The implementation of technology is not an option. It was the implementation of technology that created the free enterprise system and it will be the implementation of technology that maintains it.
- Continuous improvement is the only sustainable competitive advantage.
- Product development is the catalyst and fuel for continuous improvement.

Secondary fiber usage and market impediments

Doris Zacher

Abstract

The raw material supply for the manufacture of paper and paperboard products is changing, with continued increases in the use of recovered paper. This substitution is driven by the increased demand for paper and paperboard products combined with a potential decrease in available virgin fiber, environmental concern for conservation and preservation, and consumer demand for recycled products.

This paper will describe changes in the fiber supply, the challenges in meeting future demand, the economic consideration of collection and processing, and the concept of the recycling loop and its true market implications.

We are in a period of changing paradigms — an expectation that high quality can be achieved economically with both product "in" and product "out." Are we ready to broaden our mindset to meet these paradigm shifts and to develop diversified systems to meet the global challenges of tomorrow?

Secondary fiber usage and market impediments

During the last 10 years, total paper and paperboard production in the United States has grown by approximately 30 percent and the raw material feedstock for this production has changed with an increased substitution of recovered paper or secondary fiber. During this same period, the total recovered paper stock consumption has grown by nearly 86 percent (1). This substitution and growth is due primarily to decreased availability of virgin wood fiber, consumer demand for recycled content, and landfill diversion.

The author is Mill Buyer, Northwest Region, Weyerhaeuser Paper Co., Recycling Div., Tacoma, Wash.

The U.S. paper industry has committed to a national goal of 40 percent paper recovery for recycling by the end of 1995. Some analysts expect the industry to exceed that goal, particularly in the western United States. Achieving this goal means the collection and reuse of approximately 40 million tons of paper, a 50 percent increase over 1988 levels. Most of the growth will be seen domestically (4). The challenge ahead is how to collect the volume required to meet the demand, in the form and quality that is needed.

The history of secondary fiber usage in itself has changed as supply, demand, and substitution have come into play. Mill broke, roll stock, and pulp substitutes (unprinted converting trim) have traditionally been a viable feedback, requiring little processing and having minimal contamination. Newsprint and paperboard products have also had a long history of secondary utilization as these products are typically abundant in separated streams to be efficiently collected and sorted. However, as the demand increases, the supply becomes increasingly more difficult to obtain economically.

There are four key variables that lead toward determining the economic viability of secondary fiber utilization: collection, fiber composition, processing, and technology. Although we can address each one separately, in the final evaluation each variable impacts the other.

In the discussion that follows, remember that the end product produced, and the quality demanded, determines the specification of the secondary fiber used. Further, the fiber within each group generally must be consistent for maximum utilization (referring to fiber separation between chemical and mechanical pulping processes and further grouped between bleached and non-bleached grades).

Given that the "easily collected" material is already being used, we are concerned with finding new incremental volume. The next level of opportunity for this volume will come from medium to small businesses and from residential areas. Thus, collection costs become significant in the evaluation. Most recovered materials are of low value and, therefore, the collection and transportation is a large component in determining value and viability. The elements to be considered in transportation and collection are route density, volumes per route, numbers of separated commodities, and equipment and manpower. There are several studies that have been conducted to evaluate the economic tradeoffs within these elements and the results are dependent upon local or regional conditions. However, there is general agreement that more truck compartments (material source separation) means less volume per route. The cost consideration then becomes one of whether source-separated collection can be achieved at a lower cost than collecting "commingled" material and sorting at a processing facility.

There are more than 50 grades of wastepaper. Although there are industry standards for grade categories, wastepaper is usually sorted and packed to customer specifications. The customer also determines what contaminants will be detrimental to process and equipment capability and end product quality. There is nothing magical about sorting paper — this is usually accomplished manually, with or without the aid of a sort belt. The process may take 1.5 hours per ton for clean material to 0.3 hours per ton for heavily contaminated material; the cost can range from $10 to $70 per ton. However, as the level of collection opportunities drop, the amount of mixed paper increases along with the amount of contamination, increasing cost and decreasing quality.

Current focus seems to be headed towards accepting the limited capacity of source separation and focusing on process technology to maximize upgrading of mixed paper. This does create a challenge in that the composition may vary considerably. The generalized composition for both commercial and residential mixed paper is presented in Table 1 (2); however, any collector or processor will advise that the deviation and spectrum is very wide. Thus, we are faced once again with making choices and look towards technology to provide greater flexibility.

Table 1. — Sample composition of residential and commercial mixed waste paper.[a]

	Commercial	Residential
	- - - - - - - - (%) - - - - - - - -	
Magazine	20	25
Old newspaper	20	25
Computer printout	15	- -
White ledger	10	10
Colored ledger	10	- -
Directory	10	5
Old corrugated container	5	25
Manila file folder	5	- -
Boxboard	5	20

[a] Sources: Weyerhauser Co. (1990); Recycling Advisory Council, Task Force Report (1991).

Technology has indeed provided the industry with greater flexibility to handle new fiber streams such as mixed office papers, milk and juice cartons, moisture-tolerant containers (using a new recyclable wax), etc. Product modifications are being developed for items such as coatings, glues, dyes, and inks. But other environmental changes also impact systems, such as sludge disposal, closed water systems, dioxin free bleaching, greater responsibility for clean water discharge, and post-industrial/post-consumer content definitions. Our challenge here is to develop improved repulping, decontamination, and deinking technology both for equipment and chemistry (3).

And now we have come full circle to look at the products: How are products designed to be fully recyclable? How will the product design impact other material streams, collection, or processing systems? Are there end use markets currently available for a given product within a region of distribution? What effect will changes in products have on material substitution? We are working towards evaluating manufacturing processes that are "environmentally responsible" where we make products that are reusable, recyclable, compostable, burnable, and disposable (3).

Considering the entire recycling loop will lead us towards meeting future demand. When we evaluate products and processes as a whole, we will be able to develop the systems necessary that will allow us to meet all elements that lead towards quality products that can compete globally. However, many times we find ourselves thinking too narrowly, focusing on what has been and is now instead of what can be.

Literature cited

1. American Forest & Paper Association. 1993. Statistical Rept.
2. Oregon Recycling Markets Development Council. 1992. Summary of Report and Recommendations of the Paper Div. (Sept.)
3. Friberg, T. 1993. Alternative uses for recovered paper. Resource Recycling (Jan.) pp. 26-33
4. ________. 1993. Recycling. Pulp and Paper Mill of the Future; An Information Exchange. Univ. of Maine. (Sept.)

Technology, efficient resource use, and extending markets: New Zealand's experience with radiata pine

Owen Cox

Abstract

New Zealand has a forestry sector which is based on plantations comprised primarily of radiata pine, a fast-growing exotic conifer. The development of plantation forestry has been accompanied by the increasing use of radiata pine in substitution for other woods and materials. This market penetration occurred first in New Zealand and more recently in a number of export markets. The role of technology is discussed.

The level of the sustainable harvest is increasing. New markets are developing in Asia and North America. Other factors affecting patterns of wood use and the New Zealand processing industry have been changes in resource ownership, economic deregulation, and changes in world wood markets. The possible future directions for the sector and the role of technology in those are discussed.

Introduction

New Zealand has established a forestry sector based on fast-growing exotic conifers, particularly radiata pine. This has been achieved by the local development of the technology of plantation forestry and by a change in patterns of local wood use. In New Zealand, plantation softwoods have displaced, and in instances have been superior to, local indigenous species. Much of this has been through the development of knowledge and new utilization systems. Export markets have developed and are now rapidly expanding for New Zealand radiata pine.

In this paper an economist's view comes through. Technology has played a role in allowing radiata pine to gain wider use and in extending the new wood resource through more efficient utilization. In this sense technology has allowed the increasing scarcity of traditional wood sources to be augmented by fast-growing plantations. This provides an alternative to very high wood prices and difficult supply conditions as well as giving suppliers such as New Zealand a favorable economic environment to further develop renewable plantations to supply world markets.

Radiata pine: important attributes

New Zealand's plantation forests cover 1.3 million hectares (3.2 million acres). The plantations are generally even-aged stands of a single species. Radiata pine (*Pinus radiata*) comprises 90 percent of the plantation area. Radiata pine is a Californian pine that was introduced to New Zealand in the middle of the 19th century and has since proved to be ideally suited to New Zealand conditions.

Radiata pine is a species with good growth rates, disease resistance and tolerates a wide range of sites in New Zealand. It is well adapted to intensive silviculture and out-performs other species in growing usable wood on most sites. This has given radiata pine a dominance in the plantation resource that has enhanced its value. The concentration on radiata pine has given New Zealand processors a resource that is large enough to enable processing on a medium to large scale while avoiding many of the problems that occur when more than one species is processed.

Radiata pine is readily processed into a range of products from solid wood to pulp and paper. This versatility is a major advantage making the entire

The author is Scientist, Wood Processing Div., New Zealand Forest Res. Inst., Rotorua, New Zealand.

stem saleable and has the potential to allow growers to grow for a large number of potential uses and markets.

It is a softwood timber of medium strength and stiffness, and has excellent properties in most machining and gluing applications (3). Compared to many species it is easy to dry and easy to treat with preservatives. The branching of the tree has major impact on grade recoveries of timber for appearance and structural uses. Branching can be influenced by tree breeding, stand spacing, and pruning.

Radiata pine produces good quality medium-density fiberboard (MDF). It also yields reasonable quality kraft and mechanical pulps.

New Zealand forestry: a quick overview of development and the resource

Development and ownership

Large-scale plantings of exotic plantations began in the 1920s with both the State and the private sector establishing plantations. The plantings followed a careful review of New Zealand experience with a variety of species, by a Royal Commission which reported to Parliament in 1913. The Royal Commission recommendations involved the establishment of fast-growing plantations to supply wood once indigenous forests had been largely overcut. It was recognized that the economics of plantation forestry and experience dictated the choice of species, which was to be predominantly radiata pine. The First World War prevented earlier action on the Commission's recommendations.

Planting dropped to virtually nothing during the Second World War. Modest planting resumed following the war but the focus of attention was on the utilization of the now rapidly maturing plantations. By 1950/51 indigenous forests provided only 59 percent of the total wood harvest of 2.824 million m^3 (6). By 1960/61 the percentage had dropped to 33 percent of a harvest of 4.769 million m^3 and by 1970/71 it was only 12 percent. The harvest of indigenous forests is declining in absolute terms and in 1992/93 was only 194,000 m^3 (1.3% of the total). During the 1950s and 1960s, radiata pine displaced indigenous species as the main source of building lumber. With the development of major pulp and paper plants built in the 1950s, radiata pine also became established as raw material for pulp and paper production.

By the 1960s, experience with utilizing the plantations led to renewed interest in expanding the plantation area. Unlike the earlier planting which was intended to supply the local market, the second wave of planting was intended to supply export markets. The need for new export products was heightened as the prospects for pastoral products became more uncertain with New Zealand's traditional market, the United Kingdom, entering the European Economic Community. High levels of new plantation establishment were achieved in the late 1970s and early 1980s with a peak at 56,000 ha in 1984/85.

Planting was by both the State and the private sector. The State developed plantations for reasons other than purely commercial return. Objectives included land stabilization and protection, regional development and employment, as well as commercial returns. The State provided considerable encouragement to the private sector through a variety of mechanisms.

As part of deregulation of the economy, most of the incentives to forest growing were removed in the mid-1980s. The State's administration of forests was reorganized in 1987 with commercial forests being separated from noncommercial forests to be managed by a profit-oriented, Government-owned corporation: the New Zealand Forestry Corporation. At this point in 1988 the State owned about half the plantations in New Zealand. In 1989 the State decided to quit its holdings in commercial forests as part of a move to away from State involvement in the economy and increasing use of competitive markets to determine economic decisions (2). A first price, sealed-bid mechanism was selected for the sale, with all the State's holdings supposed to be sold in a single day (4 July 1990). The sale was not a simple freehold sale, since much of the land could be subject to land claims by Maori tribes under the Treaty of Waitangi (the 1840 Treaty between many Maori tribes and the British Crown that brought New Zealand into the British Empire). The result was that the forests were sold under long-term rights that allow the buyer (and subsequent buyers) to cut and manage the land for a minimum period. The period generally allows the current rotation and the subsequent one to be completed and it will be extended if no successful claims under the Treaty are made.

Despite the intention to sell cutting rights on 4 July 1990, very little was sold on that date. The cutting rights to most of the State's forests were sold in negotiated sales between 1990 and 1992. Today only some 18 percent of plantation area is owned by the State. The State has an interest in a further 4 percent due to earlier leasing arrangements. The major area of plantation not sold, in the Bay of Plenty, is managed by a State-owned company, the Forestry Corporation of New Zealand Ltd. Forests on the west coast of the South Island were also not sold and are managed by a separate government-owned company.

Ownership is now dominated by a limited number of major New Zealand corporations and foriegn companies. During the period of the sales of State forests there were major changes in the ownership in the private sector. Ownership of the plantation resource is shown in Table 1.

Planting of new forests slowed down during to the asset sale. There are a variety of reasons for this but among them were because existing forests, both in New Zealand and overseas, provided a quicker return than unplanted ones at the time and because of changes in the tax system that underwent major change in 1989. The level of new planting in 1989/90 was only 20,000 ha. Following a change in government in 1990 the tax laws were again changed, this time making forest planting more attractive. Planting rates have since increased and it is estimated that 48,000 ha of new plantation area was established in 1992/93.

The resource

Figures on the New Zealand plantation forest estate are given in Table 2. The mean current annual increment is estimated to be 16 m^3/ha per year (228 ft.^{3}/acre per year) which reflects the young age of much of the resource (6). On reasonable sites, mean annual increments for the normal length rotations are generally expected to be near 20 m^3/ha per year (286 ft.^{3}/acre per year). On the best sites yields well in excess of this can be achieved.

A significant feature of New Zealand plantations is the extent to which silvicultural tending occurs. Pruning is common with 52 percent of the resource (almost exclusively radiata pine) being pruned to at least 4-meter butt logs by age 12 (8). The most common height to prune to is 6 meters with pruning being commonly carried out in two or three lifts. Much of the rest of the resource is also pruned, either later than age 12 or to lower heights. Thinning at an early age, to ensure rapid diameter growth on the lower logs, is also carried out. Most thinning is to waste. Only 26 percent of forest area is thinned with the wood being recovered, and even for these areas this is often the case for only the second of two thinnings.

The harvest

The harvest of New Zealand plantation grown wood since 1950 is shown in Figure 1. This figure shows the increase in the harvest which has been a result of the increasing size of the mature resource. Initially, almost all of the harvest from both natural and plantation forests was sawn in New Zealand for domestic consumption. During the 1950s, the pulp and paper industry rapidly increased in size and became a significant user of wood with the opening of two integrated pulp- and papermills which utilized plantation wood. In the late 1960s and 1970s, further pulp capacity was opened in smaller stand-alone mechanical mills and the larger mills also increased capacity. A small amount of panel production developed in the 1950s and 1960s with the production of plywood, particleboard, hardboard, and soft fiberboard for the local market. Over the 1960s particleboard has replaced sawn timber in many applications, especially flooring. In the last

Table 1. — Ownership of New Zealand plantation forests.

Before the State forest asset sales (October 1989)	(%)
Crown owned or controlled	
New Zealand Forestry Corp.	47
Crown leases	5
Large companies	
Fletcher Challenge	12
Elders Resources	13
Carter Holt Harvey	5
Others	18
After the two sales rounds (May 1992)	
Crown owned or controlled	
Timberlands Bay of Plenty	14
Timberlands West Coast	2
Unsold	1
Crown Leases	5
Large Companies	
Fletcher Challenge	16
Carter Holt Harvey	24
ITT Rayonier	8
Juken Nissho	4
Oji Kokusaku Pan Pacific	2
Ernslaw One	2
Wenita Forestry	2
Others	20

Table 2. — New Zealand plantation forests area (1 April 1991).

	Area	
	(ha)	(%)
Radiata pine	1,156,757	90
Douglas-fir	65,478	5
Other softwoods	41,567	3
Hardwoods	25,581	2
TOTAL	1,289,382	100

Source: (8).

10 years the growth portion of the panels sector has been MDF, which is produced for both the domestic and export markets.

There has been a rapid increase in the volume of wood being exported as logs over the last 5 years. There was an earlier period in the early 1970s when a large proportion of the harvest was exported as logs. The difference between the current increase in log exports and the earlier peak was that the earlier peak included a large volume salvaged from windthrown forests in the South Island that was not able to be utilized locally. The current boom in log exports is of wood from mature forests that are being harvested in all parts of the country, but especially from the main forestry processing region in the central North Island.

In the future, harvests are projected to rise still further. The most recent projections of future harvest from the existing plantation area and domestic of consumption of wood (in roundwood equivalents) are shown in Figure 2. The two harvest projections are for two scenarios: the first is for harvest age to average 30 years for radiata pine and the second is for harvest age to average 25 years for radiata pine. The average is currently between these two ages and the 1993 actual harvest figure reflects this. The projections have been derived from figures for an estimated 90.6 percent of the forest and are based on only the existing forest area. Given expected additional planting of between 20,000 to 100,000 ha per year, which increases the area, the future potential harvest will be above the 22 to 23 million m^3 shown for early in the next century. Domestic consumption of timber and wood products (in

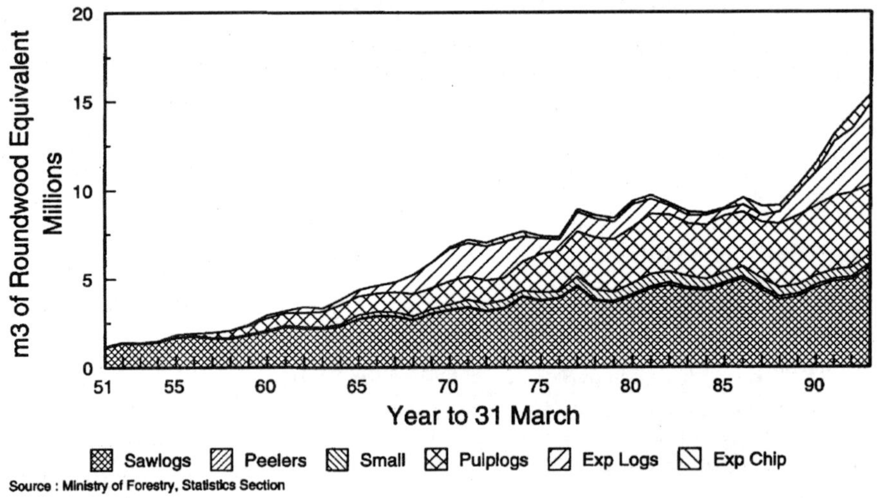

Figure 1. — New Zealand exotic wood harvest (roundwood equivalent).

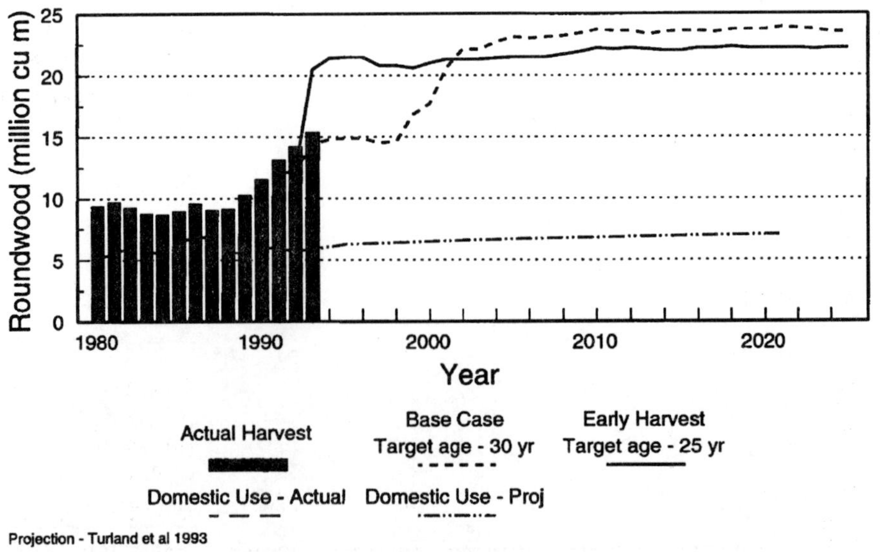

Figure 2. — New Zealand plantation forests, actual and projected harvest.

roundwood equivalents) has fluctuated around 6 million m^3 over the last 20 years and this is expected to continue. Falling per capita consumption has been balanced by increases in population resulting in constant consumption once the normal fluctuations associated with short-term economic cycles is taken into account.

Developing markets for radiata pine

The first market in which radiata pine was widely used was the domestic market. The replacement of indigenous timbers by exotics is covered by Abel (1). Radiata pine rapidly gained acceptance in New Zealand as a general-purpose building timber with advantages over the previously used indigenous species, including its ease of working and handling. The ease with which radiata pine is treated meant that with the development of simple treatment with boron, radiata pine is not prone to borer attack, unlike the main indigenous timber species rimu. The development of other treatments, particularly chromated copper arsenic (for which radiata pine is ideally suited), extended the use of the timber to areas where durability is required. Radiata pine is now routinely used for marine piles, fence posts, and house piles. The development of the domestic sawmilling industry is shown in Table 3.

Radiata pine has been widely used in packaging, especially of New Zealand's primary products. An early example was butter packaging, where kahikatea was used and was displaced by cardboard made from radiata pine. Later exports of fruit were in wooden radiata pine boxes prior to cardboard taking the market.

Radiata pine dominates the panel industry in New Zealand. The main market for New Zealand plywood and particleboard is New Zealand, although significant volumes of particleboard have been exported to Australia. The New Zealand plywood industry has been operating at less than full capacity and its ability to compete on export markets is hampered by high costs of production. There is increasing interest in New Zealand plywood with Japanese companies, particularly Juken Nissho, investing in plywood and veneer capacity since 1991. Juken Nissho are building a small plywood plant and opened a veneer/LVL plant in 1992 which operates in conjunction with a new sawmill. The growth in panel production has been in MDF, with three plants built and later expanded (Table 4). This growth has occurred due to the quality of New Zealand light-colored MDF, which is a result of the raw material being used (young radiata pine), opening good export markets. MDF is frequently used as a substitute for solid wood or plywood.

The New Zealand pulp and paper industry produces a range of pulps and papers (Tables 5 and 6). The sector has shown little growth in the last 5 years owing to constraints on the available resource. The growing MDF industry has used a large proportion of the suitable resource.

Major restructuring of the pulp and paper industry in recent years has been aimed at increasing efficiency and specialization and is in response to the market requiring greater competitiveness as New Zealand became less protected and more integrated into the world economy.

The development of export markets mirrors the use of radiata pine in New Zealand industry. The main export markets in terms of value are shown in Figures 3 and 4. Exports in roundwood equivalents (the volume of roundwood used to produce the exports) is shown in Table 7.

Impact of the increased integration with world market on efficient utilization

In this section the sawmilling industry is discussed in more detail. The sawmilling industry and its health have been seen as being pivotal to the whole processing sector (4). This is because of the constraints on developing new pulp and paper capacity imposed by the size of the resource regionally and by the

Table 3. — Production of sawn lumber in New Zealand.

	Indigenous	Exotic	Total	Treated
	(1,000 m^3)			
1960/61	752	934	1,686	507
1970/71	442	1,407	1,849	717
1975/76	374	1,629	2,003	930
1980/81	175	2,007	2,182	898
1985/86	132	2,265	2,398	1,068
1990/91	83	2,199	2,283	NA
1992/93	58	2,548	2,606	NA

Table 4. — Production of veneer and wood panels in New Zealand.

	Veneer	Plywood	Particleboard	Fiberboard
	(m^3)			
1960	25,042	19,050	1,416	47,037
1970	30,118	24,354	38,940	66,886
1975	35,433	27,047	112,866	79,470
1980	76,259	53,711	146,914	114,499
1985	77,493	55,283	188,302	145,682
1990	89,373	68,479	171,621	442,603
1993	121,241	86,426	164,933	537,602

economics of wood growing, which in New Zealand leads growers to use sawlog regimes. Integrated processing, with sawmills passing residues on to pulp, is essential in this environment, and if sawlogs are exported, so are large volumes of residues.

In the last 10 years the New Zealand economy has experienced considerable change in policy and operating environment. Major deregulation has occurred. In addition, large areas of economic activity which were previously carried out by trading arms of the government are now in the private sector following asset sales. This includes most plantation forests. Trade has been liberalized and New Zealand now has a very open economy.

Table 5. — Production of pulp.

	Mechanical pulp	Chemical pulp	Total pulp
	(air dry tonnes)		
1960/61	85,211	154,441	239,652
1970/71	215,398	317,058	532,456
1975/76	348,010	565,376	913,386
1980/81	565,479	640,583	1,206,062
1985/86	577,897	532,250	1,110,147
1990/91	615,520	733,240	1,348,760
1992/93	660,825	619,711	1,280,536

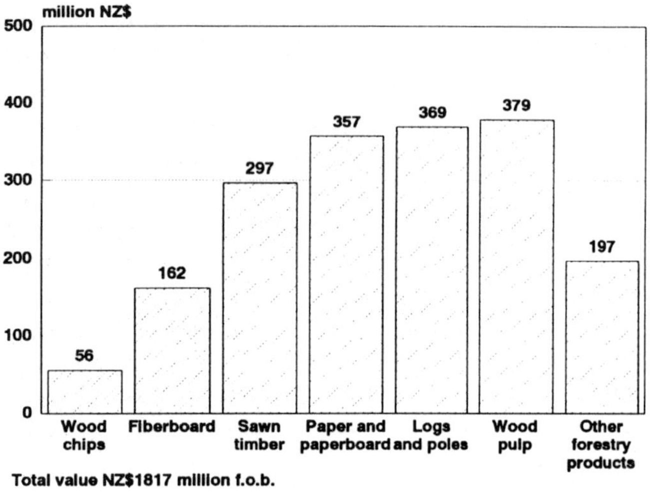

Figure 3. — New Zealand's exports of forest products (year ending 30 June 1992).

Table 6. — Production of paper and paperboard.

	Newsprint	All other paper and paperboard	Total paper and paperboard
	(tonnes)		
1960/61	90,766	93,430	184,196
1970/71	213,907	247,952	461,859
1975/76	219,134	325,580	544,714
1980/81	322,565	400,970	723,535
1985/86	212,148	458,819	670,967
1990/91	321,286	501,054	822,340
1992/93	376,105	370,674	746,779

Table 7. — New Zealand forestry exports by product in roundwood equivalents for the year ending 30 June 1993.

Product	Volume
	(m^3)
Logs and chips	4,519,000
Sawn timber	1,871,000
Pulp	2,362,000
Paper	1,378,000
Panels and other	756,000
TOTAL	9,444,000

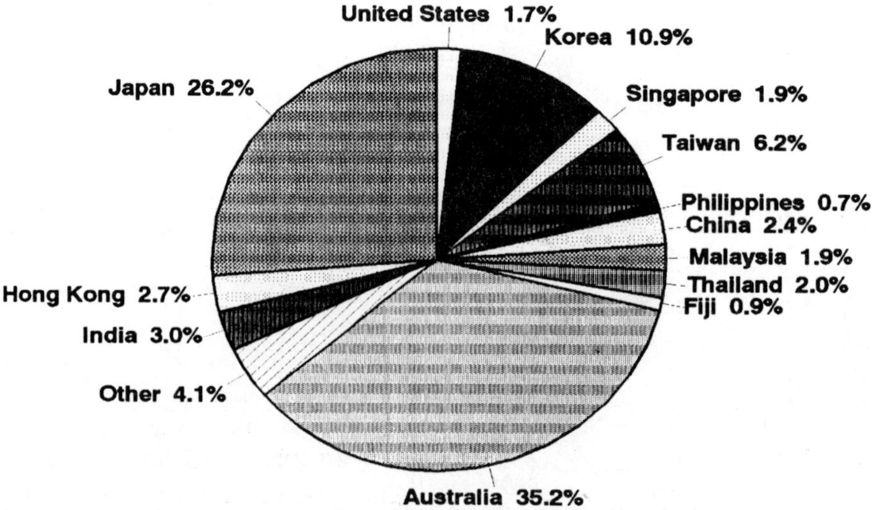

Figure 4. — Export destinations of New Zealand's forest products by value (year ending 30 June 1992).

The changes have had a major impact on the forestry sector. A feature of the sector has been the major increase in the export of logs for processing elsewhere, which has occurred despite increases in processed exports since 1987. The export of logs in relation to domestic processing for local use and export is shown in Figure 5.

Following the election of a new government in 1984, change was rapid. Initially the change concerned a greater reliance on prices and competitive markets for determining resource allocation. The New Zealand Forestry Corporation was to manage the plantations in a commercial manner for commercial goals, which lead to significant changes in the local log market. The processors that first felt the impact were small to medium processors, usually sawmillers, with only short-term sales contracts. Under previous policy these local processors had had much of the output from State forests allocated them and they were isolated and insulated from the world log market. The price of logs to domestic processors increased in the period between 1985 and 1990 as the price of logs on the domestic market adjusted to export parities. During this time there was increasing demand for radiata pine logs on international markets and the volume of logs being sold overseas increased. The volume being processed by New Zealand processors remained constant. Since then, the output of the sawmilling and panels industries has once again begun to increase.

The increases in wood costs to international levels coincided with difficult conditions in the New Zealand market for sawn timber and in Australia the major export market. The problems in Australia were partly due to exchange rate movements that made New Zealand sawn timber less competitive in Australia. One result of the problems was a rationalization of the New Zealand sawmilling industry. The larger producers, who often had some security of supply either through their parent groups owning forests or through longer term legal contracts with the State, concentrated their production on either new mills or on upgraded mills. Within the independent sawmills, however, there were a large number of shutdowns. Total output dropped in 1987 and 1988. Production has increased since then as result of improving conditions in some export markets and as others have been developed.

The export of logs, both pruned and unpruned, has been a feature of New Zealand forestry since 1986. Much of the increase in the harvest has been exported as logs, as is shown in Figures 1 and 5. Initially log exports increased, owing to the higher prices on world markets compared to those on the previously protected local market. Financial deregulation in New Zealand also contributed. The restrictions previously faced by New Zealand companies when investing overseas were reduced and major growers made substantial overseas investments, especially in Chile and Australia. Corporate debt levels increased in the mid-1980s owing to these and other asset acquisitions. Deregulation also allowed interest rates to rise, so that the low and sometimes negative real interest rates of the past were eliminated, and as inflation dropped real interest rates increased for a period, to a real rate of around 10 to 12 percent in 1986-89, owing to a lag in nominal rates dropping. The combined effect of all this was an increase in the cut. As can be seen in Figure 2, the cut increased to near the 25-year rotation projection. Prior to this it was closer to the 30-year rotation projection.

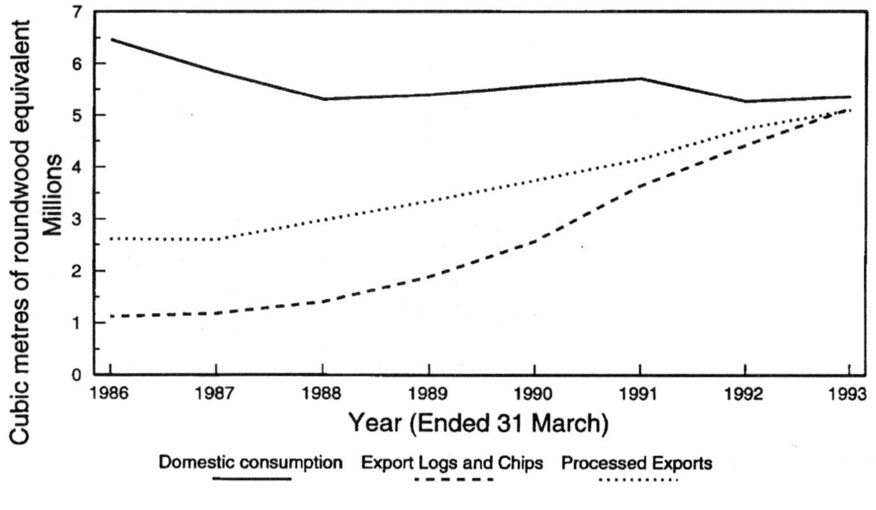

Figure 5. — New Zealand wood harvest, destination, and use in roundwood equivalent.

This shorter rotation effect is the expected response from increasing real discount rates.

The log export trade has led to some dramatic changes in the New Zealand logging industry. With the development of log exports, the industry began to supply logs to overseas processors who paid only for logs to grade specification. There are considerable variations in price between different log specifications, and each processor or market has a particular specification. With the variability of radiata pine trees there is considerable scope to lose (or gain) value according to how logs are cut. The potential to increase the value of stands by cutting for maximum value has lead to a multiplicity of log grades being cut on each skid. Local mills lacking clear knowledge of their requirements, and of the value of getting the right logs, have quickly found they are supplied with logs that are the residual after more lucrative markets have been supplied.

Prior to 1985 only a relatively few log grades were cut. These usually included peelers (to one or two specifications), pruned sawlogs, unpruned sawlogs, and pulpwood. The development of grades to recognize the differences in utilization potential and processing costs of logs with varied characteristics, such as diameter, branch size, and frequency of whorls, was occurring slowly. The New Zealand Forest Research Institute (FRI) had developed log grade specifications in the early 1980s (11) so that the variability of the radiata pine crop which was coming up for harvest was described and processed to maximum benefit. With log grading, the resource can be segregated and processed according to the match between log type, product, and process. The importance of matching log to product and process is critical and was emphasized by the Conversion Planning Project of FRI (5). Despite this, it was the export market that really boosted the use of detailed log grades.

Sawmilling developments

The sawmilling industry has become increasingly dependent on international markets. The domestic market for sawn timber has not grown since the 1970s, indeed since 1987 the domestic market has been particularly depressed and increasing production has gone to overseas markets. The first significant overseas market for lumber was Australia, although Japan also took significant volumes of large green squares for further sawing. Following depressed conditions in New Zealand and difficult conditions in Australia, in 1987/88 other markets have been developed especially in Korea, the United States, and Taiwan (Fig. 6). Buoyant demand has helped.

Unlike Australia, which imports large volumes of structural lumber from New Zealand, these other markets take timber for uses where appearance is more important. A further development has been in exporting more processed lumber products to Japan with a number of joint ventures and Japanese-owned plants in New Zealand. These export packaging timber and components.

Sawmills have increasingly matched the log grades being sawn with their markets and their process requirements.

With the development of new markets, kiln-drying has become more significant. A major part of the investment in new plants by existing sawmillers has been in increased drying capacity. FRI has surveyed the numbers and types of dryers in use and has estimated the volume being dried. The results are shown in Table 8. The almost 100 percent increase in throughput has been achieved with only a 21

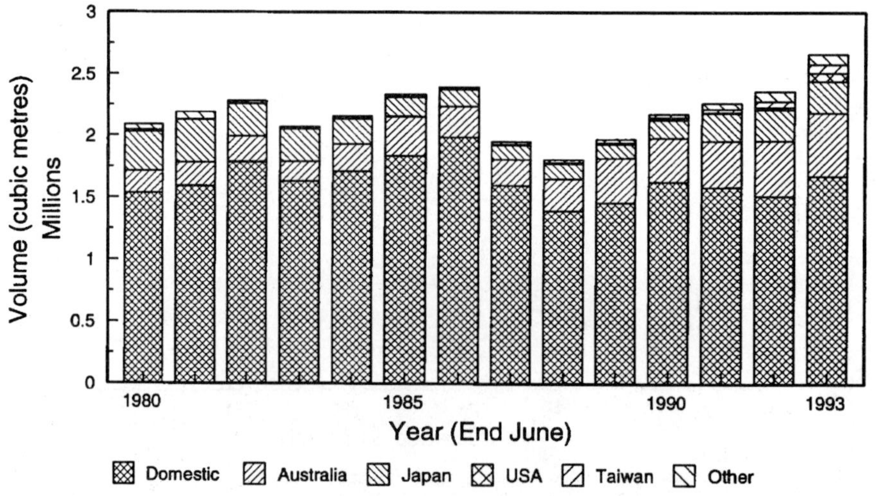

Figure 6. — New Zealand lumber production, domestic use and export destinations.

Table 8. — New Zealand kiln-drying.

	1990	1992	1993
Number of kiln chambers	341	384	413
Annual throughput (m^3)	861,780	1,134,731	1,607,052

percent increase in the number of chambers due to the development of improved kiln schedules and control, particularly in high-temperature drying.

The development of new markets requires the New Zealand industry to adopt new technology. The changes in products and in sawing technologies required can be seen by considering the main markets in turn.

Australia. — Australia has traditionally taken lumber for a variety of uses including structural, finishing, and treated lumber. The market has favored Douglas-fir for structural uses, and radiata pine for all uses. With the changes in forest management there is likely to be less Douglas-fir available for the market. Radiata pine faces competition from Australian plantations of the same species which are grown at higher stockings and for longer rotations. This means the competition from other wood is most acute for structural lumber. Steel studs are a force in the Australian market which, while only having a market share of some 8 to 10 percent of the housing market, puts a check on timber prices. In order to penetrate the Australian market, New Zealand sawmillers have had to kiln-dry significant volumes of lumber, which is not done to structural lumber in New Zealand. In addition, machine stress-grading has proved necessary to meet Australian codes and standards.

Japan. — Exports of lumber to Japan were initially almost exclusively of large squares cut from the center of logs. These were exported rough-sawn and green to Japan where they were further sawn to provide largely packaging lumber. This trade fitted the pattern of New Zealand sawmilling in the 1970s. The center of large logs provide relatively low recoveries of the higher grades of structural timber owing to the pith, which is unacceptable in these grades under the New Zealand visual grading rules. Mills were able to cut for the local building market from the outside of the log and produce a relatively low value but saleable export product from the center of the log. The disadvantage of this, however, was that the product helped to confirm Japanese prejudices about radiata pine and its suitability for any uses other than packaging.

This pattern is now changing. The relative economics of sawing in Japan as opposed to other countries is changing, and the Japanese are showing increasing interest in operating or having joint-venture processing operations overseas. The security offered by plantation resources is also an attraction. Mills have been established in New Zealand to provide sawn lumber for direct use in Japan. This has lead to a major change in the New Zealand sawmilling industry in sawing accuracy. Although mills had been aware of the potential of improving conversion by reducing sawing variation and kerf size, it wasn't until a Japanese-owned sawmill in Rotorua began sawing boards for the Japanese packaging to fine size tolerances, without planing, that the potential gains from accurate sawing sank home. All of the large sawmills are now improving conversion and product quality by paying attention to sawing variation.

United States. — Despite taking only relatively small volumes of lumber, the U.S. market is having a major impact on sawmilling systems in New Zealand. The product being produced for the U.S. markets is dried random-width lumber sawn for further processing by the millwork industry in the United States. The other area of growth is in clear components and lumber. The major change is in carrying out drying earlier in the processing sequence. In the past, lumber was sawn almost to final dimensions prior to drying so that only planing was needed after drying. The use of U.S. grading rules has helped lead to a greater appreciation of the potential of radiata pine to produce clears: instead of grading defects as with the New Zealand visual grading rules, the focus of attention is now on the potential for producing clear lengths. Further development of drying techniques and schedules has been necessary for the drying of large sizes, especially from young trees, without internal checking.

Role of technology and product development: opening new markets and improving utilization

Technology has had a major impact on the use of radiata pine and on the development of the New Zealand forestry sector.

The history of wood utilization in New Zealand has been one of radiata pine substituting for other species. This occurred first in the local market and now it is increasingly occurring in world markets. There are many other examples of substitution by one wood product for others and for wood use to become more efficient. In many respects the New Zealand experience is shared with the rest of the world. Innovation has been important elsewhere;

Sedjo and Lyon (9) suggest that for the most part "innovations have involved the substitution of one wood product or input for another" rather than the complete replacement of other products.

Radiata pine first replaced indigenous timber in New Zealand. Until the 1950s, indigenous timbers were used for all structural and finishing timber. Many, if not most, buildings were also clad in wooden boards. A variety of species were used with some species being favored until supplies ran short, then another species would be favored. In the 1950s the most commonly used species was the slow-growing indigenous species rimu.

As the size of the radiata pine resource has grown, so has the development of new markets. The development of new markets has been aided by two important characteristics of the plantation resource:

1. Its uniformity, largely being of one species (radiata pine) which has allowed technical and marketing resources to be concentrated on this species.
2. The basic versatility of this particular species.

There has been a large amount of innovation and technical development in New Zealand that has enabled the forest sector to develop new markets. Some of the more important areas include:

1. The matching of silviculture to perceived market demands. This has resulted in the adoption of regimes which have helped to match the finishing properties of radiata pine with demands for long-length clear timber and which have provided large-diameter logs quickly and economically.
2. The development of technical knowledge on the properties of radiata pine and the products produced from it. This has allowed potential buyers to know the strengths of the species and has allowed New Zealand producers to concentrate on particular markets and have the technical back-up necessary to tackle these markets. This has included codes and standards work both within New Zealand and in export markets to ensure radiata pine meets market requirements.
3. The development of processes that allow the species to widen its range of uses (e.g., preservation treatments and glue lamination).
4. The improvement of processes and their adaption to radiata pine (e.g., high-temperature drying for sawn timber, MDF production, and developments in pulp and paper manufacture including sorting the raw material into differing qualities).

Further developments are extending the range of uses for radiata pine. New processes are expanding the versatility of pine in pulp and paper and these are being adapted for radiata pine. In the solid wood area, new developments include improving the surface hardness of radiata pine and improving recovery of structural and clear timber by allowing green gluing.

Future trends

The New Zealand industry has a species which is in increasing demand on world markets because of its utility and its ability to substitute for woods which are in short supply. Increasing use of radiata pine will involve further innovation and the development of new technologies. Increased environmental awareness is generally favorable to the competitive position of New Zealand-grown plantation wood, especially radiata pine, because this is a renewable resource. Technology which enables radiata pine to substitute for long-rotation, old-growth species increases the ability of New Zealand to take the opportunities presented by the increasing scarcity of the species. Radiata pine is a relatively cheap renewable resource. This means the future questions concerning radiata pine relate more to how large the increase in area will be than to any fundamental change in the pattern of plantation forestry.

The questions that remain concern where the wood is to be processed. Deregulation and other changes in the New Zealand economy have affected the processing sector at the same time as the harvest is beginning to increase as earlier plantings come to maturity. Prior to these changes, much of the processing sector was either oriented to the domestic market or to Australia, to which New Zealand producers had preferential access. The old sales policy of the government meant that domestic log prices were lower than world prices. With the changes of the last 8 years, New Zealand processors have had to adapt to log prices based on world log prices and to both find and adapt to new markets. This occurred during difficult market conditions in New Zealand and at a time when the ownership of forests was changing. The lack of certainty over log supplies to some processors during this period inhibited investment in new processing in New Zealand.

However, change has occurred which does give some pointers to the future. New Zealand processors have become more specialized and more efficient. This has occurred in most parts of the processing sector. In pulp and paper, New Zealand mills have concentrated on fewer products to enable plants to be competitive on world markets rather than on regional markets. In sawmilling, those mills that specialize on specific products and log types are now more common. In all parts of the sector, more attention is being paid to recovery and to improving processing efficiencies.

In times of high unemployment, a major issue of concern to New Zealand is whether the opportunities offered by increasing use of radiata pine bring the employment and other benefits associated with processing to New Zealand or other parts of the world. Whatever happens, log exports will continue to have an influence on the New Zealand sector and will be an integral part of the utilization of radiata pine. The extent to which New Zealand can capitalize on the growing resource of radiata pine and provide additional wealth in New Zealand based on profitably processing the resource depends on how readily technology suited to new markets is adopted. Without the competitive adoption of new technology, the New Zealand sector cannot compete with foreign processors.

Literature cited

1. Abel, D.B. 1977. Sawn timber in New Zealand : a market and industry structure analysis. Ph.D. thesis. Victoria Univ., Wellington, N.Z.
2. Bilek, E.M. and G.P. Horgan. 1992. The challanges of privatisation: New Zealand's experience with forestry. *In:* Proc. IUFRO Inter. Conf. "Integrated Sustainable Multiple-Use Forest Management Under the Market System." Pushkino, Moscow, Russia. (IUFRO, Copenhagen, Denmark.)
3. Cown, D.J. 1992. New Zealand radiata pine and Douglas fir suitability for processing. FRI Bull. 168. New Zealand Forest Res. Inst., Rotorua, N.Z.
4. Edgar, M.J., D. Lee, and B.P. Quinn. 1992. New Zealand Forest Industries Stretegy Study. New Zealand Forest Industries Council.
5. Kininmonth, J.A. (compiler) 1987. Proc. Conversion Planning Conf. FRI Bull. 128. Ministry of Forestry, New Zealand.
6. Ministry of Forestry. 1993. New Zealand Forestry Statistics 1993. Ministry of Forestry, New Zealand.
7. Montrey, H.M. and J.A. Johnson. 1988. The role of technology in improving the competitive position of the U.S. forest products industry. *In:* Forest products trade: market trends and technical developments. J.A. Johnson and W.R. Smith, eds. Univ. of Washington Press.
8. Neumann, A. 1992. A National Exotic Forest Description as at 1 April 1991. Ministry of Forestry, New Zealand.
9. Sedjo and Lyon. 1990. The Long-Term Adequacy of World Timber Supply. Resources for the Future, Washington D.C. 230 pp.
10. Turland, J., S. Wakelin, and P. Lane. 1993. National Exotic Forest Description — 1992 National and Regional Wood Supply Forecasts. Ministry of Forestry, New Zealand.
11. Whiteside, I.D. and B.R. Manley. 1985. Log grading: FRI's answer to new-crop log size problems. Forest Industries 16(3):27-32.

SESSION V:

The Globalization of Markets and Standards

Policy and standards harmonization to open global markets for wood products

C. Michael Hicks

Abstract

U.S. exports of wood products have more than doubled over the past decade, reaching a record $6.7 billion in 1992. Log and chip exports, as a percentage of total exports, have declined in recent years. Nevertheless, log and chip exports still accounted for nearly 40 percent of total exports, and over 70 percent of our exports to Japan, in 1992. This imbalance in the mix of products exported is caused by a lack of familiarity in overseas markets of U.S. value-added wood products and a web of foreign tariff and nontariff trade barriers, many of them standards-related, that prevent value-added wood products from entering the foreign markets. It will take a cooperative effort of government and industry to correct this imbalance. The U.S.-Japan Wood Products Agreement was such an effort; U.S. exports of value-added wood products to Japan are expected to increase substantially over the next few years as a result of this agreement. U.S. exports worldwide also could increase significantly in the coming years if the U.S. wood products zero-for-zero proposal in the Uruguay Round of the General Agreement on Tariffs and Trade (GATT) negotiations is successful. Declining tariffs and rising global competition, however, will increase the temptation for governments to use technical regulations and conformity assessment programs (e.g., testing, certification, and accreditation) as protectionist measures to erect unnecessary barriers to international trade.

The Foreign Agricultural Service (FAS) is the international arm of the U.S. Department of Agriculture. The mission of the FAS is to increase exports of U.S. agricultural products, including wood products. The FAS programs combine the technical and marketing skills of the private sector with the recognized authority and resources of the U.S. government. In 1979, the FAS and the wood products industry joined forces to initiate an overseas market development program for U.S. wood products; the Forest Products Division (FPD) was established in 1982. The FPD, with a staff of 12 full-time employees, administers two overseas programs for the wood products industry: the Market Promotion Program and the Foreign Market Development Program. These programs are effected on a cost-sharing basis with the U.S. wood products industry. Funding for the two programs totaled $17.5 million in FY93, including funds carried over from FY92.

The FAS overseas activities, including those under the Market Promotion Program and the Foreign Market Development Program, focus on overcoming constraints that limit U.S. exports of wood products or impede their competitiveness in the world market. These constraints include tariffs, restrictive phytosanitary standards, discriminatory product standards, restrictive building codes, and, equally important, a lack of familiarity in overseas markets of U.S. value-added wood products. FAS activities are designed to address these constraints and can take the form of government-to-government negotiations and/or industry demonstration projects, educational seminars, trade fairs, etc.

U.S. exports of wood products have more than doubled over the past decade, reaching a record $6.7 billion in 1992. Log and chip exports accounted for nearly 40 percent of total exports and over 70 percent of our exports to Japan in 1992. This im-

The author is Coordinator, Trade Policy, Forest Prod. Div., USDA/Foreign Agric. Serv., Washington, D.C.

balance in the mix of products exported is caused by a lack of familiarity in overseas markets of U.S. value-added wood products and a web of foreign tariff and nontariff trade barriers, many of them standards related, that prevent U.S. value-added wood products from entering foreign markets. Exports through the first 8 months of this year are up 12 percent over the same period last year.

Declining tariffs and rising global competition has increased the temptation, worldwide, for governments to use technical regulations and conformity assessment programs (e.g., testing, certification, and accreditation) as protectionist measures to erect unnecessary barriers to international trade. Our ability to compete in foreign markets depends, in large part, on our ability to meet prevailing product standards, regulations, and conformity assessment requirements.

Japan

Japanese technical barriers in the wood products sector (e.g., prescriptive building codes, discriminatory product standards, and testing and certification procedures) were designated a priority trade practice under Super 301 in May, 1989; some of these barriers were arguably in violation of the GATT's Agreement on Technical Barriers to Trade (Standards Code). Negotiations between the United States (USG) and Japan (GOJ) were initiated later that year, and were subsequently expanded to cover other areas, including tariffs, subsidies, and tariff classification. Following months of intensive negotiations, an agreement ("Measures To Be Taken By The Government Of Japan Relating To Wood Products") was announced on June 15, 1990.

Implementation of the agreement is on track and has proceeded according to the schedule set forth in the agreement. Significant portions of the agreement have been implemented. The Ministry of Construction has revised the Building Standards Law to allow the construction of three-story wooden apartment buildings (to date, 90 three-story wooden apartments have been approved for construction or have built in Japan); revised the standard for interior finishings to allow increased use of wood finishings; and revised and/or established technical standards to allow the use of exterior wooden siding, wooden fire doors, and machine stress-rated lumber. The Ministry of Agriculture, Forestry, and Fisheries has revised the Japanese Agricultural Standards (JAS) for oriented strandboard and laminated veneer lumber; and established JAS standards for finger-jointed and machine stress-rated lumber.

Implementation, to date, has been marked by a high degree of cooperation, due in no large part to the framework that was set up to oversee implementation, and the diligence of U.S. government and industry. Two committees, the Building Experts Committee (BEC) and the JAS Technical Committee (JTC), were set up to oversee implementation of specific portions of the agreement and further expand the use of wood in Japan. A third committee, the Wood Products Subcommittee of the U.S.-Japan Trade Committee, was also set up to resolve any disputes or problems arising out of the implementation of the agreement. The BEC and the JTC have met at least once annually since 1990. The BEC and JTC committees, made up of government and industry experts in the area of building codes and product standards from the United States, Japan, and Canada, were also given the responsibility of making recommendations with respect to the harmonization of standards, and, in particular, the future development of common standards and testing methodologies. (The "Measures" were announced as a unilateral action by Japan; therefore, the committees are open to any country that wishes to participate.) As a result of the work of these two committees, a number of Japanese standards have been revised to make them more consistent with North American and/or international standards.

Satisfactory market access in Japan, however, ultimately hinges on the elimination or reduction of tariffs on products of interest to the United States. Under the terms of the U.S.-Japan Wood Products Agreement, the GOJ will, in the context of the Uruguay Round, reduce tariff rates on wood products of interest to the United States by an amount at least as ambitious as that achieved by formula participants in the Tokyo Round, or about one-third. The United States also received assurances that this action would result in low tariffs, and that initial tariff reductions would be greater than subsequent reductions to provide for immediate improved market access.

The United States' Uruguay Round wood products zero-for-zero initiative would take this process one step further; it proposes the elimination of tariff and nontariff barriers on wood products worldwide, including those in Japan. A significant number of countries, including Canada, Switzerland, Thailand, Finland, and New Zealand, have agreed, in principle, to eliminate some or all tariffs in the wood products sector. The European Community, Australia, South Korea, the Association of Southeast Asian Nations, and the European Free Trade Association countries also have expressed interest in the U.S. proposal.

Their support, however, is predicated upon the final Uruguay Round package. Japan has come out strongly opposed to the initiative, arguing that it goes beyond the terms of the agreement. This is not the case; the agreement was not, either explicitly or implicitly, a prenegotiation of the Uruguay Round. The outcome of the wood products zero-for-zero initiative could affect other sectoral initiatives, such as scientific equipment, electronics, paper, furniture, and construction equipment. Canada and New Zealand have stated that it cannot support any outcome on sectoral initiatives which does not respond to their primary concern, forest products.

Europe

The European Community (EC) is continuing with its effort, begun in 1985, to create a single internal market for goods and services by developing EC-wide regulations (directives) to replace those of individual member states. In 1990 the EC announced its Construction Products Directive (CPD), which covers wood products but, to date, only Denmark and the United Kingdom have incorporated it into their national standards. The purpose of the CPD is to allow building products from one EC member state to be sold and freely used in another member state as long as they are appropriate for the intended use; those products that can demonstrate compliance with the relevant essential requirements are entitled to carry the CE mark. The CPD only provides the framework for harmonizing member states' technical requirements for construction products. The interpretive documents, which are issued separately, actually spell out how the requirements can be met and provide specific guidance to European standards bodies on the development of product standards. The interpretive documents for the CPD were completed only recently, and are still awaiting translation into French and German.

Product standards, the final link in the process, also are not yet complete. The European Committee for Normalization (CEN) has been assigned the task of developing the product standards; the actual drafting of the technical standard document is done by technical committees established by the CEN. The CEN does not allow U.S. or other third-party participation on either of these committees and, as a result, U.S. exporters are concerned that the resulting standards may discriminate against U.S. products vis-à-vis those of European origin. The FAS and the American Forest and Paper Association are monitoring this process very closely.

FAS has been in discussions with the EC on the pinewood nematode issue for several years. The pinewood nematode (*Burasaphelenchus xylophilus*) is a microscopic organism, transmitted by wood borers (*Monochamus spp.*), and has been identified as the probable causal agent in the spread of pine wilt disease. It is a relatively insignificant forest pest in North America, but not so in other countries, most notably Japan. Concern over the possible spread of the disease has led an increasing number of countries to limit the import of coniferous wood products from countries known to have the pest (United States, Canada, Korea, Japan, China, Mexico, and Taiwan). In 1985, Finland was the first country to ban the import of green coniferous wood after the pinewood nematode was discovered in a shipment of coniferous wood chips from the southern United States. Norway and Sweden quickly followed after they also intercepted infested shipments. In 1990, the European Community amended its Plant Health Law to require that all coniferous lumber be kiln-dried; however, the United States was granted a series of derogations beginning in 1990 and continuing through September 1993 which allowed the United States to continue shipping some green lumber to the EC under an APHIS-approved mill certification program, pending further research. However, effective October 1, 1993, all coniferous lumber with the exception of cedar lumber originating in the United States must be heat-treated to a core temperature of 56°C for a period of 30 minutes. Shipments also must be accompanied by a phytosanitary certificate until such time as the industry gets its APHIS-supervised "Certificate of Heat Treatment Using Kiln Facility" program up and running. The EC's phytosanitary requirements for coniferous lumber (56°C for 30 min.) is quickly becoming the international standard. The United States is expected to adopt similar requirements in the coming year for at least some portion of the lumber entering the United States.

Mexico

Closer to home, the United States, Canada, and Mexico concluded negotiations on the North American Free Trade Agreement (NAFTA) in August 1992. Under the terms of the proposed agreement, tariffs on some wood products, most notably framing lumber, will be eliminated immediately; others will be eliminated over 5 years (hardwood lumber, fiberboard, doors, and windows) or 10 years (softwood plywood). Mexico also agreed to terminate its drawback programs which provide for the refund or waiver of customs duties on materials used in the production of goods subsequently exported to another NAFTA country by January 1, 2001. The

NAFTA also prohibits any new performance-based customs duty waiver or duty remission programs. The best known of the Mexican programs is the maquiladora program. Maquiladora operators are required to export 50 percent of their production. Beginning January 1, 1994, this requirement will be phased out (5% annually), and eliminated totally by January 1, 2001.

In the area of standards, each country must ensure that goods from the other two countries are treated no less favorably than like goods or services of national origin. NAFTA also provides for greater transparency in the standards setting process by requiring advanced notification of proposed changes to standards-related measures. There also are provisions to ensure that interested parties (government and industry) have an opportunity to comment in writing on any proposed standards-related measure. NAFTA also encourages the use of international standards as a basis for its standards-related measures if those standards are an effective and appropriate means to fulfill a country's objectives, as well as the acceptance of technical regulation, adopted or maintained by another NAFTA country under appropriate circumstances. The countries shall, to the greatest extent practicable, also make compatible their respective conformity assessment procedures. All of these provisions will become increasingly important in the future as Mexico begins to develop national standards in the wood products and building sectors.

Conclusion

U.S. exports of value-added wood products to Japan are expected to increase substantially over the next few years as a result of the U.S.-Japan Wood Products Agreement, as are exports to the world, particularly if the U.S. zero-for-zero proposal in the Uruguay Round is successful. Wood products exports could easily exceed $12 billion by the turn of the century. Declining tariffs and rising global competition, however, will increase the temptation for governments to use technical regulations and conformity assessment programs as trade barriers.

ISO 9000: implementation and impact on the timber industries in Europe

Robert J. Foster

Abstract

References to ISO 9000 are increasingly being made in North America, but what the standard means and what companies need to do to satisfy its requirements can be easily misunderstood. This paper will show how one company's role has changed towards an increased interest in quality systems, together with a summary of certification and accreditation in Europe. There will also be a brief guide to ISO 9000 and the steps companies need to take to be in a position to be certificated. The paper will conclude with mention of other companies that are adopting the standard in the United Kingdom, in other European countries, and further afield.

TRADA certification

The Timber Research and Development Association (TRADA), has carried out tests and issued test certificates for timber and timber products for many years. In 1977 it started its involvement in quality assurance by starting a scheme for the visual stress-grading of timber. This was followed shortly after by its first product certification scheme, which was for trussed rafters.

Other schemes followed, but after discussions with the British government we were advised that to obtain their formal approval we should set up a separate company to operate quality assurance schemes and this was formed in 1987.

This company is now known as TRADA Certification Ltd. and operates all of the quality assurance schemes formerly run as TRADA, as well as a number of new schemes. One of these schemes is the certification of companies whose quality systems satisfy ISO 9000.

The author is Chief Assessor, TRADA Certification Ltd., High Wycombe, Buckinghamshire, United Kingdom.

What is ISO 9000?

Rather than talking of ISO 9000, one should really talk of the ISO 9000 *series*, since this is made up of a number of guidance standards with ISO 9001-3 detailing specifications for the distinct forms of company organizations.

Of these three standards, ISO 9001 is appropriate if design and development are to be dealt with, but the most widely used of the standards is ISO 9002.

ISO 9002 gives 18 separate quality system requirements for the various elements of a company's operations. However, it does not cover all operations since there is no reference to marketing or financial matters. It is a system that should give your clients confidence that their product, to whatever specification is agreed, will satisfy their requirements. This means that the standard is not a product standard although the quality of the delivered service or product is of prime importance.

Of the 18 quality system elements, many will be familiar to all companies, since they form part of any organized company. The elements that may be new are those of "Corrective Action" and "Internal Quality Audits," which require a company to check if all of its system are being followed correctly, to record any problems, to take action to solve the immediate problem, and to take suitable actions to reduce the long-term chance of the problem happening again.

Most companies will already be carrying out 80 percent of what is needed, but many will not comply with the requirement that says that the quality system shall be documented and shall include documented quality system procedures and instructions.

How companies implement the system

The main path to be followed when a quality system is to be implemented is to listen to the work force and ensure that they feel part of what is going on.

Top management needs to be committed to the introduction of the system and the work force must know this. The introduction of the system should not all be left to one person, although one person will be needed to coordinate all of the activities.

Before any changes are considered a company should first look at its existing operations and compare them against the requirements of ISO 9002. Changes should not be made for the sake of change, but only if needed.

When procedures are to be written they should at least be discussed, and preferably written by those people responsible for carrying out the relevant activity. Once the procedure is written, all staff will need to ensure that they follow it, although procedures can be written so as to allow a degree of flexibility.

A company should not try and introduce everything on the same day — it is better to do it by stages, review and audit the new elements, and then amend or improve the procedure if needed.

Certification

Once a company is satisfied that its system is fully implemented, it may apply for a certification body to assess and approve its system. Typically this would start with a review of the company's documented system, followed by a visit to the company's site to check that the systems are being followed. The site visit will last at least 2 days and will typically reveal some minor problems. Once these have been fixed, a registration certificate will be issued containing a schedule that specifies the service that has been approved (e.g., manufacture of trussed rafters).

After registration the certification body will carry out at least two visits each year to the company to ensure that the system is maintained.

Certification bodies and accreditation

In the United Kingdom there are 30 to 40 certification bodies. While some of these only operate in particular industries, many operate in a wide range of industries. Certification bodies can themselves be approved or "accredited" by an accreditation authority if they satisfy the requirements of EN 45012. In the United Kingdom this is done by the National Accreditation Council for Certification Bodies which advises the Department of Trade and Industry, which in turn issues the scope of approval certificate.

The system of certification bodies and accreditation also operates in many European countries and further afield, including the United States. Although each country's national standard for quality systems may have its own number (e.g., BS 5750 in the United Kingdom), they will all be identical to ISO 9000 so that the same system is being adopted internationally.

Within Europe, building products, which includes many timber products, need to satisfy the requirements of the Construction Products Directive (CPD), which has a requirement that products should be manufactured using a system of "factory production control." This is not the same as an ISO 9000 system, but companies who have ISO 9000 are deemed to satisfy the requirement for factory production control, so long as the technical needs are also satisfied.

Companies need to appreciate that the CPD and its associated CE marking will be a regulatory requirement for construction products, whereas the need for certification to ISO 9000 is a "voluntary" requirement of the marketplace.

What is the impact on the timber industries?

Within the United Kingdom the adoption of ISO 9000 has been driven by national and local purchasing organizations and large companies insisting that they will only use suppliers who are approved to ISO 9000. These include many major timber outlets and construction companies which find that such companies give them a high level of service and product. Also, the companies who are themselves certificated will often then require their own suppliers to introduce a quality system.

Companies who are registered with TRADA Certification include those who manufacture building products, those who machine profiles, timber merchants, furniture manufacturers, and sawmillers.

In conclusion, there is great interest in ISO 9000 in the timber industries within Europe as a whole and, in particular, many sawmills and board manufacturers in Scandinavia are working towards ISO 9000 registration. As many members of the North American timber industry know, the European countries purchase timber from worldwide sources and, as seen in recent months, the pattern of trade can alter dramatically in a short space of time. Although these changes are largely driven by cost, consideration must also be given to the confidence that purchasers may have in the quality of material and service. The registration of companies to ISO 9000 can give that confidence to purchasers, while at the same time giving the supplier a cost-effective means of controlling and improving their service to customers.

JAS grading and customization: routes to success in Japan

Martin Hertrich

Abstract

An industry viewpoint is presented on possible strategic routes for American manufacturers who want to successfully supply construction lumber in Japan. The premise is established in a brief discussion of: 1) global standardization; 2) reasons why the Japanese market is a main attraction; and 3) two market niches for construction lumber in Japan. The Japanese Agricultural Standard (JAS) for lumber is introduced and Vanport Manufacturing's experience in obtaining JAS certification is summarized to provide insight into how involved the JAS certification process can become. The JAS is used in both market niches, but there are pros and cons between supplying the two in Japan. Lastly, on the topic of customization, is a list of ways to become more competitive in Japan.

The conclusion addresses the two market niches, recommending as first choice customization of the business to be more like a Japanese sawmill (producing and grading lumber to traditional Japanese construction standards used in a large market segment) or, as second choice, producing JAS-stamped 2 by 4s for a market segment that is still small.

Introduction

Standardization

In the standardization of forest products in the global market, construction lumber specifications and grades have been slow to standardize because of cultural differences between nations and their ingrained methods in building construction. The two biggest markets, the United States and Japan, are slow to standardize their grading systems to become one single standard. North American sawmills that recognize this are already accustomed to routinely switching between U.S. grading bureau rules and the Japanese Agricultural Standard (JAS) rules established and administered by the Japanese Ministry of Agriculture.

The author is Vice President, Vanport Manufacturing, Inc., Boring, Oregon.

Why export to the Japanese lumber market?

Western sawmills examining the global economy often pick Japan as a target market. There are two attractions for North American sawmills considering entry into the Japanese lumber market. One is that lumber prices are among the best. The second is that the market is large: at 1.4 million housing starts per year, it rivals the size of the U.S. market. Until the recent recession, another reason to enter the Japanese market was because it is more stable than the American market where lumber prices are subject to wide fluctuations.

Why does the Japanese market offer higher prices for lumber than other lumber markets — including the United States? In Japan, the consumer views lumber as not just a construction material, but also as a finished product that must have a fine appearance as the house is being framed, even if the wood is eventually covered. The combination of this attitude for lumber with the spending power of the Japanese economy and the strength of the yen results in the highest prices in the global market.

The consumer desire for a particular quality of lumber is a result of a cultural style that has developed over a period of 1,000 years. In a traditional Japanese house, many of the structural members are exposed in a post-and-beam architecture and are therefore valued and graded according to where it will be placed in a specific location in the home. Japanese consumers also have an appreciation for old growth and will pay a premium for fine-grained lumber.

In the past, Americans have used the same fine-grain wood to make a 2 by 4 which is ultimately covered with sheetrock. Now, however, the American standard of 2 by 4 construction has also been accepted in Japan.

The two lumber markets in Japan

Since the introduction of 2 by 4 construction methods, a second market has emerged in Japan. Although 2 by 4 construction techniques were developed in the United States, 2 by 4 grades in Japan must use JAS standards of the Japanese Ministry of Agriculture. Just as construction lumber in the United States must have a grade stamp for building inspection, 2 by 4 construction lumber must have a JAS grade stamp in Japan. Although the same agency administers the grading rules for the age-old traditional construction lumber used in the first market, the new market has its own set of JAS rules.

Therefore, although Japan has relented to U.S. pressures to open their market to 2 by 4 construction methods, North American sawmills must still meet foreign requirements such as using an unfamiliar set of grading rules. And just because Japan has developed new standards and modified building codes to allow construction lumber produced to American sizes to be used in Western-style housing in Japan, North American sawmills cannot take this to mean a new easily accessible market. North American manufacturers and their lumber graders must pass the tests and requirements set forth by the Japanese Ministry of Agriculture.

JAS certification process

Unlike in North America, where grading bureaus certify individuals to be graders for any mill, JAS certification is a process in which both the individual and the sawmill are certified. In the process, the graders receive certification individually to use the JAS stamp but only at sawmills certified by the Japanese Ministry of Agriculture. Any new graders that the sawmill acquires would have to go through the same process in order to use the JAS stamp.

The process of JAS certification is a lesson in government red tape. Mills will waste much time and effort unless they have a good reason for becoming JAS certified. As an example, Vanport Manufacturing saw an opportunity for relatively inexpensive Japanese publicity and received JAS certification for traditional as well as 2 by 4 construction lumber. Table 1 illustrates most of the types of information required when Vanport applied for certification.

Table 1. — Information required for JAS certification.

Grader Certification

- Two seminars, 3 days each taken six months apart
- Written exam at the end of the second seminar

Sawmill Certification

Sawmill and planer performance:

- Test cut to determine if mill meets minimum accuracy requirements
- Production capacity daily/yearly
- Production flow rates
- Log species cut

Sawmill layout:

- Flow charts and flow diagrams of product flow
- Machine center specification
  - — size of saws
  - — size and number of knives
  - — size and specifications of all electric motors
  - — brands and makes
  - — dimensions of machinery
- Maintenance programs and schedules

Dry kilns:

- Drying schedules for the various sizes produced
- Electric motor specifications
- Boiler capacity
- Sticker stacking machine specifications and capacities

Sawmill infrastructure:

- Log yard
  - — storage capacity
  - — log handling equipment brands, specifications, dimensions
- Lumber yard
  - — number of forklifts and their sizes, capacities, makes, models
  - — inventory capacity
- Shipping/transportation
  - — number of trucks
- Work force
  - — size
  - — management structure
- Order processing system, paper flow
- Customer base
- Measuring instruments
  - — number of: tape measures, calipers, scales, slope-of-grain measuring devices, moisture meters

Other

- All the paperwork was done in Japanese
- Process required at least 1 year from start to finish
- Direct costs incurred: Vanport covered all expenses incurred such as providing the two officials from the Japanese Ministry of Agriculture with air flight from Japan to the United States, hotel accommodations, and ground transportation.

2 by 4 vs. traditional — pros and cons

Even though a JAS stamp is required on 2 by 4 lumber in Japan, the process of obtaining JAS certification for American-sized lumber (which is mostly graded on the basis of strength characteristics) is easier because the grading rules are closer to those used in the United States. Of course, once the sawmill can put a JAS stamp on the lumber it produces, it avoids the necessity of changing to metric cutting patterns. Recently, the Japanese Ministry of Agriculture awarded the U.S. grading bureau Western Wood Products Association with the honor of becoming a certifying agent of 2 by 4 JAS grades. This makes it even easier for western sawmills to obtain JAS certification for 2 by 4 construction lumber now that contacting the Japanese Ministry of Agriculture in Japan is no longer necessary and the paperwork is in English.

Once a sawmill and its graders pass the tests and are certified to stamp 2 by 4 lumber with a JAS stamp, the manufacturer with logs suitable for producing 2 by 4 construction lumber can enjoy selling to the 2 by 4 market niche in Japan — a small segment that, although still growing in size, totals only about 7 percent of the construction market for residential housing.

But 7 percent of even a large market can easily be flooded by the large North American lumber production capacity. Therefore, the potential for any significant diversion of 2 by 4 lumber into the Japanese market is quite small in the immediate future. Rather, the mature market for traditional Japanese lumber still shows the greatest potential at this time. That is, sawmills that do not want to wait for the segment to grow in size and that have better quality logs should consider the alternative of entering the old Japanese market by adapting to the standards for metric-sized lumber used in the traditional housing construction in Japan. This market is many times bigger than that of the 2 by 4 market, making up the bulk of the housing starts per annum in Japan. But it requires the added complexity of grading lumber for appearance. The up-side is that it is not necessary to be certified to use JAS grading rules! Certification is needed only if the sawmill plans to qualify its lumber for use in the construction of public buildings where a JAS grade stamp is required in Japan.

Unlike the United States, where lumber quality standards are actively enforced by the grading bureaus, it is market forces and the consumer that plays the major role of controlling the quality standards in Japan. In other words, if the lumber grading is not strict enough, customers will shy away from buying or demand lower prices. And if the quality starts to decline, the customer will let you know and will probably make a claim, entitling him to some reimbursement.

Why do Japanese mills become JAS certified if residential houses in Japan do not require JAS-stamped lumber? The answer is that in fact many are not JAS certified, but those who are certified only use the JAS stamp for lumber orders going into the construction of public buildings where JAS-stamped lumber is a requirement. Since the market for public wooden buildings is relatively small, there is little incentive for a Japanese sawmill to endure the process to become certified. Neither is it the intent of the Japanese Ministry of Agriculture to have every Japanese sawmill certified.

The JAS grading rules in traditional construction of residential homes was created to serve two purposes: to improve quality and to assist consumers in product selection. This goal is being achieved because the industry is using it as a minimum quality standard that applies when the lumber received seems to be a lower grade than what was specified by the consumer. The JAS grading rules are then used to determine if the lumber quality does not at least meet the specifications described by the JAS rules. When they don't, the consumer has a legal right to some compensation and the mill's image becomes tarnished.

It is therefore in the sawmill's best interests to grade its lumber to exceed the JAS standards. The result is a good reputation and a premium price for the mill's product. But the process of obtaining JAS certification for traditional building construction of residential houses is simply not required.

If JAS certification is not required on traditional lumber products, why would a North American company want JAS certification? For Vanport, it provided an effective method of advertising because Vanport was the first sawmill outside Japan to receive JAS certification. Vanport became news in all the Japanese lumber journals. Another reason is promotion: JAS certification enhances the reputation of the manufacturer in the Japanese marketplace. The process of certifying the graders can also provide them with improved understanding of the Japanese grading system and its minimum quality standards. Finally, if the target market is not in traditional lumber but is for the new 2 by 4 construction design, then the JAS stamp for 2 by 4s is required.

Customization

The ultimate goal of a sawmill that wants to take advantage of the lucrative Japanese market is to become like its main competitor, a Japanese sawmill. This involves more than just adapting the machinery to cut to metric sizes and being able to grade to Japanese rules. The way a sawmill does business also influences its competitiveness.

One way to compete in Japan with Japanese sawmills is to sell lumber in "yen per cubic meter" instead of "dollars per thousand board feet." Not only does this remove the customer's risk of fluctuations in the exchange rate, but it makes the American sawmill's prices directly comparable to the lumber prices in the Japanese market to give the sawmill leverage in price negotiations. (The customer can no longer make the argument that prices need to be reduced to compensate for the change in exchange rate.)

Ordering lumber from a North American sawmill becomes more favorable if the Japanese customer's responsibilities regarding transportation are limited to arrangements for truck shipments from ports in Japan to their final destination. This means that the North American sawmill should make all the shipping and vessel arrangements to get the lumber to the Japanese port as specified by the customer.

There are other business practices that not only make it easier for a Japanese firm to do business in North America, they also reduce costs to the North American sawmill. For example, just because letters of credit are a standard instrument of payment in foreign trade, this does not mean one must use it. Letters of credit merely add to the cost of doing business if the customer is a credit-worthy foreign corporation. A better practice is to offer open accounts to customers who pass credit checks. Open accounts offer the customer most of the same advantages they have when ordering from a sawmill in Japan: ordering is simple, usually done over the phone or facsimile, and order changes do not require a massive amount of paperwork. And when a customer's order is invoiced, it is helpful to have a bank account in Japan where Japanese customers make yen payments directly to your cash account. The bank is also a source of credit reports on potential customers.

Avoid other unnecessary costs related to foreign trade. Becoming self-insured on shipments by ocean vessels avoids paying insurance coverage for the unlikely loss of lumber at sea. (How often is a cargo ship lost at sea?) A North American sawmill, which regularly and frequently sends lumber to Japan distributed on various ships bound for different ports, will incur an insignificant loss with a single shipment aboard a sunk ship as compared to the amount in expenses incurred over time in insurance premiums. (Besides, ships are covered by their own insurance.)

As the North American sawmill sells further down the distribution channel, closer to the retailers and end users, the mill will need salesmen who are fluent in Japanese to market the lumber and negotiate volume and prices. The large trading companies and wholesalers have English-speaking employees on their purchasing staff but the smaller companies often communicate only in Japanese.

Conclusion

There are two ways a sawmill can adjust to supply the Japanese market: 1) become JAS certified to supply 2 by 4 construction lumber; and 2) do business more like a Japanese sawmill which can grade and cut to the traditional Japanese grades and dimensions. The first route (2 by 4 JAS) is the easiest. The second route, although it requires more time and customization, is still the favored choice because it has a greater payback.

The globalization of markets: a European perspective

Roger J. Cooper

Abstract

Key trends in the European Community (EC) forest products market are identified and the extent of globalization of the market is assessed. The EC accounts for approximately 25 percent of the global forest products market and 52 percent of world forest products imports. The region has become increasingly self-sufficient in sawn softwoods, particleboard, fiberboard, and wood pulp, but external import dependence has increased with sawn hardwoods, plywood, and papers. Growth areas for North American supplies are with sawn softwoods and hardwoods, softwood plywood, and pulp. The extent of "globalization" in the market is low with solid wood products, greater with wood panels, and highest with pulp and paper. These differences will continue, but in all product categories there will be a gradual strengthening of globalization in the future.

Introduction

Over the last 20 years, world trade in forest products has consistently outpaced production growth. Comparative world growth rates for sawn timber, wood-based panels, and paper from 1970 to 1980 are shown on Table 1.

In terms of regional export growth, the Pacific Rim has become a key growth area for North American exports. Over the last decade, imports of processed products by Japan, Korea, and other newly industrialized countries in Asia increased in value by over 10 percent per year. By contrast, European imports have grown more slowly, with annual growth averaging 6.5 percent during the 1980s. Despite this slower growth, Europe still plays a key role in the world forest products market and in world trade.

This paper first reviews key trends in the European market over the last 20 years. This is followed by an assessment of the past and future globalization of the market. The paper concentrates mainly on the European Community (EC) but first the overall European market is examined.

Review of European market trends

The European market in a global context

Europe is the world's second largest consumer of forest products after North America. The continent accounts for about 25 percent of the world market. Table 2 shows a regional breakdown of consumption in 1990. It will be seen that Europe accounted for one-fifth of the sawn timber market, one-third of the wood panels consumption, and over one-quarter of paper and board use. However, it can be seen in Table 3 that per capita consumption in Europe is well below North American and Japanese levels. European sawn timber consumption per capita is one-third of the North American level while wood based panels and paper and board figures are 50 percent and 40 percent, respectively, of the American levels.

Europe's role in the international forest products trade is dominated by the region's heavy import

Table 1. — Comparative world growth rates.

	Annual rate of growth	
	Production	Exports
	- - - - - - - - (%/yr.) - - - - - - - -	
Sawn timber	0.9	2.3
Wood-based panels	3.8	6.4
Paper and paperboard	3.2	4.5

Sources: (4,5).

The author is a Lecturer, School of Agric. and Forest Sci., Univ. of Wales, Bangor, Gwynedd, United Kingdom.

Table 2. — World consumption of forest products in 1990.

Region	Consumption			Market share		
	Sawn timber	Wood panels	Paper and board	Sawn timber	Wood panels	Paper and board
	- - - - - (million m^3) - - - - -		(million tonnes)	- - - - - - - - - - - - - (%) - - - - - - - - - - - - -		
Europe	101	42	64	20	33	27
North America	145	38	83	29	30	35
Japan	39	12	28	8	10	12
Former USSR	90	12	10	18	10	4
Other developed	8	4	5	1	3	1
Developing	114	18	47	23	14	20
World	**497**	**127**	**237**	**100**	**100**	**100**

Sources: (5,9,11,12).
Note: Figures are averages for 1989/91.

Table 3. — Per capita consumption of forest products in Europe, North America, and Japan in 1990.

Region	Sawn timber	Wood panels	Paper and board
	- - (m^3/1000 capita) - -		(tonnes/1000 capita)
Europe	181	72	126
North America	525	137	294
Japan	316	98	235

Sources: (5,9,11).
Note: Figures are averages for 1989/91.

Table 4. — World and European imports of forest products, 1970 and 1990.

Product	World imports 1970	World imports 1990	European share 1970	European share 1990
	- - - (billion $) - - -		- - - - - (%) - - - - -	
Sawn timber	3.2	20.4	59	51
Wood panels	1.2	10.4	58	53
Pulp and paper	7.2	65.6	54	57
Others	2.5	13.0	36	27
TOTAL	14.1	109.4	52	52

Sources: (3,5).

dependence. While North America is the world's largest consumer, Europe is the world's largest importer and accounts for over half of the world imports of forest products. Over the last 20 years there has been an almost eight-fold growth in the value of Europe's forest products imports while the overall percentage of world imports has remained at 52 percent. However, there has been a fall in the region's share of sawnwood and panels imports but a growth in the share of pulp and paper trade. Unlike Japan and Korea, the vast bulk of European forest products imports have for many years been of processed products rather than logs and chips (Table 4).

Table 5. — European Community's share of European imports of forest products in 1990.

	(%)
Coniferous sawn timber	86
Broad-leaved sawn timber	92
Plywood	90
Particleboard	86
Fiberboard	80
Wood pulp	84
Paper and board	89

Sources: (9,11).
Note: Percentages based on import values.

The European Community market

The European forest products market is dominated by the market in the EC. The EC accounts for 60 percent of Europe's sawnwood use and over 70 percent of its consumption of wood-based panels and paper.

One feature of the EC has an overriding influence on the forest products market — the low forest resource endowment. Whereas North America and Nordic Europe have, respectively, approximately 2.0 ha and 2.9 ha per capita of forests, the EC has only 0.12 ha per capita. Consequently, the EC is heavily dependent on imports and accounts for 85 percent of total European forest products imports. Table 5 shows the EC share of European imports on a product basis.

Key market features and trends for the main product segments are summarized in the following sections.

Sawn softwoods

- Consumption is strongly cyclical, broadly following the cyclical movements of the construction industry (Fig. 1).
- The long-term consumption trend is virtually static.

- Self-sufficiency within the EC is increasing, rising from a level of 45 percent in 1970 to 55 percent in 1990 (Fig. 1). This reflects increased production from maturing plantations particularly in the United Kingdom, France, Germany, and Ireland.
- Consumption is concentrated in the northern and central regions of the EC which account for 80 percent of softwood use.
- The import pattern is undergoing a gradual change. Table 6 shows there has been a fall in the import share of the Nordic suppliers, a rise in the North American share, and a growth of trading within the EC. Other shares have remained relatively steady. The declining share from Nordic countries is a reflection of the region's strategy of concentrating growth in the pulp and paper sector and maintaining a fairly static level of sawn wood production.

Table 6. — Import shares of major suppliers of sawn softwoods to the European Community.

Region	1970	1980	1990
		(%)	
Intertrading	3	5	10
Nordic Europe	48	43	35
North America	6	15	20
Former USSR	17	16	15
Eastern Europe	11	6	5
Austria/Switzerland	13	14	14
Chile/Brazil/other	2	1	1
Total EC imports (million m^3)	23.1	23.7	25.3

Sources: (2,4,5,10).
Note: Import quantities include trade between member states.

Sawn hardwoods

- Total consumption at between 11 and 14 million m^3 per year is one-fifth of softwood consumption (Fig. 2).
- Like softwoods, this market is also strongly cyclical, responding to similar construction industry influences.
- Self-sufficiency is falling (Fig. 2). This partly reflects falling tropical sawnwood production from declining log imports and also a gradually declining log production in some countries, e.g., Germany and the United Kingdom.
- The long-term consumption trend is static.
- There is a higher share of consumption of hardwoods (33%) in the Mediterranean countries compared with softwoods (20%).
- Up to the late 1980s there was a relatively static market share held by tropical timbers (one-third of the market) and temperate timbers (two-thirds of the market). However, over the last 5 years the tropical share has fallen partly as a result

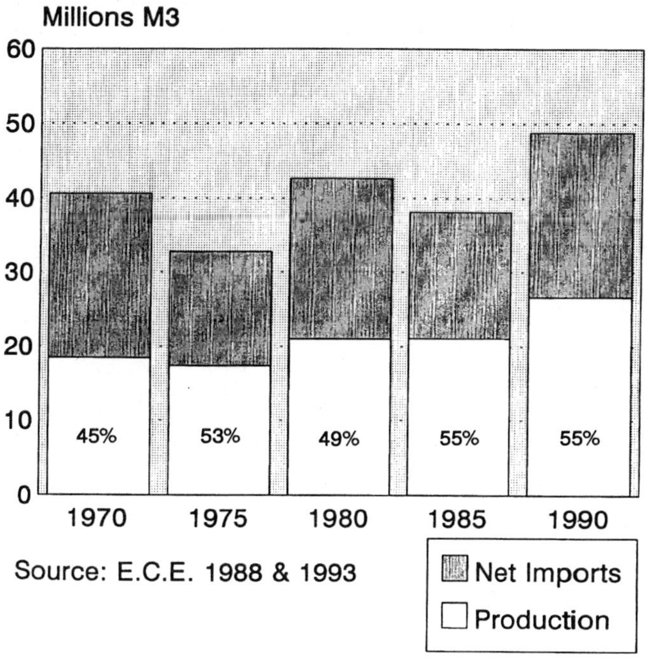

Figure 1. — EC production, net imports, and consumption, 1970-1990: sawn softwoods.

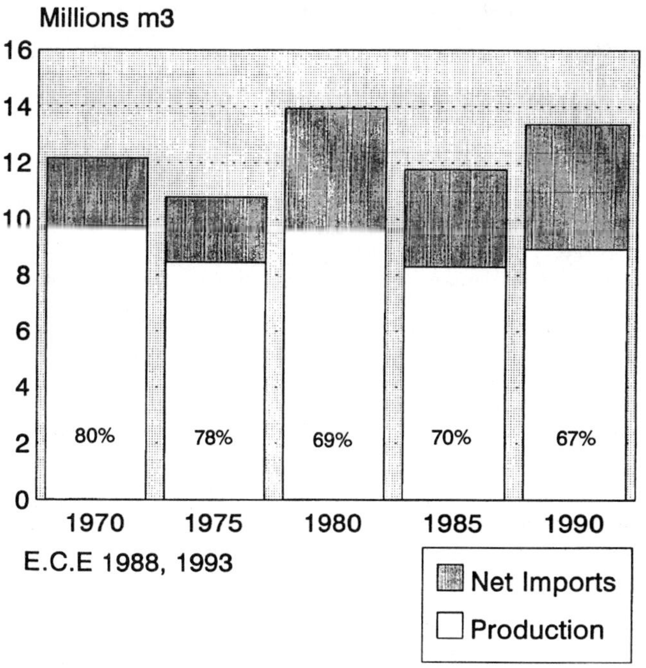

Figure 2. — EC production, net imports, and consumption, 1970-1990: sawn hardwoods.

of environmental pressures against the use of tropical timbers.

- In the tropical sector, imports of tropical sawn wood increased while log imports fell.
- In the temperate timber market, there has been a strong growth in supplies and import share from North America and a decline in the quantity and share of Eastern European suppliers (Table 7).

Plywood

- Consumption has grown from 3.5 million m^3 in 1970 to 5 million m^3 in 1990 with an average growth rate of approximately 1 percent per year (Fig. 3).
- Self-sufficiency has fallen rapidly from 60 percent in 1970 to 35 percent in 1990 (Fig. 3). This reflects the sharp decline in tropical plywood manufacture in Europe based on imported logs and the growing difficulty of competing with lower cost producers in Southeast Asia.
- Nearly half of EC plywood consumption occurs in the northern EC countries (United Kingdom, Ireland, Denmark, Belgium, Luxembourg, and Netherlands).
- The growing import supplies have been drawn mainly from Southeast Asia (Indonesia, Malaysia) and North America. These countries accounted for 75 percent of imports in 1990 compared with 47 percent in 1970. The Nordic countries (principally Finland) have lost market share. Table 8 shows changes in import shares from 1970 to 1990.

Table 7. — Import shares of major suppliers of temperate sawn hardwoods to the European Community.

Region	1970	1980	1990
		(%)	
Intertrading	29	46	39
North America	6	21	28
Eastern Europe and former USSR	61	30	32
Nordic Europe	4	3	- -
Total EC imports (million m^3)	1.7	2.6	2.8

Sources: (2,4,5,10).
Note: Import quantities include trade between member states.

Table 8. — Import shares of major suppliers of plywood to the European Community.

Region	1970	1980	1990
		(%)	
Intertrading	18	16	12
Nordic Europe	26	13	10
North America	29	29	40
Tropical suppliers	18	32	34
Eastern Europe and former USSR	9	10	5
Total EC imports (million m^3)	1.8	2.6	4.1

Sources: (2,4,5,10).
Note: Import quantities include trade between member states.

Particleboards

- While North America leads world consumption in most product sectors, in the case of particleboard, EC consumption is more than double the North American level.
- During the 1970s, particleboard consumption experienced a strong growth of 6 percent per year but this slowed to 3 percent in the 1980s as the market for conventional boards approached saturation. In the last 5 years, growth has been stimulated by the introduction of new types of boards, in particular oriented strandboard (OSB), now totalling about 200,000 m^3 in the EC (6).
- In contrast with other product categories, the EC is virtually self-sufficient in particleboards (Fig. 4) although there is a substantial level of trading within the EC accounting for 25 percent of consumption.

Fiberboards

- During the 1970s and early 1980s, there was an irregular but clear decline in fiberboard consumption as traditional boards lost markets to other materials (Fig. 5).
- This fall has been dramatically reversed by the 20 percent annual growth in medium-density fiberboard (MDF) consumption from 1982 to 1992 (Fig. 6). Production capacity now stands at 3 million m^3, nearly 40 percent above the North American level (15).

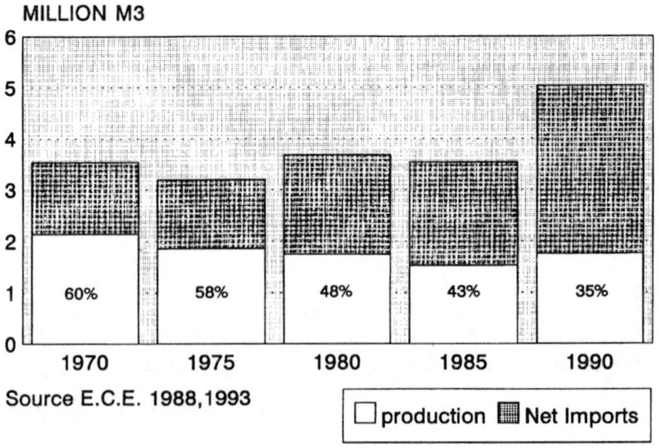

Figure 3. — EC production, net imports, and consumption, 1970-1990: plywood.

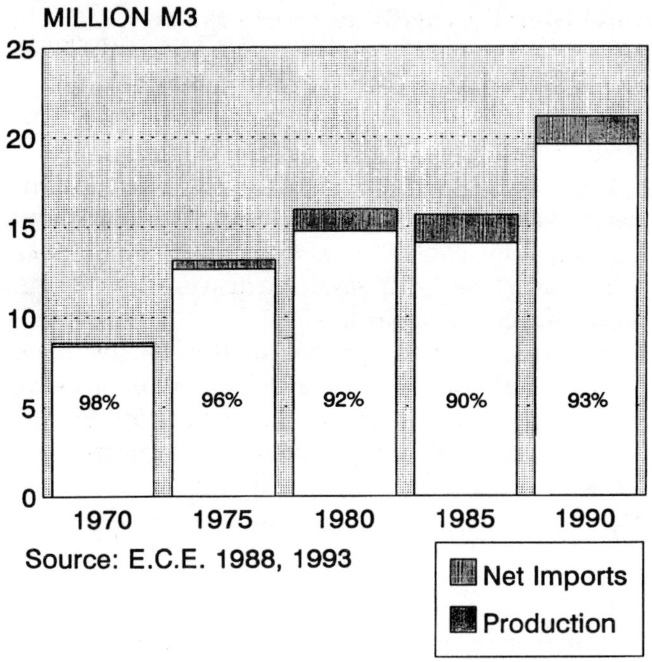

Figure 4. — EC production, net imports, and consumption, 1970-1990: particleboard.

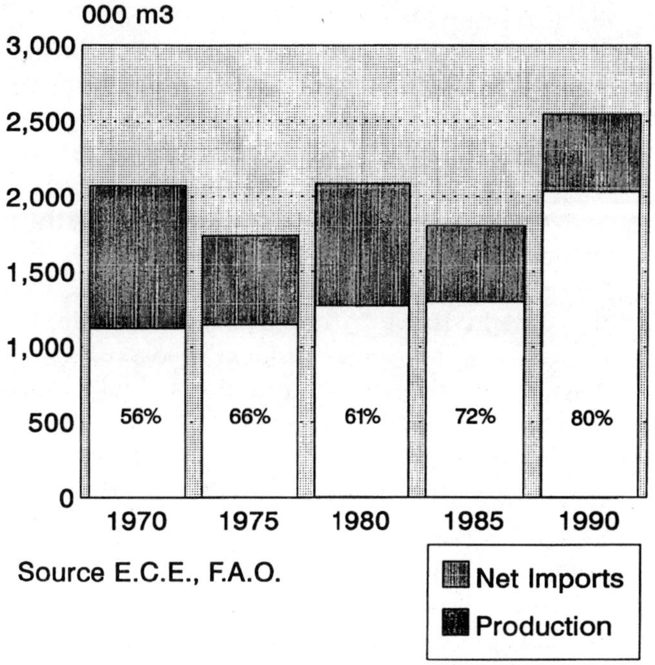

Figure 5. — EC production, net-imports, and consumption, 1970-1990: fiberboard.

- MDF replaced particleboards in the cabinet furniture industry and is now competing strongly with sawn timber in the joinery and mouldings market.
- As a result of the rapid rise in MDF production and use, the EC has become more self-sufficient in fiberboards (Fig. 5) and there has been a rapid rise in intertrading within the EC and fall in imports from traditional suppliers.

Wood pulp

- Consumption of wood pulp grew at 1.5 percent per year during the 1970s and 1980s, but increasing production caused self-sufficiency to rise from 42 percent to 54 percent over the period (Fig. 7).
- The fastest growth in production occurred in Portugal and Spain where kraft pulp production based on pine and eucalyptus plantations increased by 2 million tonnes.
- In the northern and central EC, production growth was mainly concentrated on mechanical pulps.
- The market pulp industry is relatively small. Only Portugal and Spain are significant suppliers.
- Recycled fiber has gradually increased its share of the EC paper industry's fiber use. The ratio of pulp consumption to paper production fell from 64 percent in 1970 to 51 percent in 1990 and this trend is expected to continue.

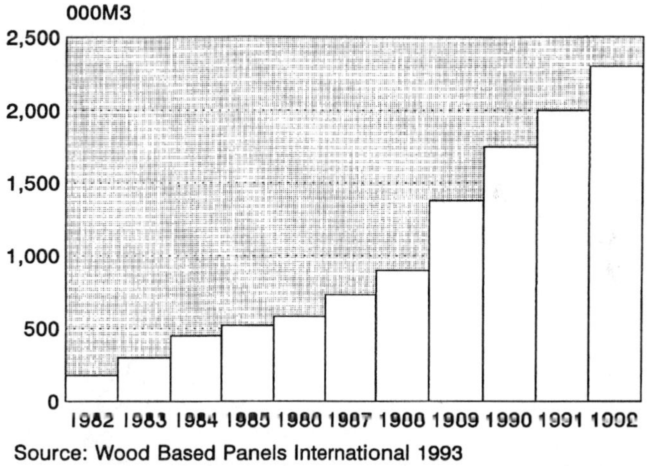

Figure 6. — European MDF production, 1982-1992.

- Sources of market pulp have also changed. The import share and shipments from Nordic Europe have declined as the Scandinavians have expanded paper production and exports. Growing import volumes and shares have come from North America, developing countries (mainly Brazil and Chile), and from intertrading (mainly from Portugal and Spain) (Table 9).

Paper and board

- Consumption grew by 3 percent per year during the 1970s and 1980s, and self-sufficiency dropped slightly from 77 percent to 73 percent (Fig. 8).
- EC imports grew at a considerably faster rate of 5.5 percent per year reflecting an increasing specialization in paper production both within Europe and globally.
- There have been substantial shifts over the last 20 years in the EC's import pattern. Table 10 shows there has been a considerable loss of import share by North America and also a smaller loss by Nordic countries. The main trade growth has been in intertrading and particularly in exports by France, Germany, and the Netherlands to other EC countries.

Conclusion on the EC market review

Key points from the above reviews are as follows:

- In all product areas, the international trade of EC states has grown at a faster rate than consumption.
- The EC has become increasingly self-sufficient in sawn softwoods, particleboard, fiberboard, and wood pulp. The EC's *external* import dependence has increased with sawn hardwoods, plywood, and paper and board.
- Increased intertrading within the EC has been particularly marked for particleboard, fiberboard, and paper where considerable benefits for EC countries arose from the removal of internal tariffs in the 1970s.
- The two major external supply areas to the EC

Table 9. — Import shares of major suppliers of wood pulp to the European Community.

Region	1970	1980	1990
		(%)	
Intertrading	6	4	17
North America	33	42	39
Nordic Europe	56	42	35
Others	4	12	9
Total EC imports (million tonnes)	9.2	9.4	11.4

Sources: (2,4,5,10).
Note: Import quantities include trade between member states.

Table 10. — Import shares of major suppliers of paper and board to the European Community.

Region	1970	1980	1990
		(%)	
Intertrading	18	31	37
North America	21	15	7
Nordic Europe	55	50	48
Others	5	4	7
Total EC imports (million tonnes)	9.1	14.0	26.4

Sources: (2,4,5,10).
Note: Import quantities include trade between member states.

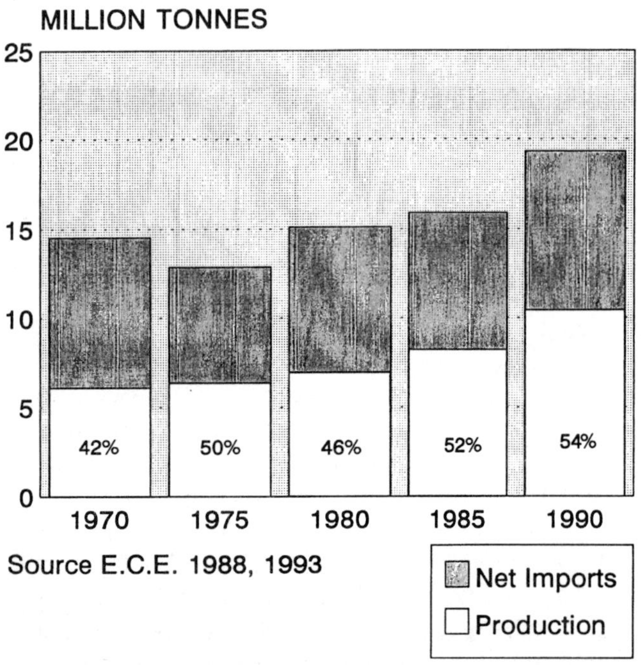

Figure 7. — EC production, net imports, and consumption, 1970-1990: wood pulp.

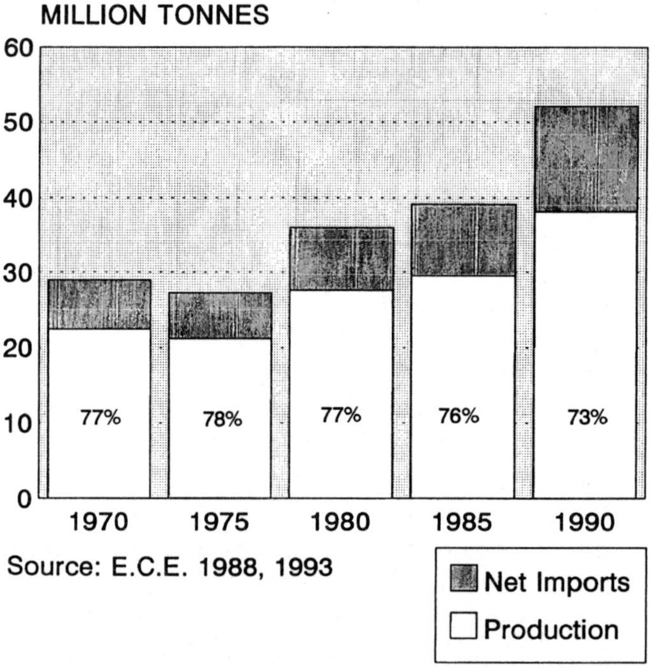

Figure 8. — EC production, net imports, and consumption, 1970-1990: paper and board.

are North America and Scandinavia, with Southeast Asia a third key area for tropical timber products. North America has increased exports and gained market share with sawn softwoods and hardwoods, softwood, plywood, and wood pulp. Scandinavia's main market growth has been in paper where it has tariff advantages over North America. Tropical suppliers in Southeast Asia have a major share of the plywood market but are losing ground in the sawnwood sector.

Globalization of markets

The EC forest products market has clearly become more international in the sense that import requirements have increased. The question now addressed is to what extent the markets have and will become more global.

First, it is necessary to define globalization. There has been considerable debate in the literature of what is meant by "global" and "international" marketing (1). In essence, global marketing involves selling standardized products in the same way to all markets. By contrast, international marketing recognizes that different countries or markets have special requirements which cannot be met by standardized products. This means that different products and marketing methods have to be employed for individual markets (1). However, it must be recognized that different interpretations have been applied to the term "globalization." For instance, there is debate about the extent to which standardization of product specifications is a necessary condition of globalization, given that technological advances make it quicker, easier, and cheaper than in the past to manufacture customized products for specific market needs (7).

An evaluation of markets in each major product sector is given below, based on the criteria summarized in Table 11.

Table 11. — Global and international market characteristics.

Global market characteristics
Standardized product lines
Minimal product modifications
Standard international product specifications
Similarity of product uses across markets
Similar distribution channels across markets
International market characteristics
Different products for different segments
Variable product specifications
Varied uses across markets
Varied distribution channels across markets

Sawn timber

At present there is a marked absence of global characteristics in the sawn wood market. The following points support this judgment.

- There are widely differing patterns of use by species between users and countries.
- There is little standardization of specifications (sizes, grades) among buyers.
- There is only limited harmonization of grading systems by suppliers (e.g., in the United Kingdom there are at least four major sets of grading systems operating in the softwood market, and a similar situation prevails with hardwoods.
- There is a very large number of independent, mainly small suppliers to the market. In the EC alone there were nearly 2,800 mills each producing over 10,000 m^3 per year in the late 1980s (13).
- New plant health regulations have reduced market access by North American softwood suppliers.

For the future there seems little likelihood of a fundamental change to the present situation. However, some movement towards standardization of specifications will result from the introduction of a common European structural timber code. Once introduced this will lead to European wide recognition of stress grading rules and harmonized criteria for the specification of structural timbers.

Wood panels and wood composites

The present European market exhibits some features which make these industries more global in character than sawmilling.

- Products are relatively homogeneous.
- Common product standards and grades are used across Europe.
- Patterns of use by material vary less than with sawn wood.
- Production is more concentrated (e.g., the EC had 156 particleboard and 42 fiberboard plants in 1990) (14).
- There is substantial multinational ownership of plants.
- There are strong Europe-wide trade associations.

However, some features offset these globalization features.

- European tariffs are a disadvantage to non-European suppliers.
- Some products, such as standard grades of particleboard, possess low "transportability."

Greater globalization of these markets is expected for the future brought about by the following:

- Continuing rapid growth in consumption, particularly of MDF and OSB.
- Increasing multinational ownership and concen-

tration into larger groups necessitated by rising investment costs.

- New composite materials will be aimed at global niche markets, e.g., new structural composite materials. This process will be assisted by harmonization of structural codes.

Pulp and paper

Currently the European pulp and paper market is judged to be the most global of the three main product sectors for the following reasons:

- Standard product specifications apply across markets.
- The industry is strongly vertically integrated from forest to paper use.
- The industry is concentrated in the hands of a relatively small number of large companies. The top 15 European producers account for 50 percent of paper production (8).
- There are similar uses and technical specifications across markets.
- A higher proportion of exports by European countries go to non-European markets than with sawn wood or panel products.
- The industry has powerful trade associations.

In the future the trend to increased globalization will continue for the following reasons:

- Increasing investment costs in the industry will lead to further mergers and acquisitions and industrial concentration.
- Trade will become more diverse with increasing European supplies drawn from producers in resource rich countries, e.g., Brazil, Chile, and Indonesia.
- The European industry will be stimulated to attain higher environmental standards (effluent treatment, recycled fiber use, minimization of the environmental impact of paper use). This will be driven by European environmental legislation, consumer pressure, and the eco-labeling of end products.

Conclusions

Seen from the European perspective there will be diverging trends in the process of globalization in the future. The sawn wood market will not develop very far or very rapidly to a more globalized state. Any movement in this direction will mainly be with structural sawn wood. Opportunities for North American suppliers will mainly lie in specific market segments (e.g., temperate hardwoods to replace tropical hardwoods, joinery quality softwoods to replace European softwood species).

The composites market will continue to become more global in nature. Rising investment costs in new plants will stimulate multinational investment and further concentration of ownership. As specialization of production proceeds, international trade will continue to grow more rapidly than production. Products currently imported from outside Europe face severe challenges. North American softwood plywoods will meet increasing competition from European OSB production and Southeast Asian plywoods will face increasing environmental pressures.

Europe's imports of further manufactured wood products (joinery products, furniture components) will continue to expand as exporters place greater emphasis on added value strategies. Suppliers will be in both the developed and developing world. However, specifications of these products will be extremely varied and strong market segmentation will be practiced by exporters.

The pulp and paper sector will continue to be the most global of the forest industries and the trend to increasing globalization will continue. Continued specialization of production on a global scale will occur, stimulated by differences in wood, energy, and environmental costs. EC imports from outside Europe will increase. Shipments by EC and Scandinavian producers to world markets particularly to the expanding markets in the Pacific Rim countries will also grow.

In conclusion, the European market will become more "global" in the sense that the region's imports will be drawn from increasingly diverse sources. But globalization in the sense of standardization of products and marketing methods will be mainly limited to the wood composites and paper sectors.

Literature cited

1. Cohen, D.H. 1990. Marketing Strategies for the Global Market, Vol. 5 of Proceedings of XIX World Congress of IUFRO. 5-11 August 1990. Montreal, Canada.
2. FAO. 1972. Yearbook of Forest Products 1970 and 1971. FAO, Rome, Italy.
3. ________. 1977. Yearbook of Forest Products 1966-1977. FAO, Rome, Italy.
4. ________. 1982. Yearbook of Forest Products 1969-1980. FAO, Rome, Italy.
5. ________. 1992. Yearbook of Forest Products 1979-1990. FAO, Rome, Italy.
6. Jaakko Poyry. 1993. Personal Communication.
7. Perry, A.C. 1989. International versus Domestic Marketing. Four Conceptual Perspectives. European J. of Marketing 24(6).
8. Pulp and Paper International. 1993. The Top 150. Pulp and Paper International 35(9). Miller Freeman, Brussels.
9. ________. 1993. The Top 150. Pulp and Paper International 35(7). Miller Freeman, Brussels, Belgium.
10. UN/ECE. 1992. Forest Products Trade Flow Data 1990-1991. Timber Bulletin XLV, No. 4, United Nations, N.Y.

11. ________. 1993. Forest Products Statistics 1988-1992. Timber Bulletin Vol. XLVI, No. 2. United Nations, N.Y.
12. ________. 1993. Annual Forest Products Market Review 1992-1993. Timber Bulletin Vol. XLVI, No. 3. United Nations, N.Y.
13. UN/ECE/FAO. 1992. Survey of Structure of the Sawmilling Industry. Timber Bulletin Vol. XLIV, No. 2, Geneva.
14. ________. 1990. Survey of Structure and Capacity of the Wood Based Panels Industries 1989. Timber Bulletin, Vol. XLIII, No. 2, N.Y.
15. Williams, W.C. 1993. MDF Wood-Based Panels International 13(4):7-11. Benn Publications, Tonbridge, U.K.

Luncheon Address and Technical Forum Presentations

Luncheon Address:

What on earth have we done to our forests?

A brief overview on the condition and trends of U.S. forests

Douglas W. MacCleery

Historical significance of U.S. forests

In our highly urbanized society, it is easy to overlook the critically important role that forests played in the history and development of the nation. The single most important event in the evolution of the American landscape as we know it today was the clearing of forests for agriculture. This clearing was essential to provide food for a rapidly growing nation.

In addition to clearing for agriculture, forests were utilized for a wide variety of products and uses. The forests were habitat for wildlife which were the basis for a lucrative export markets in furs, as well as providing an important supplement to the diet of millions of Americans.

Wood was virtually the only fuel used in this country for most of its history. It warmed its citizens, produced its iron, drove its locomotives, steamboats, and stationary engines. Lumber, timbers, and other structural products were the primary material used in houses, barns, fences, bridges, even dams and locks. Such products were essential to the development of rural economies across the nation, as well as to industry, transportation, and the building of cities. American forests — the products derived from them and the land they occupied — were, in a very real sense, the economic foundation of the nation.

In a spiritual sense, the forest, and the wilderness values it represents, has also played an important role in the identity of the nation. This was expressed in the writings of Thoreau, Emerson, Marsh, and others and was first evidenced politically in the late 1800s in efforts to address concerns over wildlife decimation and forest depletion.

There is no question that without its forests America would have had a decidedly different history, and would be a decidedly different place than it is today.

In the past few years, we have seen an increased interest in all aspects of the environment. A portion of this interest has focused on concerns over the condition of the nation's forests and associated wildlife. An enlightened perspective on the current condition and trends of our forests and wildlife should be based on a general understanding of how they came to be what they are today. This paper seeks to fill that need by briefly outlining the natural and human influences that have shaped our forests over the years.

Nature and extent of North American forests prior to European settlement

Early European settlers to America were awed by the ocean of trees which greeted them. Forests blanketed much of the eastern third of the United States, extending from the Atlantic Coast to the prairies beyond the Mississippi River. These vast forests were a sharp contrast to those of England and much of the rest of Europe, which had been severely depleted for fuel and building materials. The original American forest covered a little over one billion acres, or about half of the U.S. land area (including Alaska).

The author is Assistant Director, Forest Inventory and Planning, Timber Managment, USDA Forest Serv., Washington, D.C.

Today, about one-third of the nation is forested — about 737 million acres. This is about 70 percent of the area that was forested in 1600. About 310 million acres of forest have been converted to other uses since 1600 — primarily to agricultural lands.

Forests remained the dominant feature of the landscape in eastern North America for centuries after initial settlement. In 1796, almost two centuries after the first European settlements, a French naturalist visiting the new American nation wrote that "The most striking feature (of the country) is an almost universal forest, starting at the Atlantic and thickening and enlarging to the heart of the country." He said that in his travels into America's interior he "scarcely passed, for three miles together through a tract of unwooded or cleared land."

About three quarters of the nation's original forest was in the eastern third of the country. West of the Mississippi, as rainfall diminished, forests and woodlands gave way to vast, treeless prairies and deserts. However, in mountainous areas where rainfall was sufficient and along the Pacific Coast, extensive and often magnificent forests developed.

A popular myth is that, prior to European contact, America was dominated by impenetrable, relatively uniform ancient forests, that cloaked the landscape in a long-term static balance with the environment. The reality was quite the contrary. Presettlement forests were exceedingly dynamic, shaped by a myriad of both natural and human-caused influences, disturbances, and catastrophic events that had a profound effect on the age, plant species, and wildlife of the forest environment. Presettlement forests in both the East and West were a diverse mosaic of forest stands whose age, tree species, and wildlife varied widely and reflected the disturbance history of the area.

America's original forests were not pristine in the sense of being uninfluenced by humans. In both the East and the West, forests were strongly influenced by Native Americans. In the eastern forests, most Native Americans lived in fixed villages. Domesticated crops accounted for more than half of their diet. It was a maize-based agriculture. Population densities were at least five times that of the nomadic hunter/gatherer societies to the north and west. Hundreds of thousands of acres were cleared for fields. Tens of millions more were burned frequently to improve game habitat, facilitate travel, reduce insect pests, remove cover for potential enemies, enhance conditions for berries, and to drive game. It was a shifting type of agriculture. Fields and villages were abandoned when their natural fertility ran out, new forests were cleared, and the abandoned lands quickly reverted back to forest.

European settlers spoke of the open park-like forest that they encountered (a condition created by frequent burning), and of the frequency of Indian burning. Settlers in New England reported that Native Americans burned the woods twice a year. A general statement can be made that, wherever there were Native Americans, there was fire.

The abundant wildlife that was reported also gives an indication of the frequency of disturbance: deer, wild turkeys, and a variety of game birds abounded. Even elk and bison, normally associated with the western prairies, were common in the eastern forest. Bison were reported as far east as Massachusetts.

The south was dominated by fire-created forests, such as longleaf pine savannas on the Coastal Plain and Piedmont. The hardwood forests of the Appalachian Mountains were also burned frequently by Native Americans. In Virginia, the Shenandoah Valley — the area between the Blue Ridge Mountains and the Alleghenies — was one vast grass prairie. Native Americans burned the area annually.

On the western fringe of the forest, fire-dominated forest types, such as oak and pine savannas, covered tens of millions of acres. The forests were heavily influenced by fires sweeping in off the western prairies. Fire-created prairies extended well into Ohio. Evidence of the dominant role fire played in these forests is demonstrated by the fact that, when farms finally began to move out onto the prairies and to cut off prairie fires, millions of acres of open oak savannas and even treeless prairies to the east of these farms became dense woodlands and forests within two decades.

As we see rising interest in protecting some of our forests in their "natural" condition, the complex natural and presettlement human history of U.S. forests raises equally complex technical and policy questions over whether to allow wildfire to assume its natural role in these areas, as well as whether to seek to replicate presettlement human influences. We know that it is virtually impossible to separate natural from human-caused influences in presettlement forests — North American forests have been both occupied and influenced by humans from the time these forests advanced north before the retreating continental glaciers 8,000 years ago.

European view of the forest

There can be no doubt that the era of European settlement ushered in a vast increase in the impact of humans on the forest. The abundance of land and resources and the scarcity of labor was a defining difference between America and Europe, where the situation was reversed. This difference affected everything we did, from the stewardship we applied to our resources (or the lack thereof), to the adoption of the institution of slavery.

The seemingly endless forest was viewed as a mixed blessing by early European colonists. On one hand, it provided an abundant and readily available source of fuel and building materials. It also yielded abundant game which remained for decades after settlement as an important source of food. But the forest was also habitat for wolves, eastern panthers, and other predators that found colonial livestock easy prey and against which the colonists waged unrelenting war. It provided cover for sometimes hostile Indians. But most importantly, it occupied potential cropland that could be liberated only after intensive and back-breaking labor using the primitive hand tools of the day.

For the first three centuries of our history, most Americans were farmers. In 1800, 95 percent of the people lived on the land. Except for a relatively few people engaged in plantation agriculture in the South, most were subsistence farmers. The predominant view that emerged in the early 1600s, and that continued for almost 300 years, was that the forest was both inexhaustible and an obstacle to the much preferred agricultural use of the land.

After all, the forest was vast and cropland was scarce. The nation was much more concerned about feeding itself than it was over the spiritual value of the forest. Forest clearing became a win/win situation. It liberated cropland and pasture while providing fuel, fencing, and material for building homes, barns, mills, and factories. Indeed, often selling the fuelwood and potash that could be made from wood ashes would pay most of the cost of buying the land.

Use of wood for fuel

Energy was the dominant use of wood on a volume basis until well into the late 1800s. In the late 1700s, about two-thirds of the volume of wood harvested was for fuel. By 1850, wood still provided over 90 percent of the nation's energy BTUs.

By far the dominant use of fuelwood was for domestic heating and cooking. However, it was also the primary industrial fuel. Until after the Civil War, virtually all steamboats, railroads, and stationary engines used wood fuel. The per capita consumption of fuelwood averaged over four cords per year for most of the 19th century, and the volume of fuelwood consumed rose 15 times between 1800 and 1900.

Even before 1800, fuelwood cutting was depleting the forests around population centers. In 1759, one visitor described the area around Philadelphia as "bereft of forest." In the late 1700s, fuelwood was being hauled nearly 100 miles to several coastal towns, causing the price to rise beyond the reach of the poor.

Fences

Until the middle of the 19th century, the second most important use of wood on a volume basis was for fences. Fences were not ornamental. Because labor was scarce, herding livestock, as had been common in Europe, was generally not practiced. Instead, hogs, cattle, and other livestock were turned untended into the woods. Fences were needed to keep them out of crops and gardens. Construction of fences was one of the most labor-demanding tasks of farming, second only to forest clearing in its labor requirements.

In 1850, there were 3.2 million miles of wooden fences in the United States, enough to circle the earth over 120 times.

Early industrial use of wood

Most mechanical power for textile mills, grist mills, sawmills, and other mills and factories was provided by water. But most mills were built of wood, the dam in the river was most likely made of wood, even most of the machinery inside the mill was built of wood, tipped with iron.

The production of iron itself depended upon wood. Until 1850, the majority of U.S. iron was made from wood charcoal. Virtually every state east of the Mississippi had a number of iron-making furnaces and forges fired with wood. In the 1850s, the tonnage of coke iron finally exceeded that of charcoal iron. Even so, the tonnage of charcoal iron produced continued to rise until 1900.

Wood was also used as process heat for breweries, tanneries, salt evaporation plants, and anywhere industrial process heat was necessary.

Transportation

By the early 1800s, the United States was one of the largest nations in the world. What tied such a large group of disparate and often quarreling states together more than anything else was a transportation system. America's forests figured heavily in this task as well.

The nation's first highways were its rivers. Wooden keelboats were followed by steamboats, which proliferated after 1830. Steamboats were made of wood and, until the Civil War, used wood for fuel. In

1840, almost 900,000 cords of wood were sold for steamboat fuel, or a fifth of all fuelwood sold.

Railroads followed the steamboats. After 1850, the railroads began to expand rapidly to link the growing cities and to provide access to the cities from the agricultural and forestland upon which they depended. Although called the "iron road," railroads used far more wood than iron. Except for the engine and rails, railroads were made of wood. The cars were wood, the ties were wood, the fuel was wood until the Civil War, the bridges and trestles were wood, and station houses, fences, and telegraph poles were wood.

The mileage of U.S. railroads increased more than 35 times between 1850 and 1910, from 10 thousand miles to more than 350 thousand miles. By the late 1800s, railroads accounted for between 20 to 25 percent of the total consumption of timber in the country.

By far the most significant railroad use of wood on a volume basis was for crossties. Each mile of track required over 2500 ties. Crossties were not treated until after 1900. Because of their rapid deterioration in contact with the ground, ties had to be replaced every 5 to 7 years. Given the mileage of track in 1910, that would be equivalent to replacing the ties on over 50,000 miles of track annually. It is estimated that in 1900, just replacing railroad ties on a sustained basis required between 15 and 20 million acres of forestland.

Farm clearing

While use of wood for industrial purposes and transportation was certainly significant, especially in the latter half of the 1800s, clearing for agriculture has been the dominant cause of reduction in forest cover. Since there was no improvement in crop yields per acre throughout the 19th century and well into the 20th century, farm clearing increased at about the same rate as the growth in population. Between 1850 and 1900, the U.S. population increased over three times, from 23 to 76 million people, while the area of cropland increased four times, from 76 to 319 million acres. For every person added to the U.S. population during the 19th century, farmers were putting another 3 to 4 acres of cropland under the plow. The area of pasture and hayland increased even more than that of cropland.

In the 60 years from 1850 to 1910, American farmers cleared more forest than the total amount that had been cleared in the previous 250 years of settlement — about 190 million acres.

To put forest clearing during this period in human terms, the nation's farmers were clearing forest at an average rate of 13.5 square miles per day every day for 60 years.

Increase in lumber production after 1850

The 19th century was a period of rapid population growth. During the century the U.S. population increased by 14 times — increasing from 5.3 to 76 million people.

The America of the early 1800s was overwhelmingly a nation of largely self-sufficient farms. However, after 1850 the United States began to restructure from a rural, agrarian society to one increasingly urban and industrialized. Between 1850 and 1900, the population of the nation's cities increased at twice the rate of the general population, causing a profound change in the demands the nation placed on its forests.

In addition to increased resource demands resulting from the growing cities, farms were pushing out onto the treeless prairies. These prairie farms themselves began to demand large quantities of lumber for barns, homes, outbuildings, and fences.

Logging and sawmilling increasingly began to be organized as large-scale industrial operations. The consumers and the forests upon which they depended for their natural resources began to be separated by hundreds of miles.

Lumber production increased dramatically. It rose more than eight times between 1850 and 1910, from 5.4 billion board feet to 44.5 billion annually, or more than twice the rate of population growth. Vast areas of the midwest, south, and Pacific coast were logged, and often relogged. The tree limbs, tops, and other debris that remained after logging were often burned in the belief that the logged areas could be converted to cropland or improved pasture. These uncontrolled slash fires burned more or less continuously, and when weather conditions were right, sometimes resulted in massive wildfires, property damage, and major loss of human life.

Repeated wildfires also killed residual trees and seedlings left by logging and created tens of millions of acres of what are called "cutovers" or "stumplands" that remained idle, unstocked or poorly stocked with desirable tree species for extended periods of time.

Wildlife depletion

The nation's wildlife faced many of the same pressures as its forests. A wide variety of wildlife species, which at the time of settlement had been present in huge numbers, were being severely depleted. The most significant factors were virtually unrestricted market hunting of all kinds of wildlife for food, furs, and feathers (which in the late 1800s were in great demand for women's hats), as well as habitat modification caused by farm clearing, logging, and massive wildfires. Even many species of songbirds, such as robins, meadowlarks, and others, were being adversely affected in some areas by hunting for food.

Before the turn of the century, a growing number of people began to be increasingly concerned about what was happening to much of the nation's woodlands. Fears about future supplies of timber were mixed with apparent implications of increased flooding and watershed damage, depleted wildlife populations, loss of the beauty of the American landscape, and even concerns about how forest clearing was affecting the climate itself. All of these concerns began to call into question the myth of forest inexhaustibility.

Predictions of an impending "wood famine" were raised frequently, beginning in 1865 with Frederic Starr. Use of the term "famine" was an apt one, for wood in its various forms was one of the most widespread and essential materials in both domestic use and industry.

Bernard Fernow, a German forester, began to promote the revolutionary idea that forests are a renewable resource that can be managed on a sustainable basis over the long term.

George Perkins Marsh, who grew up in Vermont in the 1830s, began to raise concerns about the adverse effects of farm clearing on watersheds and other environmental values. In 1864, he wrote *Man and Nature*, which became a catalyst for public concern.

In 1855, Henry David Thoreau retired to Walden Pond in Massachusetts, to reflect on the intrusion of civilization and to mourn the loss of America's wild places. Thoreau's was one of the earliest voices that began to put a value on "unspoiled" places for human spiritual renewal in a society increasingly urban and artificial.

Teddy Roosevelt, a well-born but sickly New Yorker, went out to the Western frontier to test his mettle. He came back transformed and became a lifelong sportsman and conservationist. In 1887, Roosevelt founded the Boone and Crockett Club, which became a strong and effective advocate for sportsmen. Its membership was small in number, but very large in influence.

More importantly, in 1901 an assassin's bullet propelled Roosevelt into the Presidency. In that position he had a profound effect on the conservation history of the nation.

The "conservation movement," if it could be called that, was made up of a disparate of group of sportsmen, foresters, concerned citizens, and intellectuals, all concerned about what they saw going on. It was not an organized movement, in the sense that the groups strategized to work in concert. Instead they were groups with related interests which moved more or less independently, seeking to achieve ends that had a common theme.

The rapidity of the changes was a factor in the public concern. People could clearly see the changes in the forest landscape within their lifetime. In 40 to 50 years, many areas went from 80 percent forest to 10 percent or even less. Two wildlife species became the symbols of the deterioration of American wildlife. Bison, which had covered the plains in the tens of millions in 1870 were reduced to scattered remnant bands in 15 short years. The passenger pigeon, probably the most abundant bird in North America, in 1870 had darkened the skies over much of the Midwest. It was virtually extinct 20 years later in 1890.

Early responses to such concerns were expressed in setting aside land in protective categories, such as Yosemite in California (1864), Yellowstone in Wyoming (1872), and the Adirondack Preserve in New York (1885). In 1891, Congress authorized the President to designate Forest Reserves out of Public Domain lands, but made no provision for their management. The Forest Reserves, unlike the park preserves, were generally not tied to the preservation of a nationally significant unique area. These designations, which had grown to 40 million acres by 1897, generated much opposition; and rightfully so, because it was unclear how such set-asides would address society's need for grazing, wood products, and other resources.

Out of the debate emerged a new idea, one rooted in the sustainable forest management concepts practiced in Europe. Conservation, the notion that forests, wildlife, and other renewable resources can be managed under scientific principles on a sustainable basis over the long term, increasingly began to be promoted by professional foresters, such as Fernow, as well as by hunter/sportsman's groups concerned about the depletion of wildlife. As the result of these efforts, in 1897 Congress finally gave the Forest Reserves a management mandate, to "preserve and protect the forests," to "secure favorable conditions of water flows," and to "furnish a continuous supply of timber for the use and necessities of the citizens of the United States."

The idea of conservation as wise use received widespread public support under the dynamic advocacy of Gifford Pinchot, first Chief of the Forest

Service and his close friend and mentor, Theodore Roosevelt.

Thus, the Forest Service emerged as a product of the nation's first conservation movement. The agency became a major instrument for the practical application of conservation as the wise use of forest resources. That role encompassed much more than the management of the National Forests. It also included leadership in working with the states and private landowners in advocating the protection and productive management of nonfederal forestlands, as well as research in forest management and improved utilization of wood products.

Policy framework for forest and wildlife conservation

The policy framework that had emerged by the 1930s to address these issues emphasized protection of forests from wildfire and wildlife from over harvest, and the management of both forests and wildlife under scientific principles. Specific actions focused on: 1) promoting and encouraging the protection of forests, regardless of ownership, from wildfire, insects, and disease; 2) the acquisition of scientific knowledge on the management of forests and wildlife, and on the improved utilization of wood products; 3) encouraging the productive management of private forestlands through tax incentives and technical and financial assistance; 4) the adoption and enforcement of strong state and federal wildlife conservation laws; and 5) the acquisition and management of public lands for both commodity and amenity uses and values. (See the Appendix for a further description of this policy framework.)

A key element of the public policy framework was strong cooperation among federal, state, and private sector interests to achieve common goals. A stronger, more coercive federal role in the direct regulation of private forestlands was considered and debated, but ultimately was rejected.

Condition of U.S. forests and wildlife in 1990 versus 1900

It is a measure of both the inherent resilience of our forests, and of the success of the policies that were put in place in response to public concerns in the early decades of this century, that forest conditions over much of the United States have improved dramatically since 1900. The following table is a snapshot of the forest and wildlife situation that existed in the 1900s, as contrasted to 1990.

Forest wildlife today

Several species of American forest wildlife became extinct as a result of forest changes and human uses during this century, such as the passenger pigeon, heath hen, and Carolina parakeet. A larger number of subspecies and wildlife populations were substantially diminished and some disappeared altogether.

Many species, however, which were poised on the brink of extinction in 1900, have staged remarkable comebacks. Due to actions that were set in motion in the early decades of this century, most forest-wildlife species are both more abundant and more widespread than they were in 1900.

The pattern that has emerged since the 1930s is a substantial increase in forest wildlife that can tolerate a relatively broad range of habitat conditions. The numbers and distribution of the so-called "habitat generalists" have increased dramatically. Fortunately, most U.S. forest wildlife species are habitat generalists. One reason may be the natural dynamics of North American forests and the frequency of disturbance in the natural regime.

Some species abundant in North American forests prior to European settlement, particularly large predators and herbivores such as wolves, elk, and bison, have not returned to large areas of their former range. Yet even many of these species have staged comebacks in areas large enough to accommodate their needs for a large home range.

But saying that many wildlife species have staged remarkable comebacks does not imply the absence of problems. Species with specialized habitat requirements are increasingly of concern today. Examples include:

- The red-cockaded woodpecker and gopher tortoise, which are natives of fire-created southern pine savannas and woodlands.
- The Kirtland's warbler, which is native of young jack pine forests in Michigan.
- The northern spotted owl, which occupies mature and old-growth forests in the West.

Some forest wildlife species require active management of young forests for their survival, e.g., the Kirtland's warbler. Many other species, including a wide variety of both game and nongame species, need a mixture of forest and forest edge environments. Some need large, contiguous areas of habitat, such as grizzly bears, wolves, elk, and forest-interior birds. Some require old and ecologically diverse forests. Others, although needing mature forests, require specific habitat conditions, such as open savannas and woodlands which are created by frequent ground fires, e.g., red-cockaded woodpecker. Even the old-growth, Douglas-fir forests required by the northern spotted owl are subclimax forest types that

1900	1990
In the early 1900s, wildfire commonly consumed 40 to 50 million acres annually, an area the size of Virginia, West Virginia, Maryland, and Delaware combined.	The area consumed by wildfire has been reduced by more than 95 percent, from 40 to 50 million acres in the early 1900s to 24 million acres today, even in bad fire years.
Due largely to such wildfires, there were perhaps 80 million acres of cutovers which continued to remain idle and unstocked or poorly stocked with desirable tree species.	The cutovers or "stumplands" that existed in 1900 have long since been reforested. Today, many of these areas contain mature forests. Others have been harvested a second time and regenerated to young forests.
The volume of timber harvested nationally greatly exceeded that of forest growth.	Nationally, forest growth rates have exceeded harvest rates since the 1940s, with each decade generally showing a greater margin of growth over harvest than the one preceding. By 1986, the volume of tree growth nationally exceeded the volume harvested by 37 percent, and growth was more than 3-1/2 times what it had been in 1920.
No provision for reforestation was being made. In fact, no long-term forest management of any kind was being practiced.	Tree planting on all forest ownerships has increased dramatically since World War II, and was at record level throughout the 1980s. Many private forestlands are now actively being managed for tree growing.
Large quantities of wood were left after logging, sawmills were inefficient, use of wood in buildings was based on custom, rather than sound engineering, and huge volumes of wood were lost to rot and deterioration.	The efficiency of wood utilization has improved dramatically since 1900. Much less material is being left in the woods, many sawmills produce twice or more the usable lumber and other products per log input they did in 1900, engineering standards and designs have reduced the volume of wood used per square foot of building space, and preservative treatments have substantially extended the service life of wood. All of these have reduced by millions of acres the area of annual harvest that otherwise would have occurred.
Large-scale disastrous flooding in the east was tied to farm clearing and to logging and wildfires.	Eastern watersheds have been reforested. The headwaters of many American rivers are protected from over-harvesting by National Forests.
Clearing of forests for agriculture continued at very high levels. In the last decade of the century, America's farmers cleared forests at the average rate of 13.3 square miles per day. In the five decades ending in 1900, forest cover in many areas east of the Mississippi had been reduced from 60 to 70 percent of the land to 20 percent, or even less in some areas. Many of the areas being cleared were on steep slopes, were marginal for growing crops, and often were highly erodable.	For the last 70 years there has been no increase in cropland area About 1920, for the first time in American history, the increase in the area of cleared farmland abruptly stopped, rather than continuing to rise at the rate of population growth.[1] While farm clearing of forests continued after 1920 in some areas, it was offset by farmland abandonment and reversion back to forest in others. The stabilization in the area of cleared farmland had an immensely beneficial effect on U.S. forests.
By 1900, many wildlife species, which formerly were abundant, were severely depleted or were on the brink of extinction. Examples included game animals, such as white-tailed deer, wild turkey, pronghorn antelope, moose, bighorn sheep, and, of course bison. Furbearers, especially beaver, had been eliminated from significant portions of their ranges. Waterfowl were also severely impacted, including wood ducks and several other species of ducks, Canada geese, all manner of plumed wading birds, such as herons, egrets, ibis, and others. The passenger pigeon, perhaps the most abundant bird on the North American continent, was for all intents extinct in 1900. The heath hen, an eastern relative of the western prairie chicken, was also on the brink. The great auk, a flightless bird along the Northeast coast, had become extinct in 1840.	Wildlife has been a major conservation success story. Although a number of species, such as the great auk, passenger pigeon, heath hen, and several others did become extinct, many others which were severely depleted, or even on the brink of extinction in 1900, have staged remarkable comebacks. Many species which would likely have been on the endangered species list (had one existed in 1900) are today abundant. Examples include: wild turkey, beaver, egrets, herons, and many other wading birds, many species of shorebirds, wood ducks, and several other species of ducks, whistling swans, Rocky Mountain elk, pronghorn antelope, bighorn sheep, black bear, even white-tailed deer throughout most of its range. Many other species, although not actually on the brink of extinction in 1900, are today both more abundant and more widespread than they were in 1900.

[1] The two primary reasons that the area of cropland stabilized were that:
1) Rapidly increasing numbers of motor vehicles and farm tractors made it unnecessary to continue to raise large numbers of draft animals. In 1910, there were about 50 million horses, mules, and other draft animals in the United States. Fully 27 percent of all cropland was devoted to growing food for draft animals. By 1950, the numbers of draft animals had dropped so dramatically, compared to 1910, that the equivalent of some 70 million acres of cropland had been released to grow crops for human consumption.
2) After 1935, spurred by the development of genetically improved hybrid crops and by expanded use of chemical fertilizers and liming, agricultural productivity finally began to improve. Yields for corn are typical. Between 1800 and 1935, average corn yields in the United States remained virtually flat at about 25 bushels per acre. After 1935, average per acre yields began to rise, increasing to 35 bushels by 1945, to 40 bushels by 1950, and to 120 bushels by 1988. American farmers now grow as much corn on 1 acre as it took 5 acres to grow in 1920.

will eventually move toward different forest conditions without occasional, stand-replacing wildfires.

The eastern forest comes back

Nationally, the United States has about the same area of forests as it did in 1920. However, some areas have considerably more forest than existed in 1920. A growing tide of land abandonment and reversion to forest, particularly of marginal farmland east of the Mississippi, began to increase the area of forestland in some areas. Beginning gradually in the mid-1800s, marginal agricultural land in the east and south began to be abandoned as more productive farmlands in the midwest were opened up.

The reasons for reversion back to forest are complex. Two related factors were working in concert. The growth of the cities accelerated the transition of U.S. agriculture from subsistence to commercial. At the same time, the nation's progressively improving transportation system opened up more productive western lands to provide for the growing cities. The steep lands, small fields, and less productive lands of the East and southern Appalachians were unable to compete commercially with the lands of the Ohio Valley and much of the rest of the midwest. The opening of the Erie Canal in 1825 was the first major step that set the stage for cropland reversion in the northeast. Cropland depletion was also a factor in some cases, as was the reduction in the need for pasture resulting from declining numbers of draft animals.

The process of farmland reversion back to forest continued into the 1900s and was accelerated by the Great Depression. Under the Resettlement Act, a New Deal program, thousands of farmers in the Appalachians and elsewhere were relocated to more productive land. Between 1925 and 1945, almost 20 million acres of their abandoned farms and depleted woodlands were incorporated into the eastern National Forests under the Weeks Act. Millions of additional acres became state parks and forests.

In many ways, the forest and farmland landscape of many parts of the Appalachians, as well as other areas of the east and south, has come full circle. By the 1960s and 1970s, the pattern of forest, fields, and pastures was much like it was prior to 1800. In many areas the rural landscape has taken on an appearance much like it must have had prior to the American Revolution.

Increasing demands for nontimber uses and values

After World War II, steadily increasing national economic growth provided the basis for increasing personal income and leisure time. That, along with the proliferation of automobiles, revolutionized the recreational habits of the American people. Recreational visits to the National Forests increased from 27 million in 1950 to 263 million in 1990, reflecting a pattern that was common to other public lands.

As the nation's population has become more urbanized, mobile, and affluent, interest grew in setting aside land in parks, recreation areas, and reserves. In addition, the success of forest conservation practices began to demonstrate the nation's ability to meet increasing wood product needs from both private and public lands. Consequently, there has been a significant increase in the area of forestland set aside for amenity values in parks, wilderness areas, and similar designations under which timber harvest is prohibited.

Currently about 34.5 million acres of productive forestlands have been so designated — about double what was set aside in 1970. This is an area the size of the state of Florida.

Increased recreational demands came at the same time that the nation's public forests were experiencing increased demands for other uses as well. Such pressures have been felt especially in the last three decades as conflicts over the use and management of public lands has intensified.

American forests: a transformed heritage

Today, our forests represent a substantially transformed legacy — certainly in comparison to 1600, which most of us understand. But our forests have also been substantially transformed since 1900 — a dimension which is much less understood by most people.

The view of humans toward the nation's forests has changed profoundly over the years. Native Americans viewed the forest in a spiritual context, but they also took a utilitarian approach to the forest. They used and managed it to serve their own ends. European Americans initially viewed forests as an encumbrance to agriculture, or as a virtually inexhaustible resource to be mined. At first they used the forest — its wildlife, wood products, and land — to meet their subsistence needs for food and energy, much as Native Americans had done.

Later, the abundant wealth of the forests was used to build the homes, cities, and transportation infrastructure of a growing nation, and the lands previously occupied by forests were used to feed a rapidly growing population.

Scarcely more than a century ago, it became increasingly clear that old approaches were not sustainable. We began to view forests and wildlife not as products to be mined or foraged for, but as resources that could be managed over the long term on a scientific basis for both products and environmental services.

As our population has continued to urbanize, the principle of forest conservation for products and services has remained, but its role and scope have enlarged. A few decades ago we began to view forests as attractive settings for outdoor recreation and as places for human spiritual renewal.

Recently, we have begun to view forests as ecosystems supporting a complex web of life, of which humans are a part.

While it is impossible to predict how our view of forests may change in the future, it is important to understand how they came to be what they are today.

Lessons of the past

Today, the United States has almost four times the population it had in 1890, living at a substantially higher standard of living. Yet our forests and wildlife are, in most of their major dimensions, in significantly better condition today than they were a century ago.

Our forests and wildlife have demonstrated a resilience and responsiveness to management undreamed of by early conservationists at the turn of the century. These leaders were almost universally pessimistic about the future. Gifford Pinchot and others predicted an impending timber famine, coupled with significantly increased wood product prices, and consequent economic hardship and disruption. Wildlife leaders, such as William T. Hornaday and others, predicted the eminent extinction of scores of species. The timber famine never came. Most species whose extinction was prophesied have since recovered; many are even abundant today.

In their defense, the predictions of these early conservationists were logical and understandable given what they saw going on around them. Indeed, they were predicting what they felt would likely occur if past trends continued, and no actions were taken to address the concerns they raised.

But action was taken. New policies were debated and implemented. History has demonstrated that past public policies, coupled with the natural resilience of the resource, have generally served the country well.

In addition to the policy framework already discussed, there were other nonpolicy factors, largely unrecognized by early conservation leaders, that have also contributed to the improved condition of our forest and wildlife resources.

One has been the basic resilience of the forest resources of North America in recovering from disturbance and responding to management. In most of its past projections of future forest growth, the Forest Service has consistently underestimated the growth that subsequently occurred. Wildlife specialists have similarly underestimated the resiliency and rates of recovery of many species, once protected from exploitation and placed under management.

Conversion from wood energy to fossil fuels took a huge burden off American forests, particularly as population levels continued to grow. Indirectly, use of fossil fuels in internal combustion engines substantially reduced pressure for forest clearing for agriculture. It made possible the release of tens of millions of acres of cropland to grow food for humans, rather than draft animals. Petroleum also was also the feedstock for fertilizers and pesticides that substantially increased agricultural productivity after 1930.

American forests have been one of the many beneficiaries of the remarkable improvement in agricultural productivity that has occurred over the last half century. The inexorable, three century-long conversion of U.S. forests to farmland largely halted in the 1920s. Today, we have about the same area of both forests and cropland as we had in 1920. This has occurred in spite of the fact that the U.S. population has more than doubled since 1920, from 106 to 250 million, and that U.S. farmers also feed, through exports, the equivalent of more than 100 million additional people in other lands. This is a truly remarkable accomplishment. While questions have arisen in the last decade over some of the adverse consequences of intensive agricultural practices, such as pesticide and fertilizer runoff, these problems are being addressed and appear manageable. They should not overshadow the huge contribution to humanity and the environment that has resulted from improved agricultural productivity.

Another factor that was unrecognized by most conservation leaders was the effect that increasing real prices for wood products would have in encouraging both reduced consumption and increased supply. Real price increases for wood created powerful incentives of more efficient use for wood, e.g., less left in the woods after logging, better utilization by sawmills, and more efficient use in end-product applications through improved engineering, protection from rot through preservative treatment, and similar measures. Price increases also encouraged use of substitutes for wood, such as steel, concrete, and other materials. These private sector responses are the primary reason that wood consumption did not continue to rise after 1910 as it had in previous decades.

The conventional wisdom at the turn of the last century was that, due to the long time-frames and low economic returns involved in growing trees, the private sector could not be relied upon to shoulder much of the burden for producing the nation's wood, once the original forests were harvested. In addition to watershed protection, this was one of the rationales for establishing the National Forests.

Just as they had encouraged improved wood utilization, increasing wood prices spurred private sector investments in timber growing, although investments other than for fire protection were not significant until after World War II, when industrial forestlands began to be managed in earnest for tree growing. Today, private forests comprise 73 percent of U.S. productive forestland, yet supply 80 percent of the wood volume harvested.

As the condition of U.S forests improved, what had been a central conservation issue at the turn of the last century (whether the nation can assure itself a sustainable level of timber products for the economy) has diminished as a national level concern. While economic sustainability remains an issue in some areas, concerns over an impending "national famine of wood" are a relict of history. Concerns over forest sustainability have reemerged of late, but tend to find focus in local and regional level issues, such as sustainability of natural processes, plant and animal communities, and biodiversity on local and regional scales.

Challenges for the future

It is inevitable that as past issues have been successfully addressed, new issues will emerge. In recent years the growing urbanization, affluence, and mobility of the American population have caused a virtual revolution in the expectations and demands the public is placing on U.S. forests. Some of these are in direct conflict with traditional forest values and uses.

Issues such as protection of remaining old growth forests in the West, maintaining biological diversity, protecting endangered species, loss of wetlands, use of herbicides in forestry, the impact that atmospheric pollution may be having on our forests, and similar issues continue to be hotly debated. While most wildlife are in better condition today than a century ago, there are clear exceptions. Some species having specialized habitat requirements remain the focus of concern. The red-cockaded woodpecker and the northern spotted owl are examples.

Declining numbers of some neotropical migrating birds has raised concerns over forest habitat loss and fragmentation both in North America and in Central and South America. These are all issues we must seek to successfully address as demands on the resources of our planet increase.

One of the most profound changes in American society in the 20th century has been its transition from a rural, agrarian society to an urban, industrialized nation. This change has been accompanied by a corresponding physical and psychological separation of its people from the land that sustains them.

In a world of farms, forests, and small towns, the links between food and fields and between forests and home and hearth were clear and sustained by personal experience. In a world of cities and suburbs, of offices and air conditioning, those links have become more obscure and, for many people, virtually nonexistent. Yet today's urbanized society is no less dependent upon the products of its forests and fields than were the subsistence farmers of America's past.

In the last decade, we have seen the debate between those advocating the utilitarian use and management of forests for commodity products and those wanting to minimize human influences and emphasize amenity values (particularly on public forests) becoming increasingly shrill and divisive. This conflict existed in the early 1900s and found focus in the public debates between Gifford Pinchot and John Muir. As this century draws to a close and the nation's population has become increasingly affluent, mobile, and urbanized, we have seen increasing interest in the natural process/amenity side of the conservation spectrum.

Utilitarian use of the forest for commodities versus its protection for amenity and natural values are often viewed as irreconcilably in conflict — and on a persona level, or when the focus is on an individual parcel of land, they frequently are. Yet, in a larger sense and scale, they are not only compatible; indeed, they are inextricably linked.

Somewhat ironically, it has been the success of the scientifically based, utilitarian-oriented forest management concepts advocated by Fernow, Pinchot, and other forestry leaders at the turn of the century that has, to a very large degree, provided society with the forest abundance that has allowed it the capacity, unique to only a handful of other nations, to decide how much of John Muir's brand of forest conservation is appropriate. The fact that this nation now has the resource wealth to consider such choices is itself a clear and positive sign of the success of its past conservation policies.

But, as always, there are limits to such choices.

Society remains dependent upon forests for a wide variety of economic products. Indeed, utilization of forests for products has never been higher than it is today on a wood volume basis. Because of this, society's ability to continue to provide for the amenity side of the conservation spectrum will, in no small part, depend upon how much attention is also paid to the production/utilitarian side as well.

Today, the United States consumes about as much wood on a tonnage basis as the total for most other raw materials combined — steel, plastics, aluminum, other metals, and cements. Any significant substitution of these other materials for wood products could involve other environmental consequences. Alternatives to wood in most applications are both nonrenewable and use considerably more energy per unit of production than does wood.

As human population numbers and resource demands increase, the emerging challenge for society and its land managers is to find ways that both commodity products and amenity values can be realized over time from the same area of forest. This is the challenging new focus for the evolving concept of land stewardship and forest sustainability.

References

The following sources were the primary ones used in this paper:

1. American Forestry Assoc. 1990. Natural Resources for the 21st Century. Island Press. Washington, D.C.
2. Botkin, D. 1990. Discordant Harmonies: A New Ecology for the 21st Century. Oxford University Press, New York, N.Y..
3. Clawson, M. 1979. Forests in the long sweep of American history. Science 204:1168-74.
4. Clepper, H. and A.B. Meyer. 1960. American Forestry: Six Decades of Growth. Soc. of Amer. Foresters, Washington, D.C.
5. Cronon, W. 1985. Changes in the Land: Indians, Colonists, and the Ecology of New England. Hill and Wang, New York, N.Y.
6. ________. 1991. Nature's metropolis: Chicago and the Great West. W.W. Norton & Company, New York, N.Y.
7. Denevan, W.M. 1992. The pristine myth: the landscape of the Americas in 1492. Ann. Assoc. of Amer. Geographers 82(3).
8. Fedkiw, J. 1989. The evolving use and management of the nation's forests, grasslands, croplands, and related resources. Gen. Tech. Rept. RM-175. USDA Forest Serv.
9. Frederick, K.D. and R.A. Sedjo, eds. 1991. America's renewable resources: historical trends and current challenges. Resources for the Future, Washington, D.C.
10. Gruell, G.E. 1983. Fire and vegetative trends in the northern Rockies: interpretations from 1871-1982 photographs. Gen. Tech. Rept. INT-158. Intermountain Forest and Range Expt. Sta., USDA Forest Serv., Ogden, Utah.
11. Harper, R.M. 1918. Changes in the forest area of New England in three centuries. J. of Forestry 16:442-52
12. Koch, P. 1991. Wood vs. non-wood materials in U.S. residential construction: some energy-related intemational implications. Working Pap. 36. Center for Inter. Trade in Forest Prod., Univ. of Washington, Seattle, Wash.
13. Olson, S.H. 1971. The depletion myth: a history of railroad use of timber. Harvard University Press, Cambridge, Mass.
14. Perlin, J. 1989. A forest journey: the role of wood in the development of civilization. W.W. Norton & Co., New York, N.Y.
15. Raup, H.M. 1966. The view from John Sanderson's farm: a perspective for the use of land. J. of Forest History (April).
16. Rieger, J.F. 1986. American sportsmen and the origins of conservation. Oklahoma University Press, Norman, Okla.
17. Shands, W.E. 1991. The lands nobody wanted: the legacy of the eastern national forests. Presented at Symp. "Origins and Significance of the National Forests," Univ. of Montana, Missoula, Montana.
18. Smith, B.D. 1989. Origins of agriculture in eastern North America. Science 246:1566-71.
19. Steen, H.K. 1976. The U.S. Forest Service: A History. University of Washington Press, Seattle, Wash.
20. Thompson, D.Q. and R.H. Smith. 1970. The forest primeval in the Northeast — a great myth? Proc. of the Annual Tall Timbers Fire Ecology Conf. 10:255-65.
21. Trefethen, J.B. 1975. An American crusade for wildlife. Winchester Press and the Boone and Crockett Club, New York, N.Y.
22. U.S. Bureau of the Census. 1975. Historical statistics of the United States from colonial times to 1970, Bicentennial edition. Part 1. U.S. Dept. of Commerce, Washington, D.C.
23. U.S. Dept. of Agriculture. 1982. Analysis of the Timber Situation in the United States, 1952-2030. Forest Resources Rept. No. 23. USDA Forest Serv.
24. ________. 1990. RPA Assessment Update — 1992. USDA Forest Serv.
25. Van Lear, D.H. and T.A. Waldrop. 1989. History, uses, and effects of fire in the Appalachians. Gen. Tech. Rept. SE-54. Southeastern Forest Expt. Sta., USDA Forest Serv.
26. Whitney, G.G. and D.C. Davis. 1986. Thoreau and the forest history of Concord, Massachusetts. J. of Forest History (April).
27. Williams, M. 1989. Americans and their forests: an historical geography. Cambridge Univ. Press, New York, N.Y.

Appendix

Conservation policies that led to improved forest conditions

A number of forest policies emerged as national goals and priorities in response to public concerns. The success of those policies depended upon effective cooperative relationships among federal, state, and local governments, as well as private forest landowners and other private sector interests.

The policies and priorities that had the greatest effect on the improved condition of our forests are the following:

1. Focusing on fire suppression, prevention, and public education to protect the forest.
2. Establishing and enhancing the profession of forestry, and later of wildlife management, hydrology, and other natural resource disciplines, through establishment of accredited natural resource schools, professional societies, etc.
3. Improving the art and science of forest regeneration and management, including research establishment of tree nurseries and providing technical and financial assistance to forest landowners.
4. Improving the efficiency with which wood products are utilized in the woods at the mill, and in end-product applications. Such gains are the result of wood utilization research, its effective application, and the incentive created by increasing real prices for forest products. The Forest Service's Forest Products Laboratory in Madison, Wis., established in 1910, has been a significant contributor over the years to expanding the technical knowledge necessary for improving the utilization of wood products.
5. Improving the quality of forest management on private lands by improving economic incentives and removing tax and other disincentives.
6. Establishing the Forest Reserves (later the National Forests) for watershed protection irrigation, and sustained timber production.
7. While not established for forestry purposes, one policy that nonetheless had a significant beneficial impact on the nation's forest resources was the strategic decision made in USDA in the early decades of this century to emphasize agricultural research aimed at increasing crop yields. Prior to that, the USDA primarily focused on statistical reporting, soil and farm implement testing, and related activities.
8. An additional factor that has had a significant positive effect on forest conservation has been the increasing real price of wood over the decades. Between 1850 and 1950 the real price of lumber and of standing timber increased by more than five times, adjusted for inflation. This has created powerful economic incentives both for growing and managing forests and for reducing consumption of wood by using it more efficiently. The power of such economic incentives for conservation and efficient use of the resource by the private sector was largely unrecognized by early conservation leaders.

Conservation policies that led to improved forest conditions

A significant factor in the conservation of the nation's wildlife was the establishment and proliferation in the late 1800s of politically active sportsmen's organizations. These groups waged a protracted, and ultimately successful, war against market hunting. They also vigorously supported the enforcement of game laws, self-taxation to support state game management, and acquisition of habitat reserves and management areas.

The policies and other factors that had the greatest effect on the improved condition of U.S. wildlife are the following:

1. Adoption of a variety of strong state and federal wildlife conservation laws, and the establishment of the agencies effectively to enforce them. This game law framework includes the following:
   - Halting market hunting of wildlife for meat and most other products, including feathers (market hunting of furbearers has continued under state regulation).
   - Eliminating spring shooting of waterfowl and other game birds.
   - State regulation of resident game and nongame species.
   - Prohibition under federal law of: a) hunting of song birds, plume birds, and other migratory nongame birds; and b) interstate commerce in wildlife products taken in violation of state law.
   - Federal regulation of sport hunting of waterfowl and other migratory game birds.
   - Federal protection of endangered and threatened species after 1966.
2. Improving the art and science of wildlife management.
3. Establishing professional state fish and game departments devoted to scientific wildlife management and game law enforcement.
4. Improving habitat conditions, especially in the east and south, where millions of acres of agricultural land have reverted back to forest.
5. Reintroductions of species into formerly occupied range.
6. Establishing about 90 million acres of national wildlife refuges and 4 million acres of state wildlife reserves. Wildlife refuges and reserves in the contiguous 48 states were financed largely by hunting license fees and taxes on sporting arms, ammunition, and equipment.
7. Establishing the National Forests System:

- In the west, National Forests acted as wildlife reserves by providing protection for beleaguered populations of many wildlife species, especially large game, until state and federal wildlife programs and enforcement was put in place in the 1930s and beyond. The National Forests were the source of animals for a number of later reintroductions into formerly occupied habitat elsewhere.
- East of the Mississippi, millions of acres of abandoned and depleted farm and forestlands became national forestlands after 1920. After acquisition, feral cattle, dogs, and goats were eliminated and the land rehabilitated. Today, these areas provide superb habitat supporting rich populations of many wildlife species, some of which had not existed on these lands since before the American Revolution.
- The multiple use mandate of the 191-million acre National Forest System lands provides for full consideration of wildlife values and objectives in land management decisions. It has also encouraged the development of highly constructive joint efforts with state wildlife agencies in the management of wildlife habitats and populations.

Technical Forum Presentation:

Understanding customer needs: a requirement for winning and keeping customers world wide

Philip S. Auchincloss

Background and goals

In 1991, The Auchincloss Company conducted a market research study among oriented strandboard (OSB) manufacturers and suppliers to learn what issues were facing the industry. The earlier study found that understanding of customers' needs by industry participants was incomplete. The gaps resulted in manufacturers and suppliers working at cross purposes to the detriment of both groups. These findings suggested that other problems might exist further down the channel of distribution. This study is intended to follow up on earlier research and expands the scope of the prior work.

Today's competitive environment requires that companies understand their customers' needs to survive. Superior organizations will understand not only the needs of their direct customers but also the needs of others who will be using their products and services. Obtaining information like this is especially difficult when the channel of distribution is lengthy and complex like in the forest products industry. Accurate knowledge of needs means that sellers will supply only the services and product features valued by customers at each level of the channel of distribution. This information will save scarce marketing dollars and increase demand for products and services.

The goals of this study include:

1. Identifying customer needs throughout the OSB distribution channel including manufacturers, distributors, retailers, and builders.
2. Determine the extent participants understood the needs of their customers throughout the channel of distribution.
3. Recommend strategic and tactical action to be taken based upon an improved understanding of customers' needs.

This study is intended primarily for use by OSB industry suppliers, manufacturers, distributors, retailers, builders, and associations. Other forest or industrial product marketers can use this study as a guide for their own study of customers' needs. When an organization does its own research, additional information can be obtained on performance ratings of the company and competition to enhance the value of the information. Organizations should consider similar studies when searching for ways to reduce costs, target their marketing efforts, and improve their product and services.

Research objectives

This research examines three objectives. The first is to determine how customer needs varied by their location in the channel of distribution. This provides information for building targeted marketing plans and product offerings.

The second objective is to determine the measure of misunderstanding between levels. This research only examines the problem between adjoining levels. For example, suppliers were asked to report on their understanding of manufacturers' needs. They were not asked to estimate the responses of distributors,

---

The author is President, The Auchincloss Company, Winston-Salem, N.C.

retailers, and builders. Similarly, manufacturers were asked to report only on distributors' needs.

The final objective of this research is to learn what has changed in terms of customers' needs from the prior study done in 1991. Industry participants should be aware of these changes and to what extent they are changing.

Research procedure

The research plan was created to be low cost, easy to analyze, and simple to interpret. Twenty-six telephone interviews with key OSB industry participants were completed in October 1993. The names of the companies are listed by category in Table 1. All participants were promised their responses would not be linked in any way with them or their company, We have respected that request throughout this study.

Each company was asked to identify the most important business issues facing its organization in the next 12 months as they relate to OSB. Then, each participant was asked what they believed were the most important business issues that their customers must deal with in the next 12 months. The results were recorded on a structured research instrument and analyzed. The data with appropriate conclusions and recommendations are reported in this paper.

The small sample sizes used in the analysis may not be representative of the industry or a particular segment. Readers should use caution in directly applying these results to their business activities.

Table 1.

Category	Selection process	Company
Suppliers to the OSB industry	Suppliers of capital equipment or consumable supplies	Con-Vey Keystone GreCon Hoechst/Celanese Mobil Oil Corp. Siempelkamp Corp.
OSB manufacturers	Producers of OSB	Georgia-Pacific J.M. Huber International Paper Potlatch Corp. Weyerhaeuser Corp.
OSB distributors	Large distributors of wood products	Georgia-Pacific LMC North Pacific Lumber Co. Universal Companies Western International Forest Products
OSB retailers	Large building supply retailers	84 Lumber Anonymous Builder's Square Lowe's, Inc.
Home builders	Large home builders	Centex Corp. Lewis Homes of California Pulte Homes Corp. Ryan Homes Ryland Group Showcase Homes

Findings

This section of the report discusses the key findings from the survey work conducted among OSB industry participants. The presentation begins with a summary of the findings. It then proceeds to a discussion of the individual issues discovered during the investigation. This is followed by an examination of sellers' perceptions of industry issues compared to what buyers believed. Finally, the last portion discusses changes from the 1991 study.

Summary

The key business issues facing manufacturers and distributors are completely different from those reported by retailers and builders.

The three issues mentioned most frequently by manufacturers are demand for our product, environmental issues relating to plant production, and available wood supply. None of the manufacturers mentioned product acceptance, product defects, and product quality. This is somewhat surprising as these are the issues mentioned by the builders and retailers. Table 2 illustrates these differing view points.

The biggest issue for the retailers and builders is the price of the product. Their concerns were not about its absolute price, but the variability in price and the price relative to plywood. The recent price variability has hurt builders' profitability. When a builder signs a contract to build a house, there may be a 3- or 4-month delay before the lumber package is actually purchased. This past year increasing wood costs have made this a money-losing practice for builders.

Builders and retailers continue to expect to pay less for OSB than they do for plywood. They express concern about its market acceptance and mention recent problems in Dade County, Fla. where OSB is reportedly not acceptable to building inspectors.

Individual issues

Each issue listed in Table 2 is discussed in detail below.

Manufacturers and distributors are more likely to say demand is a key business issue than are retailers and builders.

Figure 1 shows the number of respondents mentioning at least one business issue relating to demand. Manufacturers were concerned about building new plants to meet the growing demand for OSB, increasing exports to overseas markets, the threat of substitute products like iron and plastic, and uncertainty of demand for OSB due to the economy.

Distributors mentioned issues concerning the uncertainty of demand for OSB due to the economy, the need for new applications for the product, and the increasing demand for structural panels.

Retailers and builders did not consider demand as a key business issue. Their concerns were of a more operational nature as shall be shown later. The issue raised by one builder was that consumers apparently held up buying homes because of the price fluctuations in lumber.

Environmental issues were mentioned by most manufacturers but not by other channel participants.

The environmental issue manufacturers mentioned most frequently was the need to handle plant emissions like volatile organic contaminants (VOC) and run-off ground water (Fig. 2). One manufacturer was also concerned about the problems of disposing of ancillary product waste like packaging materials. He indicated that customers are asking manufacturers to take back the packaging materials as this avoided waste disposal fees.

The only distributor comment was that manufacturers and environmentalists need to work together to manage the problems in the forests. Retailers and builders did not mention this business issue.

Supply is the number one business issue for distributors.

Distributors were concerned about the availability of wood for their business. A secondary issue was having a consistent supply of OSB for them to resell.

Table 2. — Business issues reported by manufacturers, distributors, retailers, and builders (the number in each column represents the number of people who mentioned each business issue).

Business issues	Manufacturers	Distributors	Retailers	Builders
Demand	5	3	0	1
Environment	4	1	0	0
Supply	3	4	1	3
Costs	2	1	0	0
Logistics	1	0	4	0
Price	1	1	5	6
Acceptance	0	2	3	2
Defects	0	0	2	4
Quality	0	0	3	4

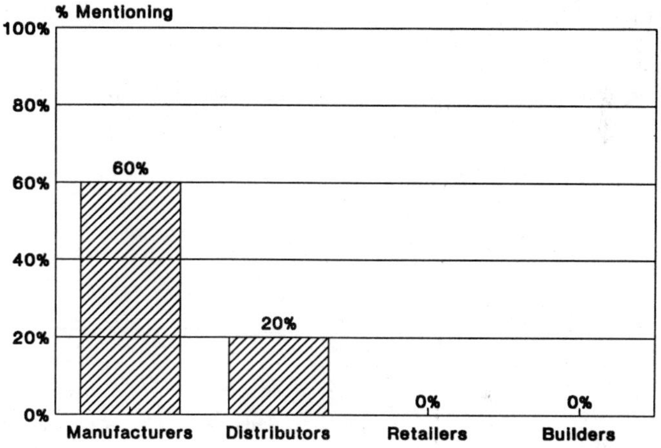

Figure 2. — Percent of repondents who mentioned **environment** as a key business issue.

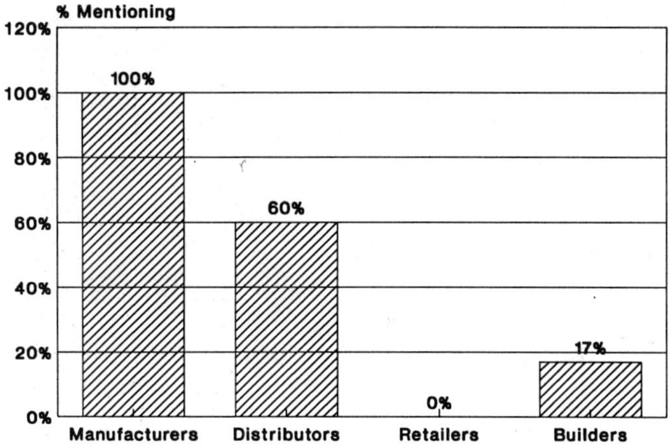

Figure 1. — Percent of repondents who mentioned **demand** as a key business issue.

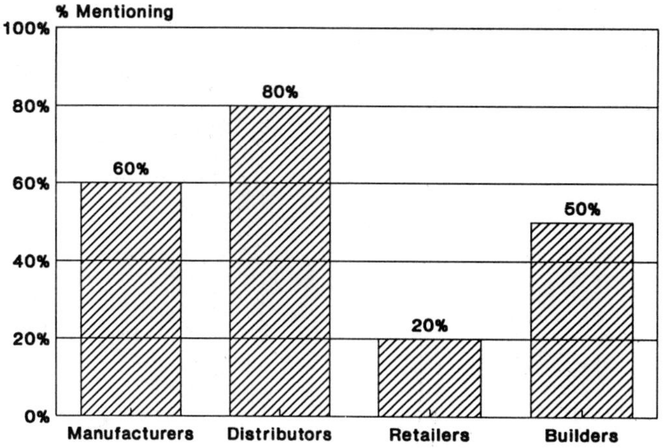

Figure 3. — Percent of repondents who mentioned **supply** as a key business issue.

Manufacturers reported obtaining a good supply of wood was an important issue to them as well.

Only one retailer mentioned supply as a key business issue. Retailers' attitudes were that they were going to get their material no matter what the circumstances were. This is likely to be true for these large volume accounts interviewed in this study. Smaller organizations are likely to be more concerned about this issue as they have less buying power.

Builders were also concerned about this issue. Many of them had experienced problems obtaining wood or, at least, obtaining it at a reasonable price.

Cost control or containment was not a key issue for those contacted.

Manufacturers who mentioned costs did so in the context of generally rising costs or retail credit risks. The higher prices recently charged for OSB increased the credit risk to retail accounts. This resulted in reduced shipments to certain retailers and the potential for an increase in bad debts when selling equal wood volumes.

Retailers mentioned logistics issues as key business issues more often than other channel members.

Retailers mentioned many logistics issues as key business concerns. The issue mentioned most often was insuring that supplies were properly delivered to all stores owned by a chain. This has become more of a problem with OSB producers selling directly to retailers instead of wholesalers stocking the stores.

One retailer was concerned about the number of pieces in each bundle of structural panels. Apparently different manufacturers are supplying product that is packaged differently. This creates a problem for the retailer in counting inventory and designing storage racks to store a complete package of structural panels.

Other retail issues included bar coding products, making space in stores for OSB, and changing channels of distribution from wholesalers to manufacturers.

The sole logistic issue mentioned by a manufacturer dealt with switching from doing business through distribution to selling direct to retailers.

Price issues are mentioned more often by retailers and builders than by manufacturers and distributors.

There are three issues that builders consider important. The first is the recent price fluctuations that have reduced their profitability. Builders pur-

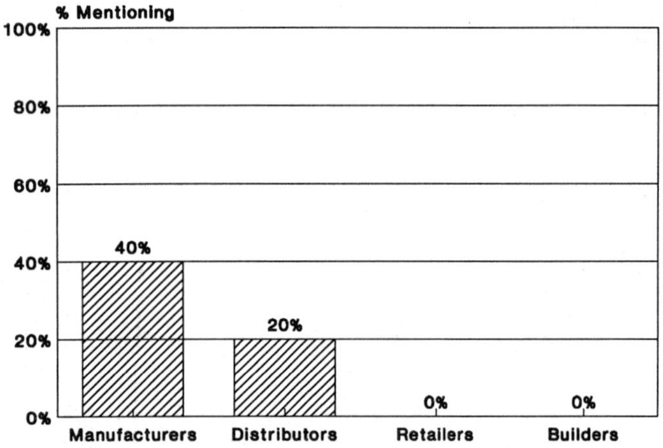

Figure 4. — Percent of repondents who mentioned **cost** as a key business issue.

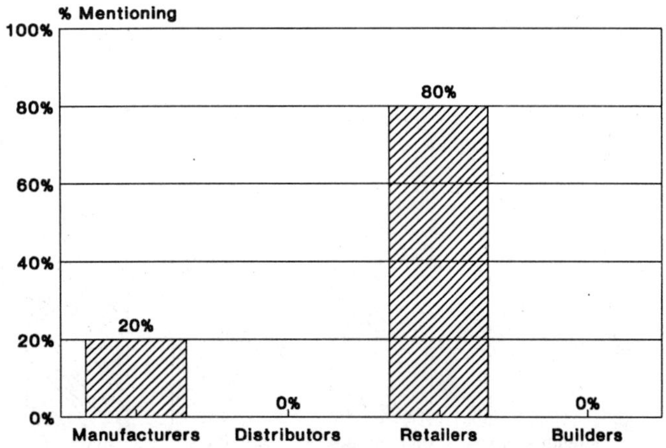

Figure 5. — Percent of repondents who mentioned **logistics** as a key business issue.

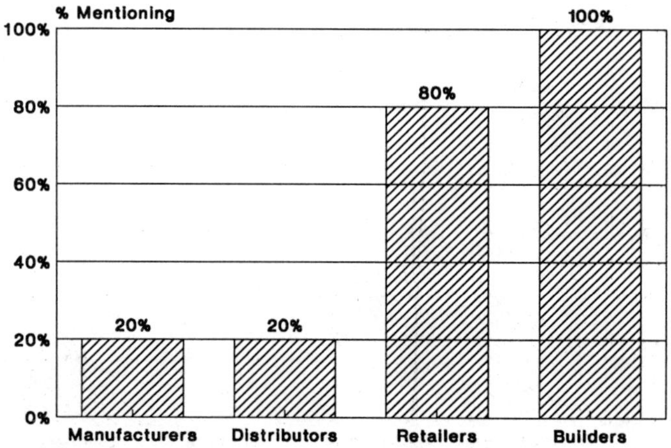

Figure 6. — Percent of repondents who mentioned **price** as a key business issue.

chased the lumber package 2 to 3 months after they contracted to build the home. They had to absorb the loss on their original package.

The second issue is the price spread between CDX plywood and OSB. Builders indicate the spread is a driving force for their preference for OSB. They like the cost savings offered by OSB.

The third issue is to achieve a competitive price in the local market area. Builders expect their retail or wholesale supplier to give them the best price. Retailers reflected this attitude as they commented that OSB is a commodity product and that builders shop the competition for the best price.

Retailers indicated that OSB must be priced below plywood to gain market acceptance. Their responses are similar to those of their customers, the builders, on the price issue.

Quality issues were important business issues only to retailers and builders.

The retailers and builders who participated in this study told us that quality of the product was important but they believed it was not a big issue. They look at OSB as a commodity product that is acceptable so long as it meets code. Builders were concerned with how well OSB would stand up over time and how consistently it will perform.

Retailers saw OSB as a commodity product. Still, they wanted a product with consistent product quality. One retailer told us that builders didn't want any call backs on homes they built with OSB or any other product.

While quality is considered a minor issue it still occupies the thoughts of retailers and builders as shown in Figures 8 and 9.

Retailers were most concerned with business issues relating to acceptance of OSB.

Some retailers consider OSB to be a new product. One retailer said, "People prefer plywood not wood chips." OSB is still viewed by some as a newcomer to the structural panel business, especially in geographic areas where plywood was the predominant structural panel. Retailers do indicate that more people are beginning to prefer OSB. Some customers have brand preferences, although this is rare.

Bad publicity arising from hurricane damage a year ago was mentioned as a factor limiting the product acceptance of OSB. On balance, there were no strong negatives that would indicate a decrease in user preference for OSB.

Manufacturers did not mention "Acceptance" as

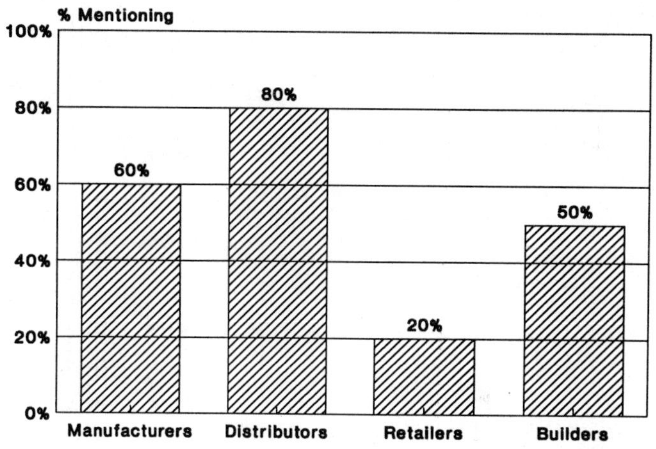

Figure 9. — Percent of repondents who mentioned **defects** as a key business issue.

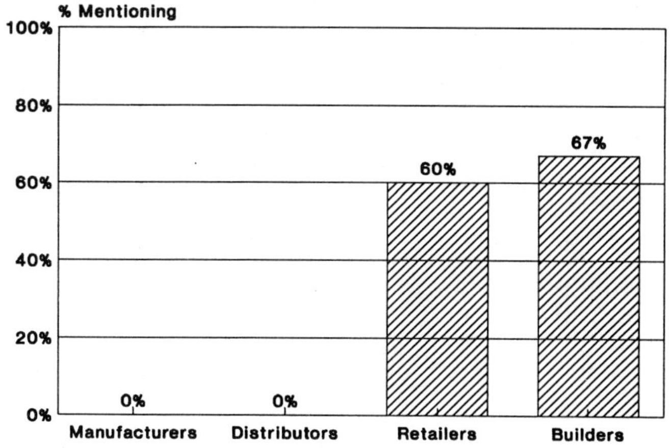

Figure 7. — Percent of repondents who mentioned **quality** as a key business issue.

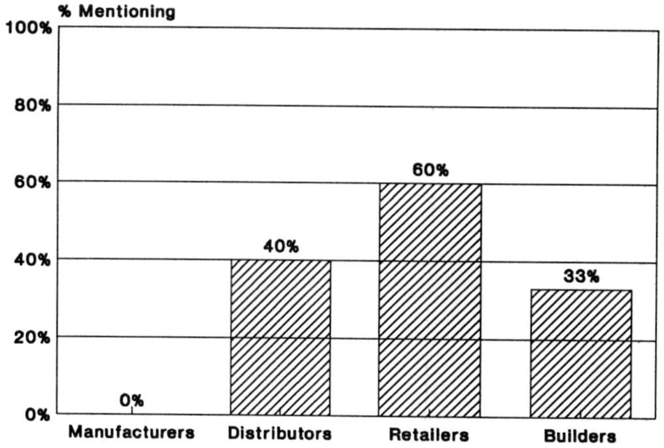

Figure 8. — Percent of repondents who mentioned **acceptance** as a key business issue.

a key business issue. They might consider that other industry participants continue to have concerns in this area.

Builders see product defects in OSB as a key business issue.

OSB defects continue to be a problem to the builder. The most common problems were edge swell and visible panel lines seen through roof shingles. Other problems mentioned include buckling, chips on edges of panels, slipper panels on pitched roofs, flaking, and peeling. One builder mentioned a problem with squeaky floors. Another builder was concerned that OSB doesn't take the wear and tear plywood does.

The issues mentioned by retailers were similar to those mentioned by builders. Manufacturers and distributors did not mention product quality as an issue.

Manufacturers should be concerned about the feedback that is not coming back up the channel of distribution about this issue. "Quality," "acceptance," and "defects" continue to be problems that bother the users of the product. Steps can be taken to reduce their level of concern through more stringent quality control or better marketing programs at the local level.

Figures 10 to 13 compare the key business issues of channel members with what suppliers believe they should be. Figure 10 compares the business issues mentioned by manufacturers to what their

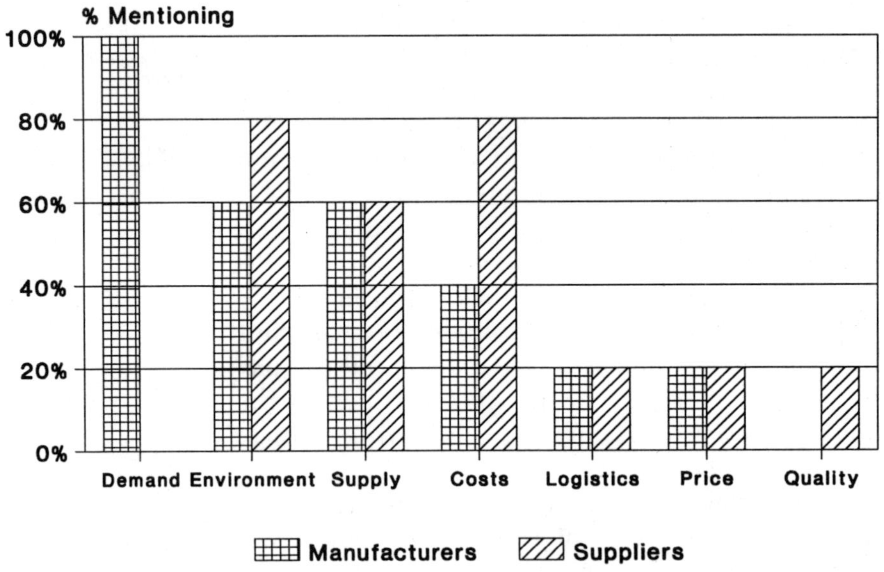

Figure 10. — OSB manufacturers' business issues as seen by manufacturers and suppliers.

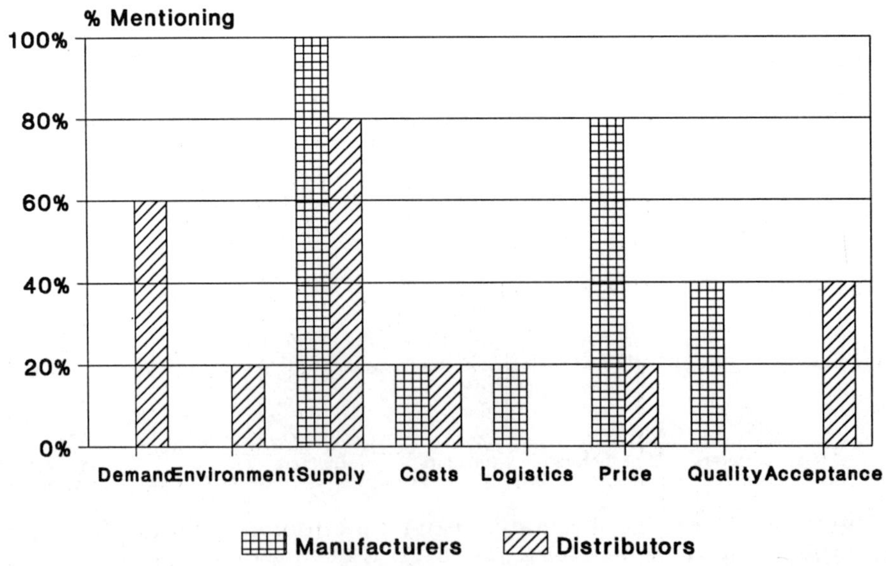

Figure 11. — OSB distributors' business issues as seen by distributors and manufacturers.

suppliers said. The areas in which there is a lack of agreement are areas of opportunity to develop a competitive advantage through creating products or services to meet costumers' needs.

Buyers and sellers view of issues

Suppliers missed manufacturers' concerns about product demand and said manufacturers would be more concerned about costs.

Figure 10 compares the answers given by suppliers about manufacturers with those given by the manufacturers of OSB themselves. Agreement between the two groups is high except for "demand" and "cost" issues. Suppliers failed to mention the importance of demand issues to manufacturers.

Manufacturer appear to lack understanding of distributors needs.

The only business issue the two agree on is "Supply." Distributors say "Demand" and "Acceptance" are two significant issues but they are not identified by manufacturers in Figure 11. Manufacturers claim distributors believe "price" is a very important issue while distributors hardly mention it.

Retailers have concerns about "logistics" and "price" which distributors do not identify.

Retailers are concerned about logistic details as shown in Figure 12. Distributors do not mention this as an issue. Distributors do mention "Supply," "Costs," and "Demand" issues which are not mentioned by retailers. The outlook here is that dis-

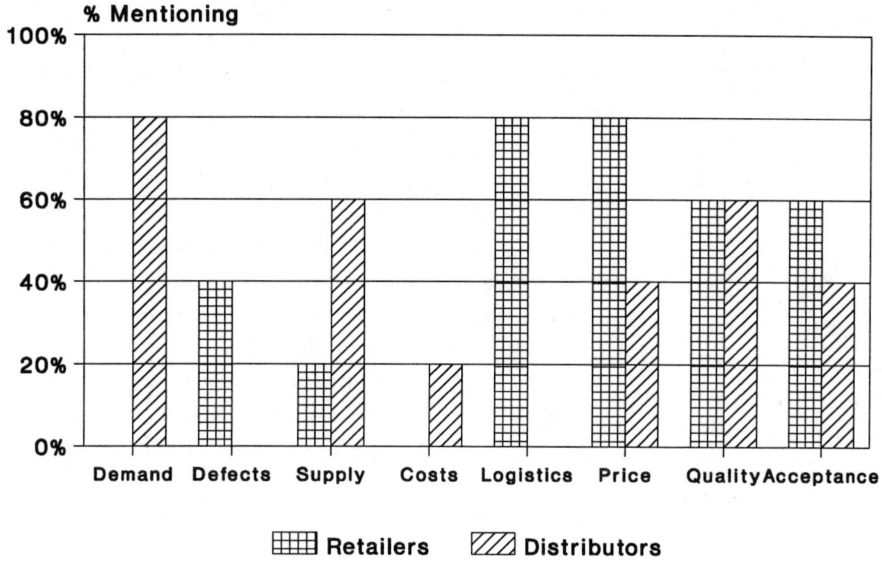

Figure 12. — OSB retailers' business issues as seen by retailers and distributors.

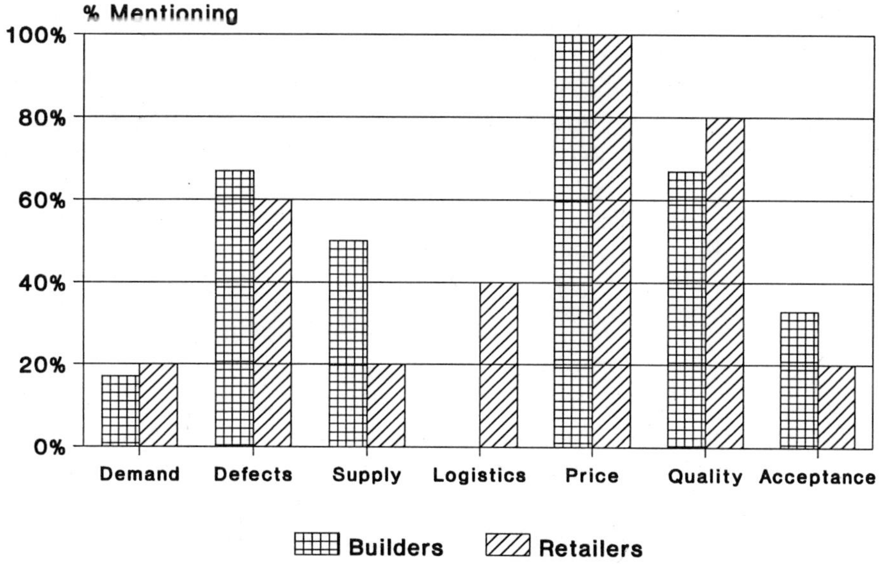

Figure 13. — OSB builders' business issues as seen by builders and retailers.

tributors need to be aware of retailers' concerns about "Logistics."

Retailers understand builders' concerns as shown in Figure 13.

Figure 13 shows just how closely aligned builders and retailers are. The figure shows very little variation from one issue to another. This understanding is driven by how closely the two must work together to survive. The builder needs a reliable local source of supply at a competitive price. The retailer depends on meeting the builders' needs to continue selling its products.

Changing issues from 1991

The focus of industry manufacturers has shifted from supply issues in 1991 to demand issues in 1993.

In 1991 we conducted a similar study on the OSB industry. Table 3 compares the responses of manufacturers then to those given in this survey. The exhibit shows that the "Demand" issue for OSB was not a concern in 1991 but has leapt to the top of the list in 1993. "Supply" has moved down to number three.

The "Environmental" issue has remained in the second position. The issue is still an industry problem.

"New product development" vanished in 1993, as has "Economic conditions." The industry has shifted focus in the intervening 2 years. We believe this reflects the intense cost pressures and restructuring the industry has undergone in the past 2 years. In addition, the diminishing supply of plywood has changed the nature of the demand structure for OSB. Thus the industry is directing its efforts to constructing new plants and gearing up to replace the plywood supply void.

Table 3. — Comparison of business issues in 1991 and 1993.

Business issue	1991 rank	1993 rank
Demand for OSB	- -[a]	1
Environemental issues like control of VOC emissions and groundwater runoff (this does not refer to restricted cuttings in protected forest areas)	2	2
Supply of raw materials	1	3
Cost control and reduction	3	4
New product development	4	- -[a]
Economic business conditions	5	- -[a]

[a] not mentioned.

Conclusions

1. Managing an enterprise that sells its product to the ultimate user through many intermediaries is a complex task. The information from this report clearly shows that the business issues at each level of the channel of distribution are unique. This suggests proper allocation of business resources must be made to adequately serve the total user base.

   Every manufacturer knows that producing a product with a minimal level of quality is a requirement for doing business in this industry. The "grade stamp" is the minimal acceptable level. As builders remind us, this is not the whole story. Manufacturers must find the correct level of quality to meet customers' needs at each level of the business. How they choose to deliver that quality and what level of quality determines their success in the market and ultimate profitability.

   These data also suggest the method of communication and message communicated must vary by position in the channel of distribution. Builders have different concerns than do distributors. Communication tools must be suited for the task, channel position, and the audience.
2. It will no longer be possible for OSB participants to survive by only understanding the needs of their adjacent customer in the channel of distribution. Industry suppliers, for example, might seek ways to solve problems for manufacturers at the builder level by reducing buckling, swelling, etc. This technique will sell their products by creating a level of demand that pulls the product through the channel of distribution.
3. The change in business issues over time is a helpful tool to managers. Manufacturers should consider what the omission of "new product development" implies from the list of 1993 business issues. An opportunity is available to obtain a competitive advantage by developing proprietary products during a time when others are focusing on different issues.
4. This study was created with a limited scope. Additional studies can probe these findings in greater detail to insure their validity over a larger number of respondents. These data should be considered as preliminary findings only. Manufacturers, suppliers, etc. should consider conducting this type of research with their specific needs in mind.

Recommendations

The following recommendations are presented by type of industry participant. They begin with supplier issues and end with the builder. Recommendations for trade associations are included, too.

1. Actions that suppliers should consider:
   - Consider developing products and services that

not only reduce manufacturing costs but provide benefits at the retail and builder level to enhance demand for OSB. For example, suppliers might find a way to reduce the alleged slipperiness of OSB on pitched roofs.

- Suppliers should develop products and services to help in solving environmental problems at plant sites. This has been a long-standing problem.
- Apparently manufacturers are not concerned with new product development. This means suppliers offering to work in this area will probably not be well received by manufacturers. These new product efforts must show real, tangible benefits before manufacturers are likely to exhibit much, if any, interest.

2. Actions that manufacturers should consider:

- Distributors are concerned about available supply. Manufacturers should work closely with their distributor customers to insure adequate supply for these key marketing partners.
- Manufacturers should develop programs to draw the product through the channel of distribution. Trade associations can be used effectively to enhance total demand for OSB.
- Manufacturers must be aware of the quality problems encountered by builders and retailers. Programs to continually improve on these problem areas will result in builders preferring OSB because of its performance qualities.
- Manufacturers need to find better ways to work with the emerging building supply retailer. They are concerned about getting enough product at the right time to their stores. Manufacturers serving this channel directly should consider ways to communicate closely with these large retailers to alleviate this problem area.
- What's missing from the issues is "new product development." Manufacturers should ask themselves if this is a strategic decision on their part or an issue that has been neglected because of the press of other issues perceived as more important.

3. Actions distributors should consider:

- Distributors selling to large building supply companies need to be concerned about their logistical requirements. Like manufacturers, distributors need to develop programs to meet the requirements of these accounts.
- Customers of distributors are concerned with quality and product defects. While distributors do not manufacture OSB they can insure that it is delivered in good condition and reject shipments that are poor quality.
- Distributors play a vital role in supplying customers with the product. They can use this fact as a tool to keep customers. Customers should be reminded that distributors obtain supplies from many sources and are likely to have material when single source suppliers may not.

4. Actions that retailers should consider:

- Retailers appear to have an excellent understanding of builders' needs. These needs, however, do not seem to be communicated up the channel of distribution. A competitive advantage could be obtained by working closely with either distributors or manufacturers to solve some product quality issues. A strategy to communicate these problems up the channel of distribution would produce improved results.

5. Actions that builders should consider:

- Quality is a key issue for builders. Builders should search for products and suppliers that provide the best quality in the long run. These may not be the lowest priced products but will save considerable monies from reduced call backs.

6. Actions that associations should consider to enhance demand for OSB:

- Develop programs that communicate to every level of the channel of distribution that there will be adequate supplies of OSB.
- Develop programs to minimize builders' concerns about "Defects" and "Quality" through promotion and education programs to builders. Associations should work with manufacturer members to develop quality standards that minimize these problems.
- Associations should develop programs for manufacturing members that increase demand for OSB by promoting the benefits of OSB to all levels of the channel of distribution.

7. This study suggests additional work to enhance a company's competitive position within the industry:

- Manufacturers should consider measuring their performance on these and other issues at each level of the channel of distribution. Comparison of their performance with competition will indicate areas for improvement and cost reduction.
- Industry participants should examine their marketing programs to determine if they are truly meeting customer needs at all levels of the channel of distribution.
- These issues can be monitored regularly to determine changes. These changes may or may not be visible to one organization in an industry. A systematic study reveals trends.

Technical Forum Presentation:

Practical experience with a fully automated line camera scanning system in a window manufacturing company

Rolf Birkeland and Helge Timland

Introduction

Norway has a cold climate. The winters are long and dark. Good, warm houses and tight, well-insulated windows are a must. We have to stay indoors for long periods of the year and it is important to have windows to look out of and to let the meager daylight in.

As a consequence, window production is a fairly advanced industry with research-based development and quality control systems and several sophisticated production units.

One of the major window producers is different from the others. The production is organized in a group of cooperating, individually owned companies. Each of these companies is responsible for its own production and sales. Functions like product development, the more public parts of the marketing effort, as well as production of hardware, wooden window components, and glass, are performed by units and factories jointly owned and governed by the group of producers.

Today the group has a market share in Norway of about 30 percent. In addition, the group has production facilities in a number of other countries: Sweden, Finland, Denmark, Faroe Islands, Iceland, United Kingdom, Ireland, Holland, and Belgium, and they have also started window production based on the same concepts in the United States.

Wood components are produced in a highly automated plant

The common product of the group is a laminated window. As shown in Figure 1, the individual parts of the casements and frames are built up by three or four pieces of softwood glued together. Those pieces that will be exposed in the finished product

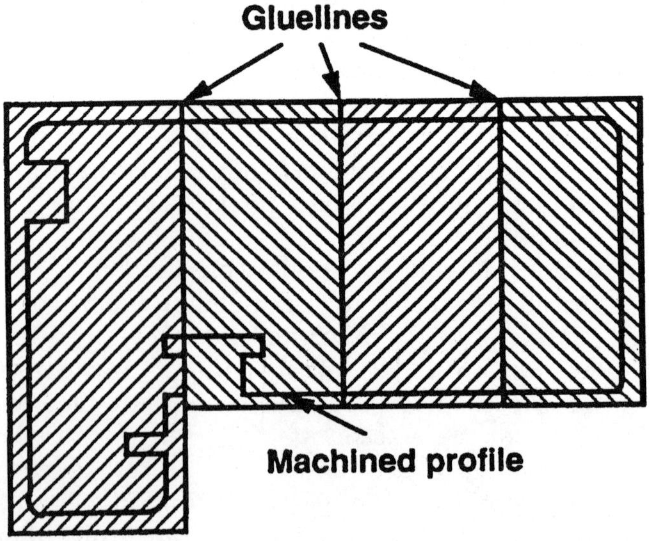

Figure 1. — Cross section of a typical glued-up part for a laminated window.

The authors are, respectively, Director, Norwegian Inst. of Wood Tech., Oslo, Norway; and Project Manager, H-Vinduets Fabrikker A/S, Kirkenaer, Norway.

are made of pine (*Pinus sylvestris*) and those not ordinarily visible are made of spruce (*Picea abies*).

This type of construction was adopted in 1972-73. In 1975, a separate production unit was established. Annually, this unit processes around 30.000 m^3 (corresponding to about 3.9 million board feet) of lumber. The result is about 3 million laminated pieces.

The ingoing material adds up to a little more than 13 million individual wood pieces. These have to be sorted according to grade and possible position within the laminated unit, making the sorting process a rather large and time-consuming enterprise. Between 16 to 20 work-years were needed to perform the sorting. Laminated windows have a higher market value than those made from ordinary wood members, and the very best quality is necessary to retain the faith of the market. The sorting function is consequently a very important part of the production process.

The fact that the logs from the Nordic forests are getting smaller all the time, and the amount of clear-select material is decreasing, makes it all the more important that the sorting function maintains a high standard.

To ascertain that the quality is good enough, all four sides of each piece of wood must be inspected. This means that about 54 to 55 million wood surfaces must be scrutinized. Decisions must be made about the wood quality and whether a piece can be used on an external surface or should be used with most of its surface hidden within the construction. In the latter case, the piece is marked to show its orientation within the laminated piece.

An automated system was needed

The company started to look for ways to automate the sorting process. In 1983, it was decided that a camera-based system should be found or developed. The system should be fast enough in both image processing and data treatment. After considerable research and a number of frustrating tests using commercially available systems, the window company established contact with a Swedish firm.

The Swedish firm's competence was with image processing in general. They had specialized in large volume handling like mail and lottery coupons, and they had developed a fast data-processing system — fast enough to suit the needs of the window production facility.

After a successful prototype in 1986, the decision was made to develop a pilot plant system in 1987, and the plant was in place in 1988.

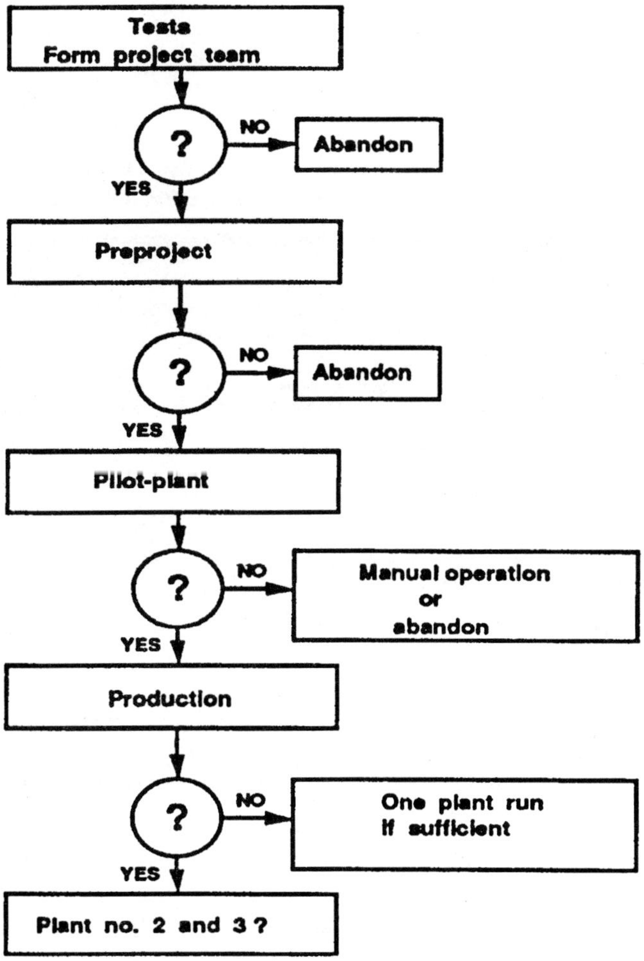

Figure 2. — Flowsheet of the project.

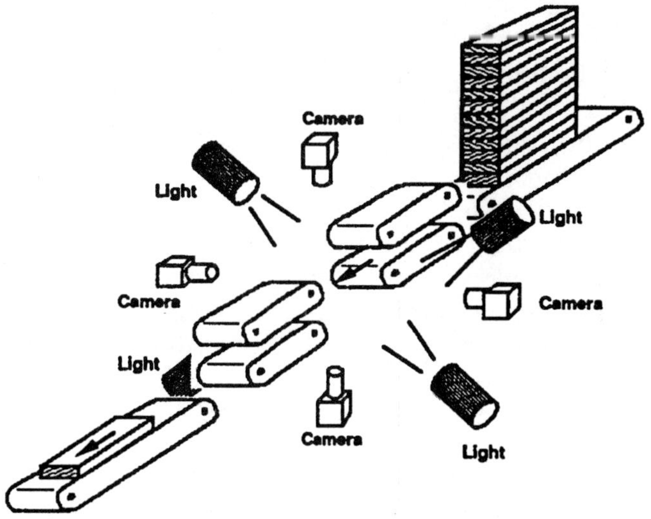

Figure 3. — Sketch of the pilot plant system.

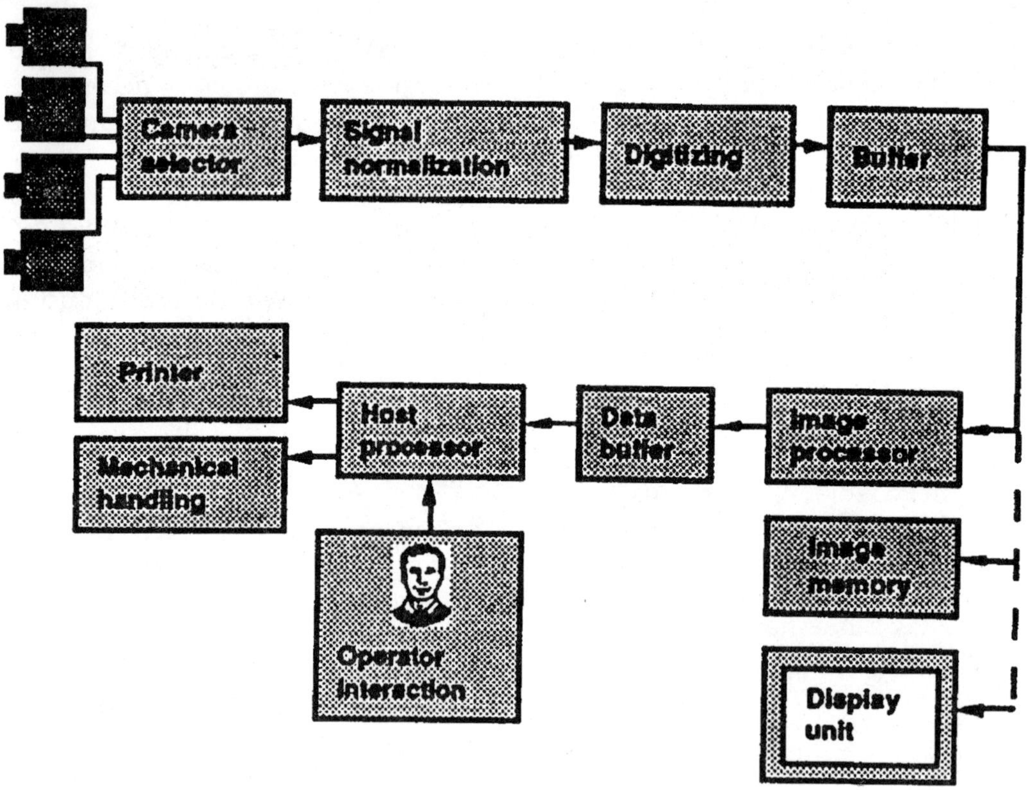

Figure 4. — Sketch of the data processing system.

The pilot plant system

The camera and light projector setting is shown in Figure 3. The system has four cameras, one for each side of the wood piece. The Reticon cameras are able to scan 1024 points per scanning line.

Problems with dust on the lenses of the cameras and light projectors were anticipated. To prevent this, a ventilation system was installed that blows filtered, oil-free compressed air in a curtain over the lenses. Despite this precaution, a layer accumulated on the lenses over a period of 18 months. The layer grew very slowly and could not be seen with the naked eye.

Because the sorting unit has a built-in ability to compensate for both variations in light and the color of the wood (up to a certain limit), the reduction in light was caused by the layer on the projector lenses went by unnoticed. However, the installation rejected more and more pieces without the wood being of noticeably lower quality. A simple cleaning of the lenses was the remedy, and is standard procedure today.

Capability of the system

The system has a resolution across the grain of about 0.07 mm. In practice, this is determined by the greatest width of a piece to be sorted, normally about 62 mm wide. With 1024 points per scanning line, the system has a small overmeasure so that the tolerance of the width is controlled. Along the grain, the resolution is in the order of 2 mm. This depends on the recording speed of the system and the feed speed of the wood pieces. In this practical case, the feed speed is about 1 meter per second. Since most defects (pitch pockets, splits, checks, etc.)

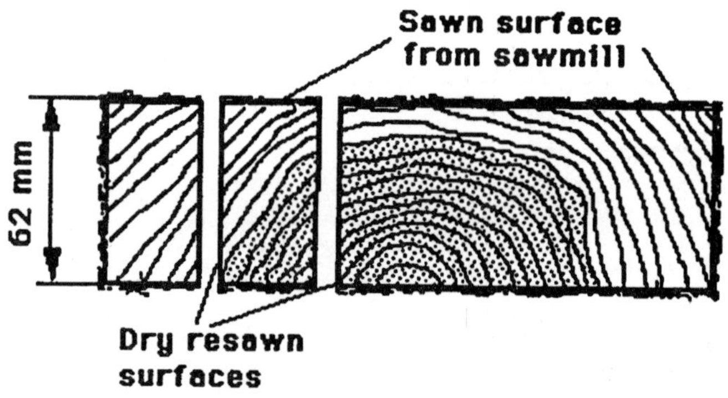

Figure 5. — How the lumber is resawn before inspection and gluing.

are larger along rather than across the grain, the lengthwise resolution has been found to be adequate.

In most cases, the pieces to be scanned have two surfaces from the sawmill and two surfaces resulting from dry resawing with tungsten carbide-tipped circular saws.

If the sawn surfaces are rough, and especially if they have long, loosely attached fibers, the system may have problems. Fibers lying parallel to the grain may be mistaken for checks.

Defects discovered by the system

In window manufacturing, the wood quality demand is not very different from many other joinery uses. Some difference do exist — the system described here was designed to detect the following defects: encased knots and knot holes, intergrown knots, checks and splits, pith, wane, dimensional defects (local and global), and dark markings (different intergrown defects of dark color).

The most important defects by this particular production are the checks. Checks make up the largest number of rejects. In window manufacturing, the finished product is painted, and even very small checks will show quite distinctly after painting. The reason for this is that the surface tension of the wet paint film tends to pull the paint away from the check (and other sharp edges).

The smallest checks may be tolerated because they will disappear in the machining operation following gluing, as it is shown in Figure 1. The limit for permissible check widths is set to be 0.2 mm. The camera and detection system will, however, see and note checks with a width down to 0.056 mm when such a check is centered within a scanning point.

The other important defect to look for are knots. At the start of the project, one of the major obstacles was believed to be distinguishing the precise size of intergrown knots. In reality this has not been a problem. It does, however, require that the operators of the system set the parameters correctly and that they adjust these if the wood character is changing.

Wane is a defect that may be difficult to detect, especially wane with no bark. If such irregularities on the edges are to be detected, a fan-shaped laser beam, at an angle, may be used (Fig. 6).

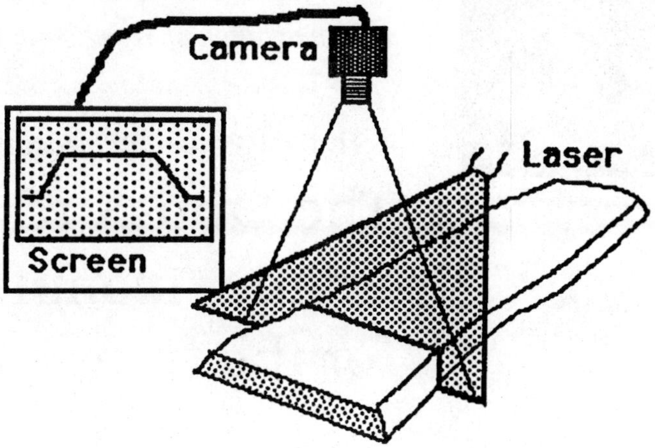

Figure 6. — Wane can be detected by a fan-shaped laser beam at an angle.

Blue stain is easy to detect because 1) its appearance is limited with the wood quality used; 2) the more pronounced blue stain spots are detected; 3) if a wood piece has blue stain on one side only and no blue stain is visible on the adjacent surfaces, the attack is normally so shallow that is it removed in the subsequent machining.

All the installation's parameter settings are accessible and can be changed fairly easily by the operators of the system by using a standard PC keyboard.

Experience and the future

The performance of the system has fulfilled the expectations. The stability of the sorting process is substantially improved. In a recent test, 98.5 percent of the approved pieces to be glued into frame material satisfied the quality limits. Compared to the manual method, this is a vast improvement.

If anything negative needs to be brought to attention, it must be that so far the system has been sorting a bit on the strict side — in the rejected heap some of the pieces could have been used!

One way to judge the results is to consider whether the project would have been repeated. In this case the answer is clearly "yes." The company is about to order a second installation, and they are thinking about investing in a third one.

Technical Forum Presentation:

The Opti-Stress System: lumber manufacturing technology for the future

Robert Czinner

The Opti-Stress System, a completely automated, computer-controlled process for manufacturing lumber, will lead the lumber industry into the future. The system's flexibility allows it to add more value to lumber than has ever before been possible, which will result in dramatically improving the utilization of the world's valuable wood resource. In fact, each plant will annually add the value equivalent to the processing of over 2 million trees. According to the USDA Forest Service, "the Opti-Stress System represents an important step in creating the forest products plant of the future." It is patented in the United States, Canada, New Zealand, and Australia, and has patents pending in 12 additional countries around the world, including Chile and Japan.

In 1992, over $6 billion of North American softwood was bought as "specialty" products: preferred lengths, cut-to-length (1/16-in. length increments), and machine stress rated lumber (MSR). These are premium-priced products that are created by processing lumber after it has been produced at the primary sawmill. There is a growing demand for these "remanufactured" products because they are cost effective when installed, though they cost more to buy. There are over 500 remanufacturers in North America. All of them are experiencing falling margins and increased competition. A low-cost remanufacturing operation that is capable of cutting all the above specialty products, during the same production run, would quickly dominate these markets.

Computer integrated manufacturing (CIM) is an automated engineering process that has been proven to reduce costs by increasing production flexibility. It utilizes a central computer to control the flow of information between MIS databases and all of a plant's automated production equipment. According to a 1992 study by the Sawmill Optimization and Control Group at the University of British Columbia's Faculty of Forestry, the installation of a CIM system in a specialty lumber plant would increase operating profits by over 500 percent. They also explain that no one has attempted to create such a system, and that if anyone did they would "gain a head start on the rest of the world in offering this technology."

Opti-Stress is a CIM system that redefines the economics of processing lumber by increasing production flexibility. CIM is the vital link enabling Opti-Stress to be the low-cost solution for producing specialty lumber products. Compared to existing remanufacturing operations, Opti-Stress reduces production costs 30 percent, increases throughput a minimum of 50 percent, and increases the value-added up to 50 percent. These factors combine to drive up operating margins over 400 percent.

One of Opti-Stress's unique features is its patented ability to electronically combine and analyze both the visual and strength characteristics of a board. This ensures that the maximum value is added when each piece is cut. Existing electronic optimizing systems are limited in their ability to add value because they can only evaluate each piece of input wood

The author is Chairman, Integrated Wood Recovery Systems, Montauk, N.Y.

based on either its visible surface defects, using optical scanning, or its internal strength profile, using MSR. By combining both systems, the number of potential cutting patterns which can be evaluated is increased. This results in higher value-added opportunity.

Used alone, optical scanning is limited in its ability to identify all visible defects, such as checks, small splits, and slope of grain. However, Opti-Stress ensures that all defects are recognized by utilizing an MSR system to locate areas of relative weakness along the length of a board. This is important because studies have proven that where there is a weak spot, there must be a defect. On this basis, Opti-Stress is programmed to cut out a weak spot located by the MSR, even if no defect was identified by the optical scanner. Adding an MSR analysis is the only way to be sure that optical scanning provide reliable, quality-controlled electronic lumber grading.

The Opti-Stress System utilizes its CIM design to achieve the reliable communication required between electronic components, to form one comprehensive intelligence system. Opti-Stress integrates four main components:

1. A computerized optical scanning system to locate defects on the lumber surface.
2. A computerized mechanical stress testing system, called an MSR machine, to test the internal strength of a piece of lumber along its entire length.
3. Computer-controlled production equipment to execute the system's handling and cutting decisions.
4. Optimization software to collect the data from the optical scanning and MSR systems and to combine it in order to create a digital "map" of each board.

While the first three components of the Opti-Stress system are proven technologies that are widely used, the fourth component, the optimization software, is our proprietary product. This software is comprised of three Expert System modules, which interact to provide the best cutting solution for each piece of lumber. An Expert System utilizes an advanced software language that allows a computer to create "intelligent" solutions to complex problems. These three patented Expert System modules are the Planner, the Cutter, and the Supervisor.

The Planner safeguards that the lumber dimensions input are the most cost effective for producing the exact specialty products required by our customers. It accomplishes this by scheduling both the raw material input (dimensions and quantities) as well as the plant output (dimension and grades) for each shift. Because input and output are interdependent variables, the Planner executes a highly complex "top/down and bottom/up" analysis of open orders, market prices, inventory, and historical production statistics. It evaluates all possible combinations of the above variables to determine which input sizes are needed to produce the required outputs, at the highest value added. The Planner automatically downloads its output decisions directly to the Cutter; it prioritizes the desired output based on value and promised delivery dates.

The Cutter produces a cutting plan for each board, at real-time production speeds. It determines its decisions on three data considerations: 1) the output production matrix developed by the Planner; 2) the board description consisting of the board dimensions and a list of the natural defects identified on the surface of the board by the optical scanner; and 3) the strength profile of the board provided by the MSR machine.

The Supervisor system controls the operations of the entire plant and all the software systems. This module links all the plant's electronic systems. It also stores all the production and defect information gathered on each board so that it can, at any time, track the system's cutting logic and execution. In turn, this data-gathering function provides the means for the Supervisor to automatically diagnose and solve and production problems. It also allows the Opti-Stress System to be modified, quickly and inexpensively, to control other wood-manufacturing applications.

All of the production equipment has been sourced. It is heavy-duty machinery designed for primary sawmill installations. This will ensure that the Opti-Stress System has reliable production with minimal maintenance.

To date, over CDN $4 million has been invested. Equipment and technical specifications are complete and patents have been issued worldwide. An engineering plan and prototype software have been developed and an international turn-key contractor has quoted on delivering the first plant. The technology and budget has been favorably reviewed by H.A. Simons Consulting Engineers, the National Research Council, the Ontario CAD/CAM Center, and the Canadian General Electric's Factory Automation Group.

Technical Forum Presentation:

An economic analysis of federal forestry cost share allocations 1985-1992

T.E. Dorrell, T.J. Straka, and R.J. Moulton

Abstract

Basic federal forestry incentive program allocation relationships are discussed. Cost share allocation funds are allocated to relatively productive forestlands and are primarily for reforestation practices. The southeastern United States receives about 85 percent of these cost share payments. Cost share allocations reflect economies of scale.

Introduction

Long-term timber supply has been and will continue to be a concern of forest managers. Nonindustrial private forestlands (NIPF) are a significant component of long-term timber supply models. NIPF land holdings comprise approximately 60 percent (260.6 million acres) of the nation's timberlands (6). Since the mid-1930s, the federal government has directly or indirectly promoted timber production on NIPF lands with cost share assistance for approved forestry practices.

In 1973, Congress authorized the Forestry Incentives Program, which was the first federal cost share program to specifically address timber production on NIPF lands (2). The Forestry Incentives Program provides up to a maximum of 65 percent cost share assistance for reforestation, timber stand improvement, and site preparation for natural regeneration (7). In 1990, the Stewardship Incentive Program was authorized to promote enhanced management of NIPF land holdings via cost sharing (9). SIP provides up to a maximum of 75 percent cost share for reforestation and timber stand improvement.

The Agricultural Conservation Program, the Soil Bank, and the Conservation Reserve Program were authorized in 1936, 1956, and 1985, respectively (4,7). These programs provided cost share assistance to promote soil and water conservation, reduce agricultural surplus production and other conservation practices. However, they indirectly promoted timber production with over 6 million acres planted into trees by 1992. These programs were not used in this study because the primary purposes of these programs was other than timber production.

This study focuses on the efficient allocation of the Forestry and Stewardship Incentives Programs cost share dollars with respect to site productivity, type of forest practice, size of acres served, and region. A positive relationship should exist between site productivity and cost share allocation, thus indicating efficient allocation of federal cost share dollars. Allocation of cost share dollars between forest practices will have profound impacts on long-term timber supply. Reforestation is expected to receive the greatest percentage of the cost share allocation. Efficient allocation of cost share dollars should reflect economies of scale related to tract size. Therefore, a negative relationship should exist between cost share allocations and acres served. Finally, regional distributions of cost share allocations should be a function of regional NIPF land holdings and forest operations intensity. Therefore, the southeastern United States should be receiving the greatest allocation of the cost share dollars.

The authors are, respectively, Graduate Research Assistant and Assoc. Professor, Dept. of Forest Resources, Clemson Univ., Clemson, S.C.; and Economist, USDA Forest Serv., Washington, D.C.

Results

The Forestry Incentives Program is offered only in designated counties where a suitable number of ownerships capable of producing at least 50 cubic feet of timber per year are available (7). However, mean regional site indices indicate allocation of Forestry Incentives Program dollars to sites with relatively high forest productivity (site indices range from the low 70s in the Northeast, high 70s in the South, and mid-140s in the Pacific Northwest). Furthermore, the data indicated no statistical relationship between site productivity and cost share allocations.

Reforestation received the greatest cost share allocation with timber stand improvement and site preparation for natural regeneration receiving smaller allocations, respectively. Timber stand improvement practices occurred on considerably fewer acres, possibly indicating a lack of enhanced forest management. However, commercial thinning by NIPF landowners may account for this difference.

The southeastern United States received the greatest cost share allocation with approximately 85 percent of total allocations for reforestation for the Forestry Incentives Program. The Northern Area and the Pacific Northwest received 6.9 and 6.6 percent, respectively. The remaining regions received less than 1 percent of the total reforestation cost share allocations. The Northern Area received the greatest cost share allocation for timber stand improvement with 53 and 59 percent for the Forestry and Stewardship Incentives Programs, respectively. Other regions received smaller cost share allocations for timber stand improvement in comparison to the Northern Area Site preparation for natural regeneration allocations was greatest for Region 8 with 60 percent. The Northern Area followed with 39 percent and the remaining regions with approximately 1 percent of total allocations.

Cost share allocations with respect to acres served has a negative relationship for all regions and forest practices. Indicating that efficient allocation of cost share dollars with respect to acres served is occurring. This relationship suggests that more dollars per acre are allocated to smaller tracts, which have higher forest practices costs. By allocating cost share dollars to these smaller tracts, the cost of practicing forestry is reduced to these NIPF landowners. This should result in an increase in timber production from NIPF lands. However, increased mechanization of forest management and harvesting activities has resulted in prohibitive costs in the management of these smaller tracts and the number of small tracts receiving cost share assistance should be minimized. The allocation of cost share dollars to larger tracts will insure that harvesting will not be prohibitive in the future due to economies of tract size.

Conclusions

Efficient allocation of cost share dollars for the Forest and Stewardship Incentive Programs seems to be occurring. Indicating that these programs are having a positive impact on the long-term timber supply. However, a continuing decline in the timberland base of NIPF landowners is expected well into the next century (6). The loss of productive timberland sites to urban development and other land uses will require more intensive utilization of the remaining NIPF lands for timber production.

Literature cited

1. Alig, R.J., T.J. Mills, and R.L. Shackelford. 1980. Most soil bank plantings in the south have been retained; some need follow-up treatments. South. J. Appl. For. 4(1):60-64.
2. Cohn, M.W., ed. 1973. Omnibus Farm Bill: New Target Price System Voted. Congressional Quarterly Almanac Vol. 28:287-310.
3. ________. 1985. Farm Bill Granted a Limited "Win" to All Sides. Congressional Quarterly Almanac. Vol. 41:517-539.
4. Dana, S.T. 1956. Forest and Range Policy. McGraw-Hill Book Co. Inc., N.Y.
5. ________ and S.K. Fairfax. 1980. Forest and Range Policy, 2nd ed. McGraw-Hill Book Co. Inc., N.Y.
6. USDA Forest Service. 1990. An Analysis of the Timber Situation in the United States 1989-2040. RM Forest & Range Expt. Sta., Gen. Tech. Rept. 199. Fort Collins, Colo. pp. 173-195.
7. USDA Agric. Stabilization and Conservation Serv. 1991. Forestry Conservation/Environmental Protection Programs. BI No. 5. Washington, D.C. 8 pp.
8. ________. 1992. Forestry Incentives Program: From Inception of Program through 1992 Fiscal Year.
9. USDA Forest Serv. 1993. Stewardship Incentives Program: 1992 Fiscal Year Statistical Summary. Washington, D.C. 17 pp.

Technical Forum Presentation:

The wood plypole: not a hollow promise

Robert W. Erickson

Abstract

It is generally accepted that 90 percent of the bending strength of a pole comes from the wood located in the outer 22 percent of the diameter. Consequently, 90 percent of the bending strength for a 10-inch-diameter pole is contributed by the 1.1-inch-thick perimeter shell. In real life, this mechanical advantage allows utility companies to continue to use poles that are hollow due to heartrot but that have sound shells of 2 inches or greater thickness. The plypole is hollow, lightweight, and economical to manufacture to the required levels of strength and durability. Use of the plypole will conserve our softwood forests, since at least two distribution line poles could be manufactured from the wood contained in one such contemporary pole. It is an attractive alternative to the contemporary wood pole as well as to poles made of concrete, steel, or fiberglass.

(full paper not available)

The author is a Professor, Dept. of Forest Prod., Univ. of Minnesota, St. Paul, Minn.

Technical Forum Presentation:

Properties of dahurian larch from the Russian Far East

Robert L. Ethington, Rakesh Gupta, and Olga Krankina

Abstract

The Russian Far East contains nearly half of the forestland in Russia, and about 9 percent of the forestland in the world. These forests are characterized by relatively slow growth, low stocking density, overmaturity, and are currently not very accessible. In development of a methodology for associating lumber design properties with National Grading Rule grades for foreign species, we have worked with dahurian larch (*Larix dahurica*), and this has given us some special insights into the physical and mechanical character of the species. According to a 1989 study by Borovikov and Ugolov, there are 670 million acres of forestland in Siberia dominated by larch. (For comparison, according to the USDA Forest Service, there are 696 million acres of forestland in the United States, representing all species). Dahurian larch dominates 56 percent of the forestland (equal to about six Oregons), and is the major larch east of 110° east longitude. Because of the abundance and general location, if a significant amount of wood is exported from Siberia to the United States, this larch is likely to be prominent. Our samples were taken from two natural stands 300 miles north of Vladivostok. They were essentially pure stands of larch, one 160 to 180 years old, the other 160 to 200 years old. Trees were about 60 feet high, with an average diameter at breast height of 10 to 12 inches, but ranging up to about 20 inches. Lumber cut from the logs had relatively small knots (was often knot free), and appeared to have an abundance of shake. It was surfaced at about 40 percent moisture content, and exuded a significant amount of sugar in that condition. Statistics summarizing some physical and mechanical properties of clear wood are provided from 231 specimens. Growth rate and percent summerwood are similar to those reported for western larch (*L. occidentalis*), but specific gravity is 10 percent higher, proportional limit in compression perpendicular to grain is 8 percent lower, and shear strength is 24 percent lower.

(full paper not available)

The authors are, respectively, Professor, Assistant Professor, and Research Associate, Dept. of Forest Prod., Forest Res. Lab., Oregon State Univ., Corvallis, Oreg.

Technical Forum Presentation:

The Asia Pacific woodchip trade: changing supply and demand trends

Robert Flynn and Dennis Neilson

Abstract

The Pacific Rim trade in both softwood and hardwood chips continues to be dominated by the demands of the Japanese pulp and paper industry, which accounts for 97 percent of the volume traded. However, the market softened in 1992 to 1993 as the oversupply of paper and paperboard in Japan impacted on raw material demand. Japanese imports of softwood chips fell from 9.1 million m^3 in 1991 to 8.0 million m^3 in 1992. Hardwood chip imports fell slightly from 13.35 to 13.25 million m^3 during the same period. Total imports in 1993 are not expected to be more than the 1992 volume, with softwood chip demand weaker than hardwood chip demand. In the United States, the Pacific Northwest continues to be the major supplier of softwood chips with about 53 percent of the total. Australia is the largest supplier of hardwood chips, supplying about 30 percent of the 1992 total. However, new and expanded operations in the southern United States, Chile, and South Africa are impacting Australia's market position. A number of other new exporting regions are expected to play an increasing role in providing chips to North Asia and the United States during the 1990s. Brazil, Russia, Vietnam, Honduras, Venezuela, Costa Rica, and Guatemala are all likely to increase or commence shipments this decade.

(full paper not available)

The authors are, respectively, President, Robert Flynn and Assoc., Tacoma, Wash.; and President, D.A. Neilson and Assoc., Rotorua, New Zealand.

Technical Forum Presentation:

Change and its effects on distribution channels in the international trade of forest products

James S. Fox and John P. Dolan

A brief review of domestic and international trade in North American forest products is in order. On the domestic front, the post-World War II condition was the dominance of West Coast species in all U.S. markets. Abundant supplies of green Douglas-fir and West Coast hemlock commons and uppers were shipped by sea and rail to distributors or, in the case of shop grades, to cut-up plants in Minnesota and Wisconsin.

Major producers like Weyerhaeuser, Dant & Russell, MacMillan Bloedel, and the partners of Seaboard, maintained cargo facilities at every coastal port. These tidewater mills shipped to their own docks in California, Gulf, and eastern ports, as well as to merchant wholesalers who negotiated with them as well as noncaptive producers.

"Portland" wholesalers bought from western mills and sold to midwest and eastern wholesalers. Soft wood lumber was king, plywood was an infant born of the millwork industry, and southern yellow pine was a regional specialty. The channels were all relatively defined.

On the international front, offshore exports meant essentially England, Europe, and Australia. Competition was among the majors only as MacMillan Bloedel, Seaboard, and EACOM battled the Puget Sound and Columbia River producers for market share. North America battled Russia and Scandinavia for the European market.

Channels followed the need for scale to succeed. To cut costs, shippers maintained their own sales offices at their own docks, served by ships under charter. Companies dealt with designated agents in each country, in effect granting a franchise. Shipping times of a month or more lent an additional strategic rather than tactical tenor to the game.

Telecommunications meant telex by wireless or transatlantic cable, or brief phone calls of tin-can quality, made more brief by system failures or operator error.

Five forces of change had gathered by the early 1980s, each with its own impact on the channels of distribution. These were: 1) emergence of new industrialized countries; 2) development of new wood technologies; 3) developments in transportation; 4) developments in telecommunications; and 5) availability of wood resources.

Emergence of new industrialized countries

Lumber trade tends to be between developed countries. The same is less true for manufactured panels but, given relative volumes, the wood products trade follows not just population growth but income growth as well. Of total trade (imports + exports) of lumber in 1989, the FAO reported North America at 45.1 percent of total trade, and western Europe with 29.3 percent of the total. Paper and paperboard trade in the same year showed western Europe with 54.3 percent of total trade, and North America with 27.9 percent.

As their economies emerge, new markets are requiring products that meet their individual cultural

---

The authors are, respectively, European Markets Manager, KMJ International, Bothell, Wash.; and Principal, Cascade/Rocky Mountain Consultants, Evans, Colo.

needs (e.g., chopstick blanks) as well as globalized products such as telephone directory papers.

The latter product is likely to be sold by a producer direct to his printer contact; the former (chopsticks) is more likely to flow through internationally specific channels, including middlemen, reflecting a degree of finishing, color preference, and similar regional varieties of the utensil. To the extent these emerging economies have fashion leaders who were educated in the United States and Canada, we could see a move towards western-style furniture and construction, so that globalized raw materials could result.

Development of new wood technologies

A good model for studying the impact of new technologies on channel selection may be observed in North America. In its initial entry into framing markets, laminated veneer lumber (LVL) was introduced as part of a total branded system with wood I-beams in carefully engineered applications. Distributors were screened, challenged with heavy training and promotional responsibilities and, in some cases, producer sales offices acted as distributors. After a 10- or 15-year period and increased capacity, LVL can now be purchased through a variety of distributors, and can be found in an impressive array of building systems. Once a product has achieved maturity in the market, different channels of distribution can be expected than were required for introduction.

Manufactured panel products are likely to be more like pulp or paper than lumber in their channel of distribution. For one thing, adoption of metric product specifications is more easily controlled. Some confusion is still possible, of course. A recent article indicated a volume of a manufactured panel as measured in *million* cubic meters (3/4-in. basis). As with paper grades, manufactured panels can be adapted to meet local requirements, including recycled material content, if so desired.

Internationally, introduction of new systems and products is likely to follow similar divisions as in domestic, North American markets. Engineered wood products will be taken through controlled agents, distributors, and contractors, during the early acceptance state; as use becomes commonplace, channels will both broaden and simplify — more outlets buying more directly.

In shelter construction, Cohen and Smith's analogy of autos and PFC construction may reach its zenith. In 2006, we might well see a PFC house in Japan built of SPF studs, SYP plates, rafters of coastal B.C. LVL flanges, and webs of eastern Canadian oriented strandboard, all looking quite conventional. As with Ford's Escort "world car," the integrator will control which materials are acceptable.

Developments in transport

In terms of the past 40 years, nothing can come close to equating the productivity gains in transport. In 1950, the post-war fleets consisted of Liberty ships and their big cousins, the C-5's. Modification for forest products meant accommodating deck loads, but little else.

Mill-to-market times were regularly 4 weeks (U.S. east coast), 6 weeks (Japan), or over 8 weeks (Europe and Australia). Besides the transoceanic voyage, there was the initial trip to a dock for staging, storage for some period, and finally a week of loading.

Today's container ship cuts the total time at least in half, and for kiln-dried material or manufactured panels, the next person to touch the material after the planer will likely be a user.

As pointed out earlier, the scale of ocean shipping dictated who would be in the distribution channel: long-term leases at shipping and receiving ports, charter or ownership of specialized forest products carriers, inventory valuation exposures in millions of dollars. All of this spoke of big middlemen and consortia dealing with reliable, stable agents. This club was the world less than 15 years ago.

Developments in telecommunications

The facsimile machine and personal computer have joined fiber optics and satellites as state-of-the-art telecommunications. This has decreased scale in forest products trade almost as much as changes in transportation.

Voice communications can be established from fixed site or remote locations, 24 hours a day. Barring language problems, I can check the competitive price and availability of species from west coast North America, Scandinavia, and Oceania, all within 20 minutes.

By fax, I can send details of profiling, ISO, or code requirements, and shipping documents for review by customers, agents, and manufacturers. Instead of an entire telex staff manning a center 'round the clock, I employ a programmable fax machine that dispatches my inquiries and quotes at times convenient for the receiving parties, at lowest cost to me. Likewise, inquiries from my clients and offers from my suppliers can arrive at times convenient for me, and in a format I am used to dealing with — not telex script.

However, the telecommunications revolution has accelerated dissemination of facts and data, not necessarily information. It is one thing to know the availability of 1/10-in. hardboard from a variety of suppliers — far different is a knowledge of their past performance in meeting commitments and fidelity to channels. The high tech cannot help increase the value of the high touch.

Shift in availability of wood resources

Our perception of availability of wood resources has shifted profoundly in the past four decades. "Available" used to mean primarily biologically available (i.e., on the stump). It meant economically available by reason of proximity to transportation and processing.

Particularly in the northern hemisphere, little attention was paid to the political side of availability. Despite the knowledge that more than two-thirds of the world's softwoods were owned by socialist organizations, (i.e., the Canadian provinces, Soviet cooperatives, and the U.S. Forest Service, users assumed access to those resources as a function of having the demand that would call for them.

The emergence and re-emergence of supplies from the southern hemisphere and southern United States in international trade has reflected the contraction of biologically and economically available supplies based upon political decisions in such diverse regions as the west coast United States, British Columbia, and Indonesia.

Availability of wood fiber for papermaking has changed as well due to similar political pressures. In the decision to export waste paper from one jurisdiction is the implicit decision to curtail the use of primary fiber from another. The specialist in bleached kraft pulps is an anachronism; the generalist in papermaking blends and their affect upon process variables is ascending.

The impacts of changing availabilities have been different depending on the perceptions of the player. The producer dedicated to a region or raw material has had to decide how much risk to bear on the raw material side. The channel middlemen have had to decide which sources will be both politically and biologically available to them, as well as those that will fulfill the economic need for profit. The user has had to analyze how much dedication to give to traditional suppliers. In some cases, as our posters show, this had meant almost complete collapse of the channel as users make direct investment in the west coast of North America to assure at least some supply.

What it means

The countervailing forces of the last 40 years have been both at odds and in concert impacting the channels of international trade. The container ship has flattened the channels of distribution by destroying scale economies in shipping. This has been true in lumber, panels, secondary fiber, etc.

Telecommunications has allowed the daily interchange of data and facts, but as yet has provided as much disinformation as information. Expert in products have lessened their importance and their hold on the market; experts in processes and substitution have grown in importance.

The politics of supply has added scale back to the market. How many small players can risk the take-or-pay contract required to secure priority treatment from a capital-intensive, risk-averse manufactured panel or specialty lumber producer?

Emerging telecommunications and transportation technologies, combined with globalized design, hold the promise of a "world consumer" of the future sitting in Telluride, Colo. watching a satellite feed from the Korean home shopping network, after a hard day of skiing. This consumer picks up the telephone and, using their BNP yen-denomination credit card, orders a KD home office workstation made by O'Sullivan from medium density fiberboard, lumber, and veneer parts, made in a half dozen different countries, and assembled in South America.

How channels ultimately serve that customer efficiently will determine which channel members survive 15 years hence. We trust that our readers will obtain useful insights into the origins and effects of change.

References

1. Cohen, D.H. and P.M. Smith. 1991. Global marketing strategies for forest products industries. Canadian J. of Forestry Res. 22:124-131.
2. Food and Agriculture Organization of the United Nations. 1992. Commodity Review and Outlook, 1990-1991.
3. International Forest Products, Ltd. 1993. Interfor Annual Information Form. March 12, 1993.
4. Knee, R. 1990. Lumber shippers brace for change. Amer. Shipper (Sept. '90) p. 30.
5. Seaboard Lumber Sales Co., Ltd. 1982. The President's Annual Review for 1981.
6. Seifert, B. and J. Ford. 1989. Export distribution channels. Columbia J. of World Business 24(2):15.
7. Shipley, C.D. and E. Barnett. 1989. Recruitment, motivation, training, and evaluation of overseas distributors. European J. of Marketing 23(2):79-93.
8. Tirschwell, P.M. 1993. Smart distribution centers help firms serve EC clients. J. of Commerce and Commercial 395(27930):1a.
9. Widman, C.R. 1993. A new era begins. Widman's World Wood Review 1(1):1-8.

Technical Forum Presentation:

Sugar cane rind: an ecological and economic alternative for OSB, laminated strand lumber, and new panelboard products

Paul Friedman

Abstract

Pulp, paper, particleboard, and fiberboard have long been manufactured from bagasse, the lignocellulosic residue resulting from the traditional process of crushing sugar cane and extracting about 95 percent of its sugar content. Depithing the bagasse is required because pith cells (about 30 to 40% of bagasse) absorb chemicals without contributing to product strength. The economics of these processes depend on many factors, including the availability and price of competing wood furnishes, cost of replacement fuel for sugar production, the cost of imported product, and market product quality requirements. Unlike traditional sugar cane processing, cane separation technology preserves the full potential of each component of the cane stalk because it produces individual streams of the juice-containing inner core, the tough fibrous structural material, and the thin outer waxy layer. This separation allows each stream to be further processed into both sugar and new sugar cane products by exploiting each stream's unique characteristics. Flakes or segments from the tough fiber are free of both pith and wax, and are 1-1/2 to 2-1/2 inches wide, 12 inches long, and 1/32 inch thick. Therefore, they provide an excellent annually renewable furnish for flake or strand-based panelboard products such as OSB, Canalam (a laminated strand-lumber product), and Ultrapanel (a proprietary, low density structural building material). Economic analyses for panelboard plants using sugar cane rind located next to an existing Florida sugar mill show that although OSB production is profitable, utilization of the rind for higher value added products such as Canalam or Ultrapanel could be more attractive.

(full paper not available)

The author is Vice President, Research and Development, SugarTree Technology, Boulder, Colo.

The Globalization of Wood: Supply, Processes, Products, and Markets

Technical Forum Presentation:

Measuring tree growth using machine vision

Mark Fuller and Timothy D. Faust

Abstract

A PC-based interactive image analysis system has been designed for the purpose of measuring wood quality factors in southern tree species. The system is capable of measuring growth rate, percent latewood, and documenting other elements such as false rings in southern yellow pine species (*Pinus* spp.), ring porous species such as southern red oak (*Quercus falcata*), and diffuse porous species such as yellow poplar (*Liriodendron tulipifera*) with varying degrees of operator intervention. The analysis is performed on a tree increment core and cross-sectional tree disks. The distinguishing characteristic of the system is the use of an X-Y table for tree cores which is used to automatically position each increment core under the camera one segment at the time. Information from each segment is then "tiled" together giving the overall growth characteristics of the core. The X-Y table makes it possible to utilize a wide range of magnifications and resolutions without greatly affecting the workload of the operator.

When completed, the system will operate in two modes. For the case of pine cores which, in general, have very distinct growth increment boundaries and transitions, software is being developed to estimate growth ring boundary locations in a semi-automatic mode. The hardwood species which in general have more subtle growth increment boundary characteristics are marked manually. In either case, the operator uses a mouse to define a scan path across each core segment. In semi-automatic mode, a luminance, or intensity vector, is extracted along each line segment defining the scan path. Luminance measurements are taken directly by measuring reflected incandescent light from the surface of the core specimens. The luminance vector is analyzed by the computer, which then places markers at probable boundary locations. Due to the extreme variability of growth increment widths and the presence of artifacts such as resin canals and pitch streaks, the software will not be smart enough to correctly mark all rings. The operator is allowed to edit mismarked boundaries. In the manual mode the operator places marker flags along the scan path segments at points where they cross growth increment boundaries.

The disk area measurement system will measure ring area and radial increment growth on four major and minor axes radiating from the pith. The disk area program has only one mode of operation.

The authors are, respectively, Graduate Assistant and Assoc. Professor, Univ. of Georgia, Warnell School of Forest Resources, Athens, Ga. This project is supported by the Southeastern Forest Expt. Sta., USDA Forest Serv., Athens, Ga.

Introduction

This paper describes a PC-based image analysis system which analyzes increment cores and cross-sectional tree disks for the purpose of measuring macroscopic wood quality factors such as growth rate and percent latewood. The project is being implemented in conjunction with the USDA Forest Service's Southeast Forest Experiment Station in Athens, Ga. Presently, a manually operated system is in use at the station for measuring various wood quality factors in southern tree species. The vision system project was conceived out of the realization that the present manually operated system by which cores are indexed under a microscope and measurements made by means of a graduated eyepiece, although having served its purpose well for many years, could be greatly improved in terms of speed and repeatability by the implementation of a computer-controlled system.

There are two major areas where the computer-aided system should alleviate limitations of human performance factors in the manual system. A computer vision system, although not able to make intuitive judgments involving the images it sees, is tireless and with a degree of human guidance will make core analysis less taxing. A second improvement comes with the motorized table by which cores are indexed under the camera. The present system utilizes a manually operated indexing stage in which the operator must follow a point across a microscope field to know when he has indexed the core by the proper distance. With the imaging system, software controls the rail table motors to precisely position cores under the camera. The operator, while viewing core images on an external monitor, defines each move by drawing a line over the core image or by instructing the computer to generate the line. While processing a scan path segment, the length of the line is extracted and used by the software to index the rail tables so that the next core segment is displayed on the monitor. In comparison to the present system, the scope of the operator's task is redefined and simplified. The operator controls the whole operation by interacting with a Windows dialog box and by observing the external monitor where he can verify that the software has accurately marked growth increment boundaries, edit computer-generated boundary markers, or place boundary markers in manual mode, if necessary.

Other approaches have been taken in automatic tree ring measurement. The WinDendro (1) represents one approach. This system performs ring width analysis on disk images which are read into computer memory by an optical scanner. By analyzing increment cores as opposed to disks, destructive sampling is avoided but at the same time less information is available for making software-generated growth boundary estimations. The WinDendro system is constrained to one degree of magnification from which scale problems could ensue (i.e., the scanner system may have trouble with very slow growth trees having many rings per inch). Another system is the Dendro 2003 (2), which utilizes X-ray densitometry along with image analysis to measure growth characteristics of increment cores. By utilizing reflected light as opposed to X-ray densitometry to generate images, there is a significant cost reduction. However, densitometry provides more information on growth characteristics, and so may be more useful for determining specific gravity variations, for instance. Yet another method of ring width analysis utilized by instruments such as the Densitomat model 1410 (3) is penetration resistance of a small needle which does not fall in the realm of image analysis but is also designed for determining growth characteristics. The penetration resistance method works well only if the needle penetrates perpendicular to the growth rings and cannot distinguish between growth rings and intra-annual density variations, whereas growth ring orientation can be accounted for in an image analysis setting and aberrant density variations are not as likely to cause problems. Therefore, our system may be differentiated from other such systems by the combined use of reflected light, computer-controlled positioning, and the option of various magnification settings.

Experimental methods

The only core variable taken into consideration is surface preparation. We have experimented with cores prepared by: 1) sanding; 2) machining (peripheral milling); and 3) slicing.

Depending on the species, different methods are appropriate. For instance, for hardwood species in which there is not a great deal of specific gravity variation from earlywood to latewood, the appropriate method is usually slicing, and sometimes machining, since this leaves vessels and other artifacts free of dust. In southern magnolia (*Magnolia grandiflora*) for instance, the thin bands of marginal parenchyma delineating growth increment boundaries may be obscured by sanding and vessels are filled with sanding dust. Southern pine species, on the other hand, exhibit a substantial specific gravity variation from earlywood to latewood and so may be sanded with good results.

During the data acquisition phase of the project, 50 core samples of each of the three species types (pine, ring porous, and diffuse porous) which have been analyzed previously by the forest service's manual system will be analyzed by the image analysis system. Within each of the three groups, cores having various ages and growth rates will be included. A comparison study between the two data sets will be carried out. The experimental design is straight-forward and consists of a comparison between methods.

Objective. — To design and implement an image analysis system which will be capable of measuring growth rate and percent latewood of southern pine, ring porous, and diffuse porous hardwood increment cores.

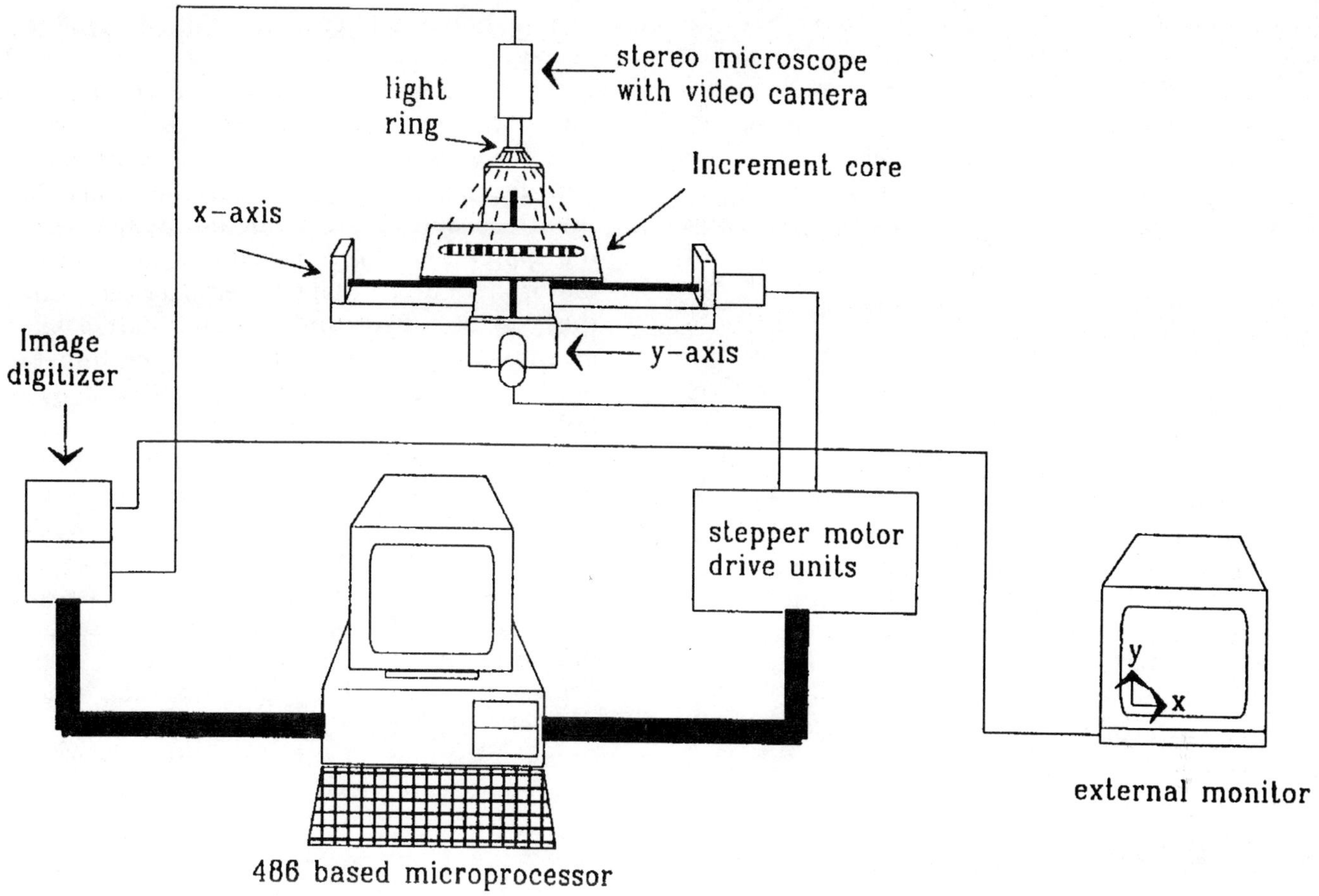

Figure 1. — Layout of image analysis system.

Supporting hypothesis. — The measurements obtained by the image analysis will be the same as those obtained from the manual method, but with substantial time savings.

Image analysis hardware and software

The system layout is illustrated in Figure 1, and its component functions and operation are described in this section.

The microprocessor serves as a link between the video hardware and X-Y table, and coordinates the two during the data acquisition process. Images are taken by a black-and-white video camera which is mounted on a stereo microscope and sent to a high resolution image digitizer, a data translation frame grabber. The frame grabber transposes the video signal to digital format and sends this information to the microprocessor. Optimas Bioscan software is used to control the hardware components that:

1. Capture, digitize, and load images into computer memory.
2. Process the image to make wood quality measurements.
3. Position the specimen under the camera.
4. Store raw data (specimen ID and data analysis) in a database for subsequent retrieval and data analysis.

Optimas Bioscan is a Windows-compatible image analysis software package with built in algorithms (macros) that perform low-level image analysis functions such as luminance and line length measurement, convolution, and image segmentation. Software platforms such as Optimas bring image analysis within reach of the "average" scientist who may not have the time, money, or programming expertise to implement all the low-level functions necessary to a functional image analysis system. This makes it possible for the programmer to concentrate on high-level, domain specific issues when tackling an image analysis problem. The power of Optimas stems also from user-defined macros which are written in ALI, an interpreted language with a structure very similar to C. Within these macros, image analysis functions may be called the same way as functions or subroutines are called in any computer language. Also, using a tool such as the Microsoft Windows software development kit, dialog boxes may be created into which the user-defined macros may be incorporated. This increases the power and utility of the software platform. Controls may be incorporated into the dialog box which are linked to Optimas macros so that complex imaging and data analysis sequences can be painlessly accomplished using push-button controls with a mouse. The dialog box which has been created for controlling the system of Figure 1 is shown in Figure 2. Following the figure is a sequence of operations which describes the steps taken to analyze a core, as well as the function of the controls in the dialog box for reading increment cores and tree disks.

Sequence of operations

Increment core analyzer

1. Clamp core onto the Y-axis stage such that the long axis of the core is oriented along the X-axis.
2. Click on the "New Core" push-button to:
  - initialize the system for a new core
  - prompt the user to type in the Core ID and year of the outermost growth increment
3. Select a "Scan Mode." The differences between Manual and Semi-automatic will be explained below.

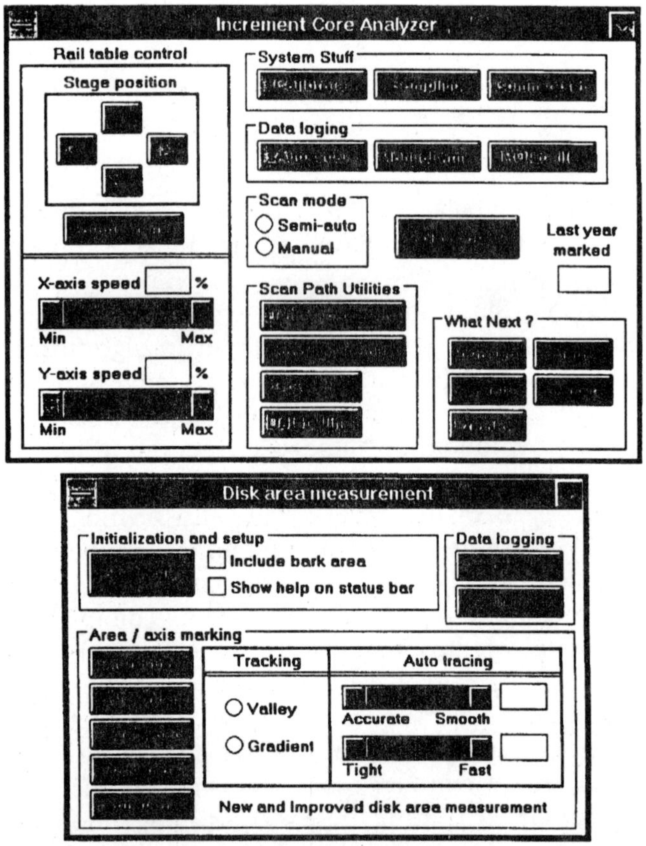

Figure 2. — Increment Core Analyzer and Disk Area Measurement dialog boxes.

4. Position the core so that the outermost ring can be seen on the external monitor. Positioning of the core is accomplished by clicking one of four directional arrows seen in the "Stage Position" portion of the dialog box which sends the appropriate run command to the rail table motor drive unit. "Stage Speed Set" allows the speed to be set on the fly to facilitate a kind of coarse-to-fine movement depending on whether the core needs to be indexed by a long or short distance, respectively.
5. Click the "Define Start Point" push-button. A mouse-controlled pointer appears on the external monitor which the user positions at the desired point along the cambial side of the outermost latewood band. Upon clicking the right mouse button, the X-table is indexed so that the start point is moved to the left edge of the external monitor. This point marks the beginning of the scan path.
6. Define a segment of the scan path by clicking the "Draw Line" or "Advance Frame" push-buttons.
  - The "Draw Line" push-button allows the user to draw a line on the external monitor. The line originates from the start point (or the end point of the previous line) and may extend to any pixel location of the monitor.
  - The "Advance Frame" push-button automatically draws a horizontal line across the entire monitor screen using the Y-coordinate of the start point or end of the previous line. In defect- or artifact-free sections of a core, the user need not define the scan path as meticulously as is possible with the "Draw Line" push-button.
7. The next step depends on whether the Manual or Semi-automatic mode is in use.
  - **Manual.** — Once the scan path segment is drawn, the user is placed in flag editing mode. Marker flags may be placed at growth increment boundaries, earlywood to latewood transitions, and at locations of cracks in the core. Upon clicking the right mouse button, these flag locations are extracted by the software and the user is prompted to make a selection from the "What Next" portion of the dialog box. In cases where the user is unsure that he has made the correct move or marked the boundaries accurately, he may view the "Previous" segment, "Preview" the next portion of the core, "Remark" (or edit) present markers, or accept the

present markers and proceed on to the next core segment. If "Proceed" is selected, the stage is indexed in the X direction by a distance equal to the length of the scan path segment. When the user chooses "Proceed," the total length of previously defined scan path segments is added to boundary marker locations on the present segment and then the present segment locations are appended onto an array. If there are cracks present in a segment, the width of each crack is subtracted from subsequent boundary marker locations so that the boundary locations are stored as if there were no cracks.

- **Semi-automatic.** — Once a scan path segment is drawn in the semi-automatic mode, the user is prompted to make a selection from the "What next" portion of the dialog box just as in the manual mode. The "Flags" push-button may be chosen, which places the user in flag editing mode in the same way as for the manual mode, except only cracks are marked in this case. The "Previous," "Preview," and "Re-mark" push-buttons may also be used and behave identically as for the manual mode. When the "Proceed" push-button is selected, a luminance profile (Fig. 3) is extracted along the line defining the scan path segment. A luminance profile is simply an array of grayscale values, between 0 and 255 in the 8-bit case, where black corresponds to 0 and white to 255. Cracks are omitted from the luminance profile in the same way that their width is omitted from the scan path length. The luminance of each path segment is appended to an overall luminance array for the whole core.

8. Repeat step 7 until the scan path is defined to the pith (or any other desired point along the core).
9. **Manual.** — In the manual mode of operation, all boundary locations have been defined when the user reaches the end of the core. At this time, the user selects the "Manual write" push-button which causes the growth increment information to be written to an Excel spreadsheet. Each line of data written to the spreadsheet contains the year of the growth ring, the width of the growth ring, and the width of the latewood band, if these were marked.

   Semi-automatic. — After selecting "Proceed" for the final scan path segment, the user selects the "Auto scan" push-button which initiates the automatic boundary location algorithms. At this point the user has defined the entire scan path and the luminance profile of the whole core has been extracted and stored in an array of bytes. This array is used to compute a "running average" array of bytes along the scan path where each point of the averaged array X_i is:

$$0 \le i \ge N/2 - 1 \;\rightarrow\; X_i = \Sigma_j L_j/j + 1$$

where: $0 \le j \ge i + N/2$

$$N/2 \le i \ge L - N/2 - 1 \;\rightarrow\; X_i = \Sigma_j L_j/(N + 1)$$

where: $i - N/2 \le j \ge i + N/2$

$$L - N/2 \le i \ge L - 1 \;\rightarrow\; X_i = \Sigma_j L_j/(L - i + N/2)$$

where: $i - N/2 \le j \ge L - 1$

where:
N = an even number of points to average
L = number of points in whole core luminance profile

   This running average array is in turn used as a threshold for determining the probable location of growth increment boundaries and transitions (that is, for every point in the luminance array there is a corresponding threshold point in the running average array). After the last scan path segment has been defined, the core is indexed back to the starting point and the first scan path segment is displayed on the monitor. Locations of boundaries are determined as follows: The luminance value of each point along the scan path is compared to the corresponding running average point value. If enough successive luminance points are found to be less than (or greater than) the corresponding threshold value and then a point is found to be greater than (or less than) the corresponding threshold value, then there is assumed to be a latewood to earlywood (earlywood to latewood) transition in the neighborhood of this sample point. After all such transition points for a path segment are located, marker flags are automatically placed on the monitor screen. If these are correct, then they are appended to an array of marker flags for subsequent writing to an Excel spreadsheet. If some markers are incorrect they may be edited by the user.
10. **Semi-automatic.** — Repeat 9 until each scan path segment has been analyzed, then write data to a spreadsheet in the same manner as was described earlier.

Other controls which are not mode-specific include:

Calibrate: used to calibrate distances when magnification settings are changed.

Sampling: used to set the length (number of pixels along a line) of a sample point and also to the

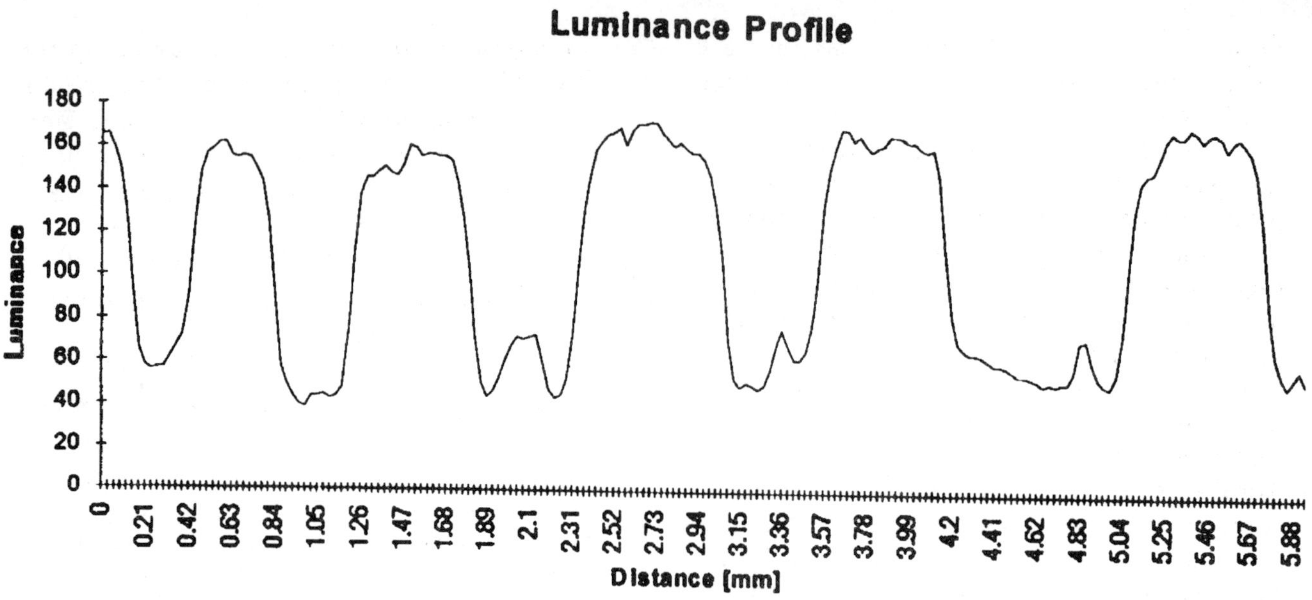

Figure 3. — Typical luminance profile.

width (number of pixels perpendicular to a line) which will be averaged for a single sample point.

"Region of interest" to file: save the image of a region of interest to a file.

Disk area measurement

1. Name specimen and declare the year of the outermost ring of growth.
2. Calibrate the image with ruler in the field of view.
3. Mark the pith.
4. Trace the annual rings in succession from pith to bark (bark may be traced optionally).
5. Press "Mark Radii" push-button to calculate radii and measure radial increments.
6. Press "Write XL" to send data through DDE to spreadsheet.
7. You may optionally save the marked image to a disk file and print the marked image.

The resulting visual marking of the disk image is illustrated in Figure 4 below. The images can be acquired from an image digitizer (frame grabber) or they may simply scanned on an inexpensive flatbed scanner.

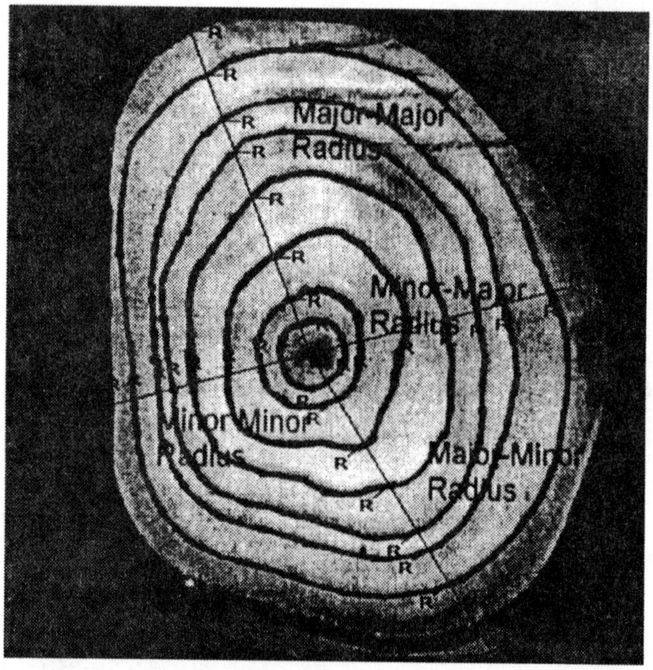

Figure 4. — Image with radii and increment boundaries marked.

Comments

Some explanation is needed for the Semi-automatic mode. This mode is workable with pine cores which have luminance profiles similar to the one in Figure 3 with pronounced jumps in grayscale from earlywood to latewood. To apply it to hardwoods would require craftier methods of boundary detection, which would soon become computationally expensive and require many hours of programming time. The boundary detection computations in the case of pine cores are performed on one-dimensional "Line" objects which are the scan path segments. To apply boundary detection to porous species would require extraction of "Area" objects, such as vessels, which would increase computation time as well as the complexity of programming by orders of magnitude.

There are obstacles associated with pine cores such as the high variability of ring widths. Latewood bands may be anywhere from several millimeters to one or two cells wide. This presents a "scale" problem which has traditionally been a dilemma in image analysis. Our eyes are able to automatically adjust to see objects at many different scales, but this is not so easily accomplished with machine vision. This is why we have incorporated the ability to change magnification into the system which helps to alleviate this problem. Another sticky point arises from the presence of artifacts such as resin ducts and growth stress residuals, either of which may appear to be growth boundaries to the software. Such artifacts can be partially weeded out by making the width of the scan path line greater than one pixel, so that a sample point intensity value is actually the average of several adjacent pixels which are oriented such that if a line were drawn through their centers it would lie perpendicular to the scan path segment. This method works well for artifacts which are small or which lie nearly parallel to the scan path. It does not work so well for larger artifacts or those which lie perpendicular to the scan line. Hence the software cannot be 100 percent accurate in its placement of growth increment markers and some operator intervention will be required.

Initially, the basic question was "How feasible is the measurement of macroscopic features of wood using reflected light as measured by an image analysis system?" Presently, the hardware development has been completed and software development is nearly complete. It has become apparent that the system is undoubtedly practical and, although will not be 100 percent automatic, will be a dramatic improvement over the manual system.

References

1. MacDendro and WinDendro Automatic tree-ring measurement system. Designed by Regent Instruments Inc. and National Optics Institute in collaboration with Rejean Gagnon and Hubert Morin of the University of Quebec.
2. Dendro 2003. Walesch Electronic. Effretikon, Switzerland.
3. Densitomat product list. DeRiTec GmbH, St Josef Strasse 14, D-8501, Rückersdorf, Germany.

Technical Forum Presentation:

Japanese hardwood product imports: current situation and historic trends

Bruce G. Hansen

Abstract

Japan represented a $3.7 billion market for world exporters of hardwood logs lumber, veneer, and plywood. Of this amount, U.S. exporters accounted for $237 million or roughly 6.5 percent. Between 1976 and 1992, imports of logs declined nearly 50 percent while imports of lumber increased 478 percent, imports of veneer increased 1492 percent, and imports of plywood increased 1993 percent. In contrast, imports of logs from the United States increased over 200 percent, imports of lumber increased over 3,100 percent, and imports of veneer increased over 11,000 percent. Imports of plywood from the United States were down nearly 40 percent.

This presentation examines the current mix, in dollar value, of products from all suppliers and the United States along with changes in the mix of products by value and changes in volume from 1976 to 1992.

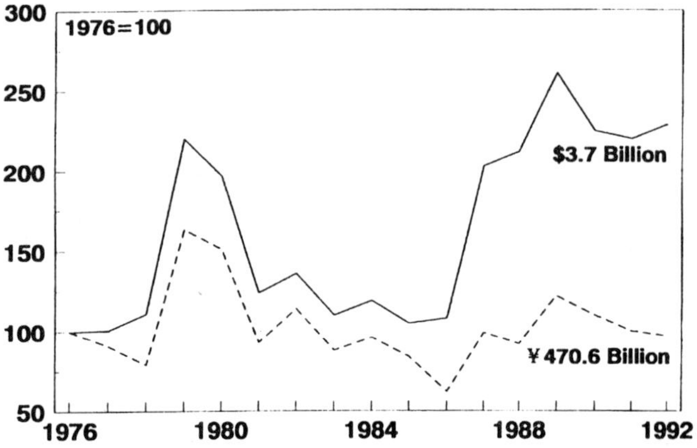

Figure 1. — Japan imported $3.7 billion worth of hardwood logs, lumber, veneer, and plywood in 1992. This was more than 2-1/4 times the value imported in 1976 and twice the average during the mid-1980s. However, due to an increase in the value of the yen, Japanese expenditures on imports in 1992 were no larger than in 1976.

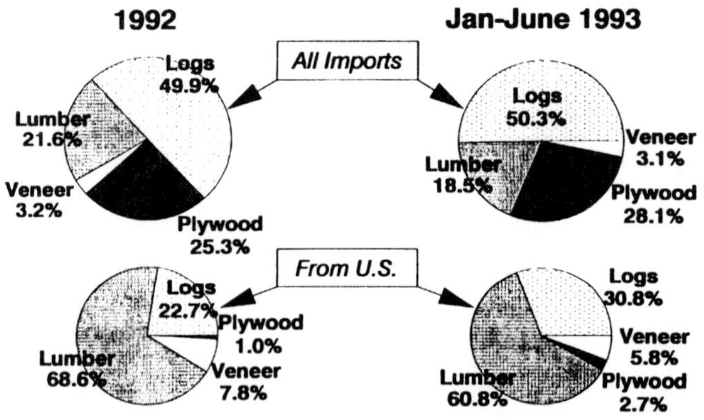

Figure 2. — Overall, logs made up half the value of all imports in 1992 and the first half of 1993. Hardwood plywood was next, then lumber and veneer. However, data for the first 6 months of 1993 indicate a slight change in the mix away from lumber to hardwood plywood. In contrast, the mix of imports from the United States was made up of more than two-thirds hardwood lumber in 1992 followed by logs, veneer, and plywood. Data for the first 6 months indicate a change in the mix to include more logs and plywood and less lumber and veneer.

The author is Principal Economist, Northeastern Forest Expt. Sta., USDA Forest Serv., Princeton W. Va.

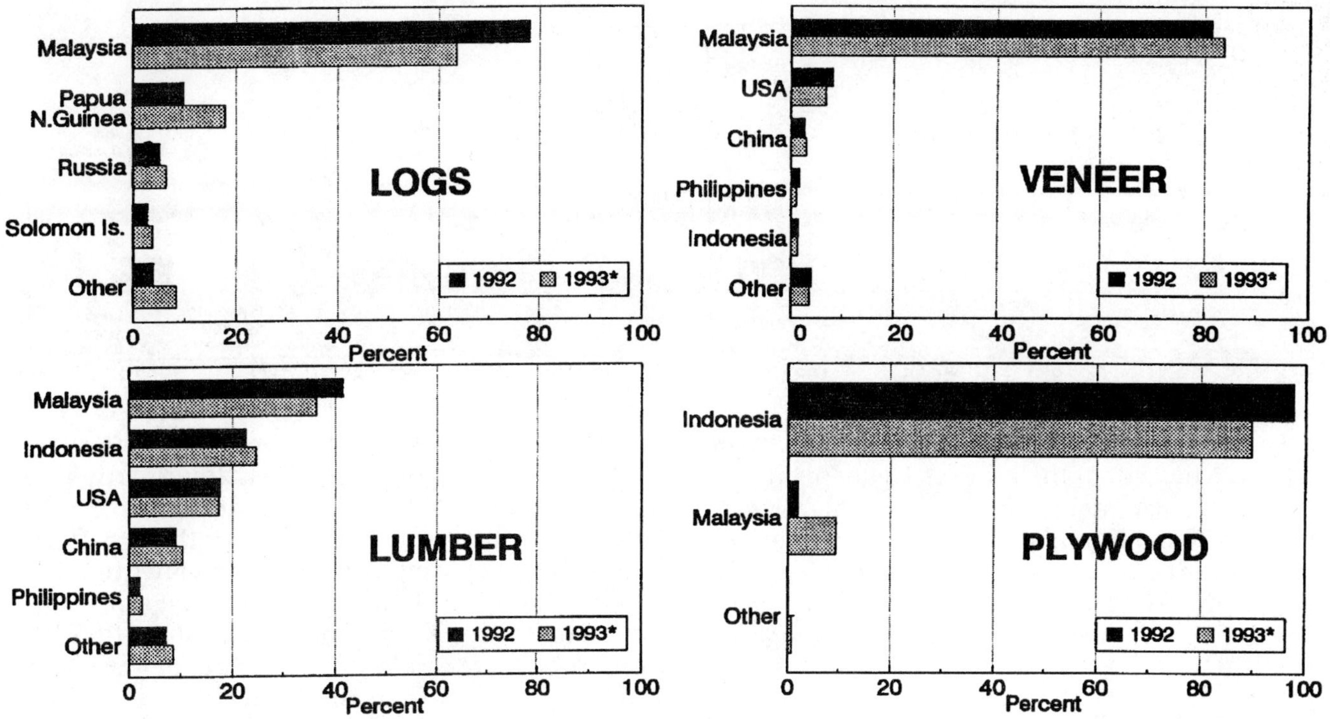

Figure 3. — About 80 percent of Japan's log and veneer imports were imported from Malaysia while nearly all plywood was imported from Indonesia in 1992. Hardwood lumber imports were the least concentrated but, even here, 90 percent of the imports were supplied by just four countries. Data for the first 6 months of 1993 suggest that Malaysia may have a reduced role in supplying logs to Japan in the future but a larger role in supplying hardwood plywood.

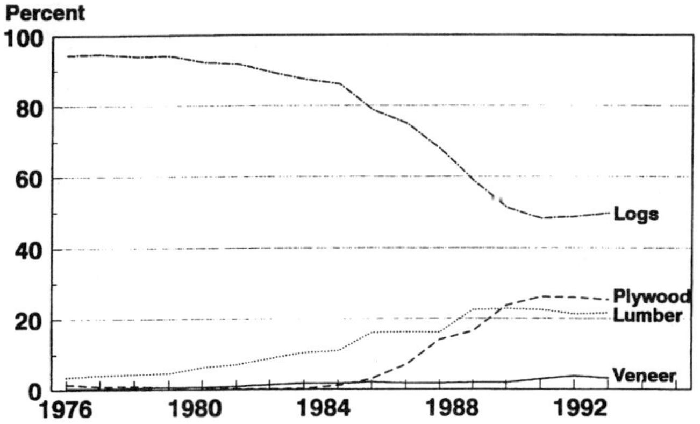

Figure 4. — Ninety-five percent of the combined value of imports in 1976 was accounted for by logs. However, their share dropped precipitously throughout the 1980s, and by 1990 logs accounted for under 50 percent of the total value. Offsetting the decline were increases in the proportion of lumber and plywood.

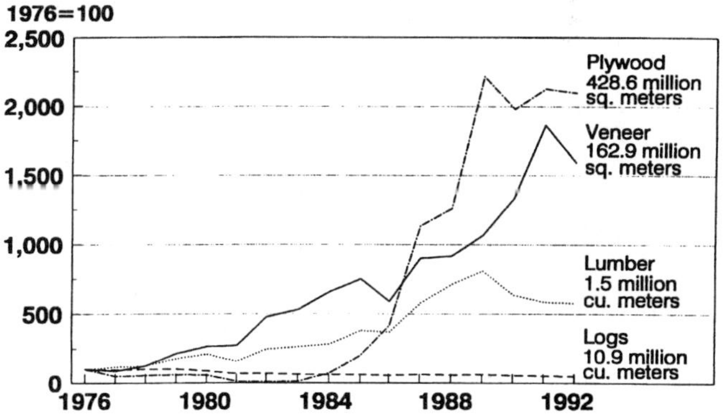

Figure 5. — Imports of hardwood plywood in 1992 were more than 22 times their early 1980s level, leading the growth in all imports. Hardwood veneer and lumber were, respectively, 16 and 6 times their level in 1976. In contrast, logs were only half their 1976 level, reflecting ever-tightening restrictions on log exports by an increasing number of Asian suppliers.

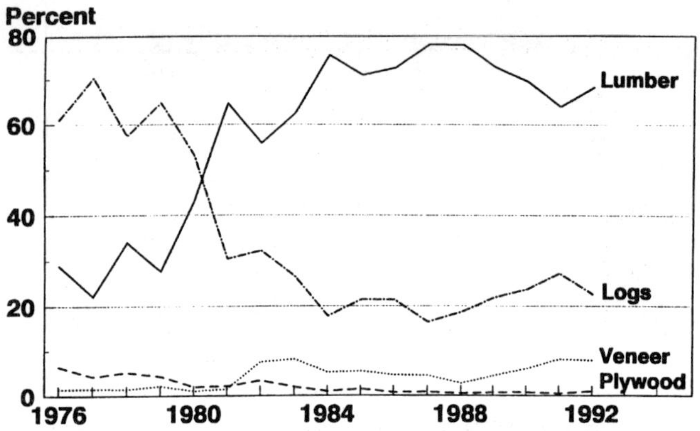

Figure 6. — Between 1976 and 1980, logs made up 60 to 70 percent of the value of Japanese imports from the United States. Since 1980, increases in the proportion of lumber mirror declines in the proportion of logs. The contribution of veneer was at an all-time high in 1992. Hardwood plywood has had a very limited role in the mix throughout the period.

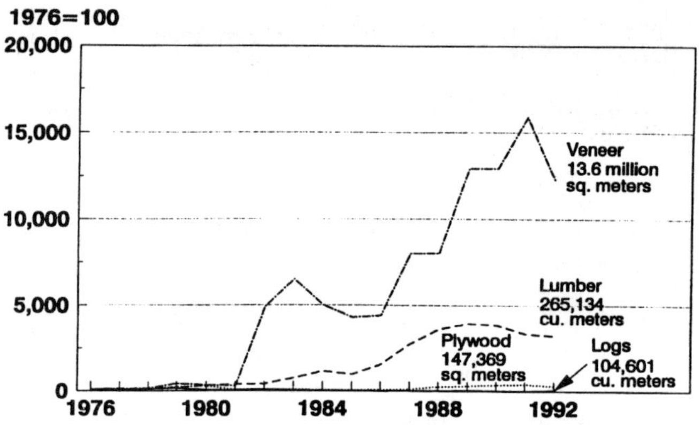

Figure 7. — Increases in Japanese imports of veneer led increases in all other products (due partly to a low base value). Veneer imports from the United States in 1992 were 120 times larger than in 1981, despite this level being down considerably from that recorded in 1991. Lumber imports have also increased dramatically since the early 1980s. Although dwarfed by increases in imports of veneer and plywood, log imports from the United States in 1991 and 1992 were 4 to 6 times the levels recorded throughout most of the 1980s.

Table 1. — Japanese hardwood log imports, 1976 to 1992.

	Quantity			Value		
Year	All imports	From U.S.	Percent U.S.	All imports	From U.S.	Percent U.S.
	- - - - (1,000 m³) - - - -			- - - - - (million ¥) - - - - -		
1976	22,042	35	0.2	455,988	2,616	0.6
1977	21,227	52	0.2	416,527	3,720	0.9
1978	22,030	40	0.2	358,998	2,821	0.8
1979	22,597	81	0.4	739,744	5,810	0.8
1980	19,202	100	0.5	670,355	5,711	0.9
1981	15,022	26	0.2	409,977	2,418	0.6
1982	15,220	26	0.2	493,234	3,131	0.6
1983	14,279	29	0.2	372,147	2,760	0.7
1984	13,386	24	0.2	397,557	2,541	0.6
1985	13,335	23	0.2	319,722	2,622	0.8
1986	12,267	25	0.2	222,297	2,711	1.2
1987	13,914	42	0.3	323,842	3,253	1.0
1988	13,061	87	0.7	259,987	5,021	1.9
1989	13,869	103	0.7	301,464	7,550	2.5
1990	12,316	127	1.0	256,515	9,009	3.5
1991	11,357	138	1.2	235,843	9,503	4.0
1992	10,910	105	1.0	233,899	6,786	2.9

Table 2. — Japanese hardwood lumber imports, 1976 to 1992.

	Quantity			Value		
Year	All imports	From U.S.	Percent U.S.	All imports	From U.S.	Percent U.S.
	- - - - (1,000 m³) - - - -			- - - - - (million ¥) - - - - -		
1976	262	8	3.1	17,191	1,240	7.2
1977	303	9	2.8	17,824	1,169	6.6
1978	326	12	3.8	16,609	1,674	10.1
1979	468	14	3.1	35,979	2,483	6.9
1980	555	28	5.0	45,256	4,641	10.3
1981	414	30	7.4	31,506	5,143	16.3
1982	646	33	5.2	49,116	5,417	11.0
1983	670	62	8.9	44,923	6,518	14.5
1984	743	95	12.9	50,926	10,780	21.2
1985	1,004	79	7.9	65,647	8,696	13.3
1986	970	127	13.2	48,290	9,187	19.0
1987	1,504	224	14.9	77,549	15,393	19.9
1988	1,854	295	15.9	100,115	20,865	20.8
1989	2,116	324	15.3	134,095	25,176	18.8
1990	1,669	316	18.9	120,117	26,536	22.1
1991	1,529	275	18.0	102,945	22,351	21.7
1992	1,512	265	17.5	101,439	20,532	20.2

Table 3. — Japanese hardwood veneer imports, 1976 to 1992.

Year	Quantity			Value		
	All imports	From U.S.	Percent U.S.	All imports	From U.S.	Percent U.S.
	- - - - (1,000 m^3) - - - -			- - - - - (million ¥) - - - - - -		
1976	10,227	111	1.1	1,961	61	3.1
1977	8,777	153	1.7	1,718	79	4.6
1978	13,005	181	1.4	2,160	70	3.2
1979	21,956	476	2.2	4,782	186	3.9
1980	27,290	387	1.4	5,327	108	2.0
1981	28,212	305	1.1	4,256	113	2.7
1982	49,042	5,470	11.2	7,608	732	9.6
1983	54,370	7,241	13.3	7,303	848	11.6
1984	67,635	5,563	8.2	8,192	739	9.0
1985	76,757	4,787	6.2	8,514	661	7.8
1986	60,368	4,898	8.1	5,101	574	11.3
1987	92,077	8,931	9.7	8,224	875	10.7
1988	93,722	8,948	9.6	8,499	724	8.5
1989	109,264	14,391	13.2	11,552	1,506	13.0
1990	136,128	13,999	10.3	15,590	2,248	14.4
1991	190,660	17,659	9.3	18,342	2,829	15.4
1992	162,862	13,559	8.3	14,958	2,343	15.7

Table 4. — Japanese hardwood plywood imports, 1976 to 1992.

Year	Quantity			Value		
	All imports	From U.S.	Percent U.S.	All imports	From U.S.	Percent U.S.
	- - - - (1,000 m^3) - - - -			- - - - - (million ¥) - - - - -		
1976	20,473	241	1.2	7,292	277	3.8
1977	10,063	211	2.1	3,983	223	5.6
1978	11,684	275	2.4	3,757	257	6.8
1979	12,432	366	2.9	5,054	388	7.7
1980	12,083	163	1.4	4,928	206	4.2
1981	2,886	121	4.2	1,051	168	16.0
1982	2,390	247	10.4	1,089	325	29.8
1983	2,899	148	5.1	1,362	194	14.2
1984	14,973	98	0.7	4,890	135	2.8
1985	41,457	90	0.2	12,271	177	1.4
1986	85,492	30	- -[a]	20,873	83	0.4
1987	232,418	79	- -	67,851	158	0.2
1988	257,618	42	- -	73,392	133	0.2
1989	453,496	70	- -	139,703	236	0.2
1990	404,978	48	- -	138,859	273	0.2
1991	434,438	59	- -	125,755	140	0.1
1992	428,604	147	- -	118,870	287	0.2

[a] Less than 0.05 percent.

Technical Forum Presentation:

Future trends for Scandinavian wood products

Timo Hartikainen

Abstract

This paper is part of a future-oriented study that aims at constructing prospective market scenarios for sawnwood, panels, and value-added products around 2015.

Motivation for this study can be found both in the information needs of marketing planning and in public decision-making related to forest sector. To understand how the future of an industry can evolve, environmental assessment must identify weak signals of emerging forces. A cross-impact based multiple scenario approach is being used (method adapted from Michel Godet's *la prospective*).

World politics, geopolitics, world economics, and social factors all make up the general climate for social and economic activities. The demand for wood products is dependent on population growth and the economic fluctuations. Problems arising from unequal division of welfare between industrial and developing countries will probably play a major part in world trade. The definition and status of sustainable development in world politics will shape the world's material production significantly. Sustainable forestry will be an important goal of international cooperation. The social aspects of sustainability will strongly affect the wood products market and position (e.g., increasing demand for food and shelter, demand for environmentally acceptable building materials, demand for better productivity in construction, and demand for nature conservation).

The preferences of both industrial end-users and final consumers will carve out a new competition environment. Conflicting realities of the environmentally concerned people (emerging "green" social paradigm) and the people relying on traditional utilization of forests (prevailing "industrial" social paradigm) will be a key question in the future. Wood products have some environmentally friendly characteristics in the substitution race between building materials, like energy efficiency, renewability, recyclability, and disposability. The wood products industry will have several strong competitive advantages, if only it can retain common sense in forest management, sensitivity in harvesting practices, technical control in manufacturing processes, and socially responsible behavior in marketing activities.

The data for these intermediate results come from literature and a small-scale survey (53 participants, both Finnish and international). The next phase of the study will result as prospective market scenarios.

The author is a Research Associate, Dept. of Forest Economics, Univ. of Helsinki, The Academy of Finland, Helsinki, Finland.

Scenario approach

Alternative futures can be described through multiple (qualitative) scenarios. Scenarios are not forecasts, but a sample of the multitude of future situations and the paths of evolving events leading there. Only after charting the unknown waters of the future one can define the desired future location of the company, and build the marketing strategies so that the path leading to this future can be brought about.

Scenarios differ from more forecasting-oriented approaches in two important ways. First, they usually provide a more qualitative and contextual description of how the present will evolve into the future, rather than one that seeks numerical precision. Second, scenarios usually try to identify a set of several possible futures, where each occurrence is plausible, but not assured. The prospective scenario method is presented as a flow chart in Figure 1.

The prospective scenario method emphasizes holistic world view with systematic analysis of phenomena of the research field. As the research object is not the future, which cannot be observed, the data for research comes from detailed analysis of present reality, historical developments, empirical results of various specialized scientific disciplines, and the intentions of people and organizations. This method was selected for the research project because of its emphasis on pro-activity and attitude of "bringing the future around" — people can affect the evolution through their own decisions. The prospective approach also provides clearly defined basis for the selection of key issues for scenarios, a systematic expert assessment survey, and it assists in abandoning the nonplausible scenarios.

Meristö (3,5) has combined scenarios to the working process of corporate management. A company can assess its environment using a "PESTE" structure

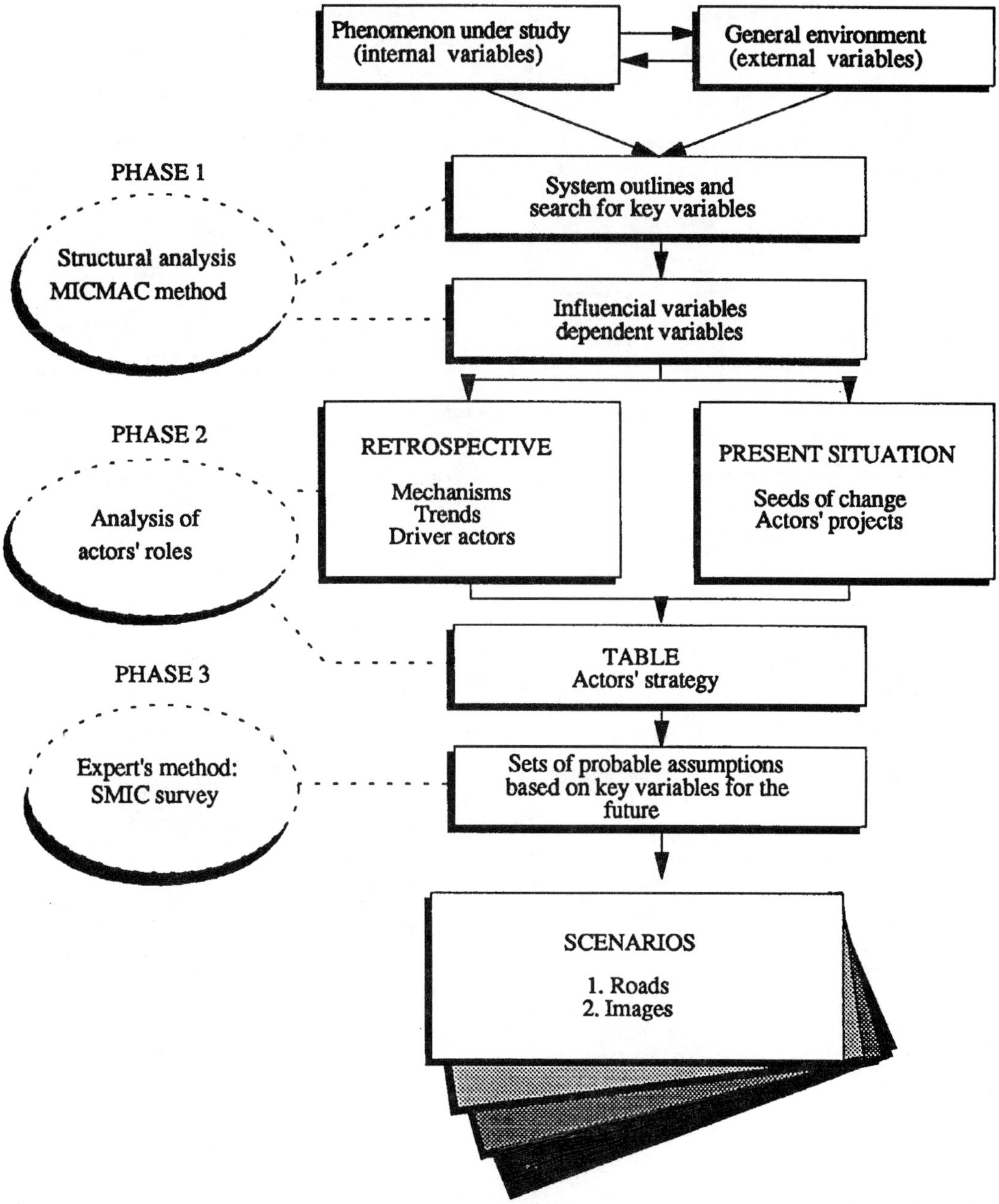

Figure 1. — The prospective scenarios method (1,2).

(political, economic, social, technological, and ecological factors) for its macro environment, and Porter's concepts (buyers, suppliers, competitors, substitutes, and structural changes of an industry) for assessing its competition environment in a more detailed manner. Uotila (7) has applied this approach in his research for Finnish sawmilling industry. Figures 2 and 3 present the visualizations of environmental scanning models by Meristö and Uotila.

Structural analysis

Factors that were considered to be important for the future evolution of the wood products market are introduced and shortly described in Tables 2 and 3. This list of factors is a synthesis based on

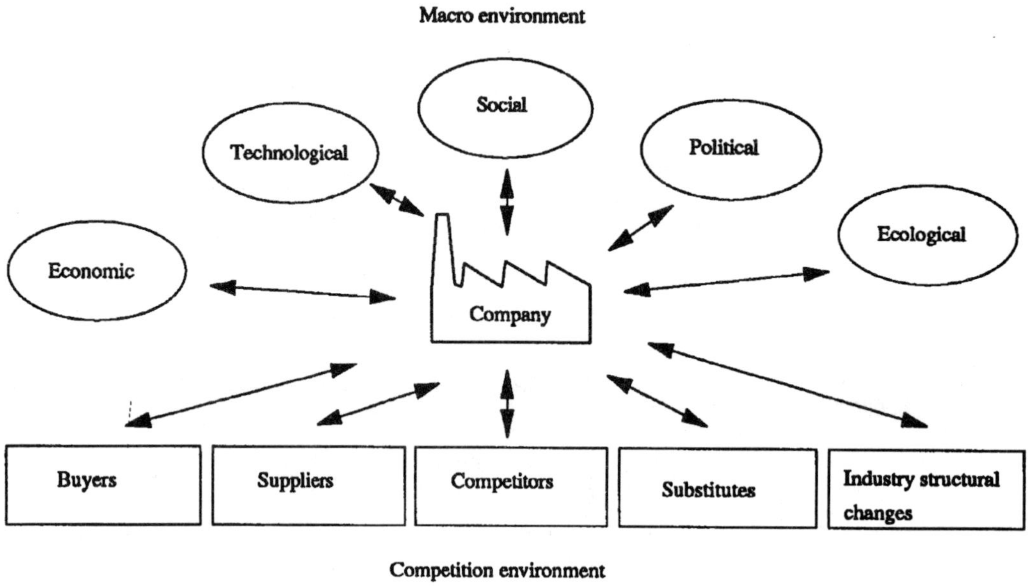

Figure 2. — Macro environment and competition environment used in environmental assessment (3,4).

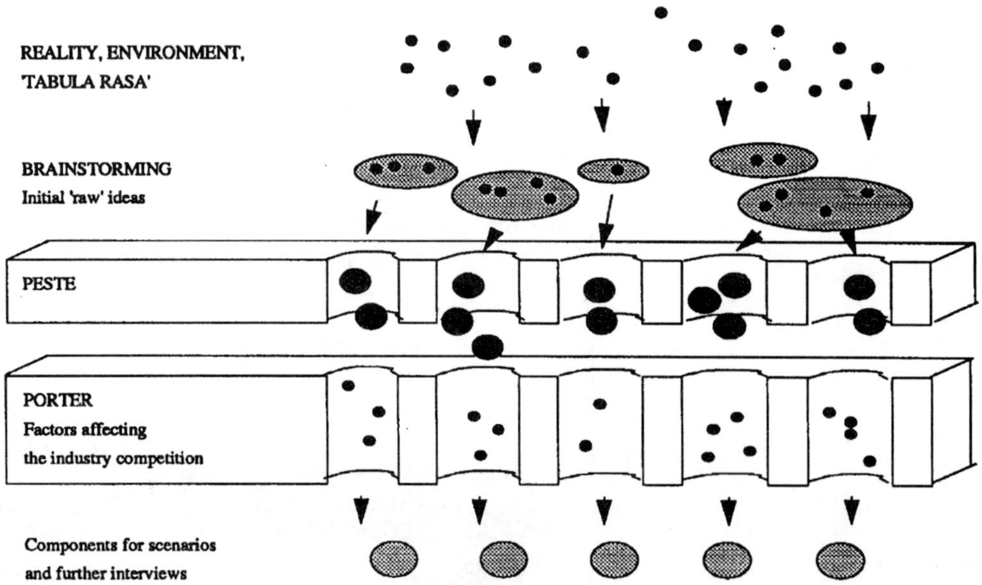

Figure 3. — PESTE analysis of macro environment and Porter's competition environment applied to identifying scenario components (7).

Table 1. — Factors in the macro environment according to the international and the Finnish panels.

Factors	Short description
Political factors	
World politics	New challenges of power politics after the cold war.
European integration	Effects of free mobility of goods, services, capital and human resources. European Community. European Economic Area.
Uncertainty in Eastern Europe	Countries enduring economic and political restructuring after collapse of the USSR.
Problems of developing countries	Unequal division of welfare between industrialized and developing countries. Rich north vs. poor south.
Strive for sustainable development	Problems of exponential growth becoming acknowledged. Future generations get a vote.
Forest policies	National/international policies over utilization and protection of forest resources. Execution of Rio resolutions.
Environmental policies	National/international policies over utilization and protection of environment. Energy sources and consumption, pollution control, taxation.
Economic factors	
Profitability of forest sector	Demand for forest products and services. Production costs. Development of productivity.
Financial integration in Europe	European Financial Area 1993, European Monetary System, Exchange Rate Mechanism, European Monetary Institution 1994, European Monetary Union 1999.
New trade practices	Environmental competitive advantages. Social responsibility. Ecological marketing.
Demand for building activity	Rebuilding in Eastern Europe, strong need for additional housing capacity.
Pulp and paper industry	Competes of fiber sources. Connections to the structure of mechanical forest industries.
Economics of developing countries	Direction of economical development in developing countries. Possibly emerging new strong economies.
Social factors	
Population growth	Estimated world population 6 milliard in 2000 and 8 milliard in 2020. After that, either stagnation or continuing growth.
Environmental concern	Publicly expressed concern for environmental damages is growing. Wasteful use of natural resources. Worldwide problematique.
Mass media	Forum for societal discussion and platform for fighting over the consumers' attitudes.
Building and housing cultures	Technical solutions spread quickly, but adapt to local culture. Trend towards natural and environmentally friendly materials.
Employment factor	Important especially in countries strongly dependent on forest sector. A determinant of political decisions.
Technical factors	
Declining quality of roundwood	Share of high quality raw material is declining. Old growth forests being replaced by second generation forests. Bulk of future timber supply is of lower quality and smaller diameter.
New production technologies	Development of more cost-effective milling methods. Technological advances in drying, sorting, veneer cutting and slicing, adhesives, and fiber techniques. Computerization.
New wood products	Increasing the wood value at the point of production. Reconstituted wood products. Technical wood. Progress in panel products.
Construction systems	Building with ready-to-install components. System products. Total service deliveries. Competition between materials. Tailor-made, customized products.
Building regulations	Standardization organs, building authorities, fire safety norms. EC building codes, ISO-9000, DIN etc.
Technical qualities of wood	Reliability, conformity, consistency, durability.
Non-technical qualities of wood	Image, attractiveness, marketing.
Ecological factors	
Natural phenomena	Greenhouse effect, ozone layer, deforestation, erosion, natural catastrophes.
Demand for sustainable forest management	Consumers express strong concern for the origins of wood they use. Wood production, productivity of woodland, multiple use, biodiversity.
Availability of roundwood	Decreasing supply of high quality timber in Pacific NorthWest. Unstable situation in Russia. Availability in Sweden and Finland stable. Tropical wood debate.
Environmental qualities of wood	Renewable, natural material. Disposal or recycling of waste wood from construction, renovation, and demolition. Aspect of sustainability possible.
Demand for environmental protection	Concern for species extinction and loss of biodiversity. Threats perceived in industrial forestry. Old-growth protection areas.

Table 2. — Factors in the competition environment according to the international and the Finnish panels.

Factors	Short description
End-users and consumers	
Green values reflected in consumption	Consumers are applying more non-material values, harmony with nature, and attach intrinsic values to natural environment.
Consumers' perception of wood	Amount of accurate information about forestry and qualities of wood. Shaped mainly by technical knowledge, social considerations, and financial resources. Reflection of tropical problems to temperate areas.
Specifiers' perception of wood	Amount of accurate information about forestry and qualities of wood. Shaped mainly by knowledge about wood as material compared to substitutes and technical performance of wood products and services.
Production and distribution systems of end-users	Tendency for smaller and more detailed deliveries by precise schedules. Demand for wider information exchange between producer and manufacturer.
Competitors	
United States and Canada	Development of North American home market. Development of roundwood supply and cost. Market share in Europe.
Sweden and Finland	Development of roundwood supply and cost. Restructuring of primary mechanical forest industries. Development of business attitudes.
Southeast Asia	Development of forestry practices. Roundwood utilization policies. Demand on home market. Especially market share in panel products.
Domestic production	Production within net importers of wood. Germany, United Kingdom, France, others.
Potential new entrants	
Russia	Political stabilization. Rebuilding of forest industry. International cooperation in forest management. Vast potential for wood products supply at least in quantity.
Plantation forests	Plantations of fast-growing species, e.g., in Portugal, Chile, and New Zealand. Potential for competition in structural wood and fiber products.
Raw material suppliers	
Goal setting of forest owners	State forest policies in North America and Russia. Private forest owners' market behavior strongly influences situation in Finland and Sweden.
Forest management practices	Environmental effects of forestry. Public demand for more sensitive methods in harvesting, reforestation, and other measures.
Competition in roundwood trade	Possibility of getting right raw material for right end use in global scale. Performance of national roundwood markets. Protectionism.
Competition in harvesting and transport	Especially in Finland and Sweden. Development from forest owners / industries oligopoly to free competition of forest owners / real marketplace / industries.
Substitutes	
Plastic substitutes	Plastics (PVC and possible new ones)) compete with wood strongly in building parts and joinery products. Comparison through life-cycle-analyses. Pressures from pollution effects.
Metal substitutes	Metals (aluminum) compete with wood strongly in structural constructions and building parts. Comparison through life-cycle-analyses.
Mineral substitutes	Concrete and tiles compete with wood in house-building. Comparison through life-cycle analyses.
Wood substitutes	Competition between conventional and engineered wood. Reconstituted wood, product systems, new technology. Pressures for sustainable forestry.
Tropical wood	Competition from tropical hardwood. Strongest in products where durability, appearance, and social prestige count. Pressures from forest protection.
The Finnish forest industry	
Customer orientation	Now underdeveloped. Pressures from industrial end-users. Customer relationships all the way from middlemen to the final consumers. Away from raw material and production orientation.
Innovative product development	Demand for new wood products to survive competition. Ecological products. Product related services. Engineering and construction know-how.
Production systems development	Development of productivity. Flexible processing and manufacturing. Demand for customized products. Decrease of bulk production.
Marketing development	Understanding of market mechanisms. Information systems development. Distribution systems.

literature and survey results. The division into macro environment (PESTE) and competition environment (Porter) was well confirmed by the respondents' opinions. Some issues had to be placed arbitrarily to one group when alternative interpretations were possible.

The structural analysis attempts to interrelate the various variables in a dual input table. The 53 factors were used to build a 53 × 53 matrix using a scale from 0 to 5 (no relation — very strong relation).

In the micmac procedure the interdependence matrix was raised to the powers of 2, 3, 4, 5, 6, and 7. The sums of each variable over the rows and columns were calculated. The hierarchy of variables became stable at the power of 6, and the classification of indirect influence was established according to those figures.

The next step in the analysis of the micmac results was to plot the variables on a chart according to their driver power and dependence after micmac classification. The individual variables have been regrouped according to their substance and location on the chart. The regrouped variables are presented in Figure 4.

The chart can be divided into five parts (2):

1. Determinants or influential, explanatory variables.
2. Relay variables. Both highly influential and highly dependent variables are unstable by nature. Any action on these variables will have repercussions on other variables. The impact will be circulated among the variables, either amplified or defused.
3. Non-related variables. Low influential, low dependent variables constitute market trends or factors which are relatively unconnected to the system and have only few relationships with it.
4. Middle cluster. Average influential and/or dependent variables.
5. Dependent variables. Low influential and high dependent variables.

The participants of the survey defined the group of actors on the wood products market quite clearly. The consumers and specifiers were the most often mentioned, then came environmental pressure groups, legislators, standardization organs, wood products industry, forest owners, research organizations, and substitute industries. Table 3 defines the actors.

Conclusions

Based on the analysis of variables and actors,

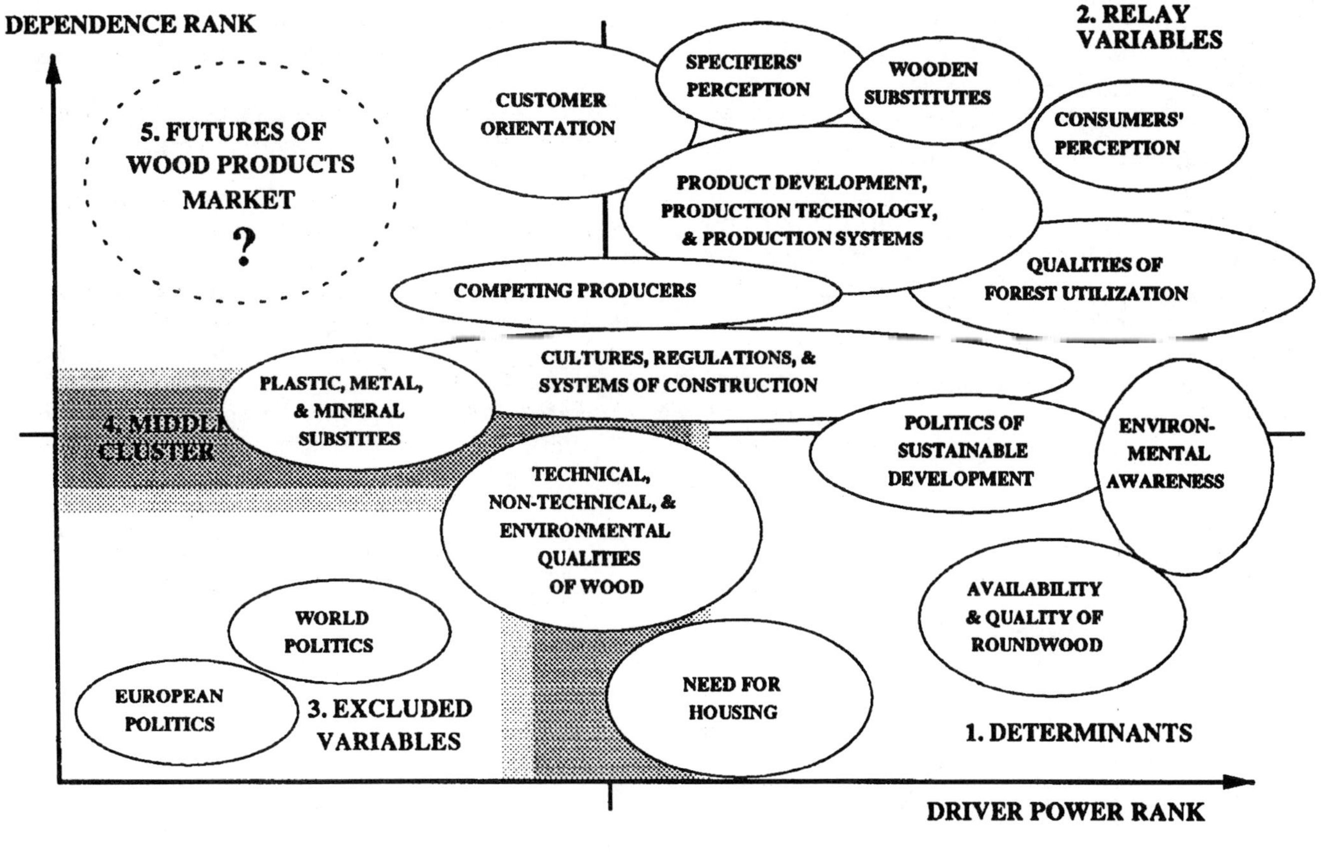

Figure 4. — The micmac results by variable group.

Table 3. — Actors in the wood products market.

Final consumers	Individuals or groups who finally make use of buildings, furniture, and other products made of wood or otherwise based on wood.
Specifiers	Architects, engineers, designers, planning offices.
Industrial end-users	Secondary manufacturers, either construction companies, producers of prefabricated components, joineries, etc.
Environmental pressure groups	Local, national or international groups that actively participate in the discussion about environment and ecology (e.g., Friends of the Earth, Greenpeace, BUND Deutschland, Kulturbund Deutschland, World Wide Fund for Nature (WWF), the greens).
Sawmilling industry	Companies that process wood from timber to lumber by sawing, cutting, planing, etc. Refers to the industries located on the boreal, coniferous forest area.
Panel industry	Companies that process wood from timber to veneer, plywood, particleboard, or fiberboard. Refers to the main producing countries in the world.
Forest owners	The institutions or individuals who own the forests. Must be noted that in Finland and Sweden majority of the commercial timber comes from nonindustrial private forest owners.
Substitute manufacturers	Companies that produce materials that compete with wood in its various applications, e.g., plastics, metal, and mineral industries.
Technical research institutions	Technical research bodies, either independent or with universities, public or private.
Legislators and policy makers	Building code authorities, politicians, and other public decision makers.
International organs	International organizations with some power, e.g., UN departments, etc.

the strategic questions for the future of wood products market were identified as:

1. What is the future of the present drive towards common definition for sustainability and its effects on reality? Will it have any significant effects on society's economic, physical, and social life?
2. What is the future of growing environmental awareness among people? Will it continue its growth and retain its strength?
3. What is the future of forestry? How will the conflict between green and industrial values will be settled?
4. What is the future of market information and market communication? Will people have access to accurate and impartial information, and will they understand and use it?
5. What is the future of product development and competitiveness of wood products? How will the various qualities of wood take the competition?
6. What is the future of physical and technical utilization of wood products? Will they find their place within construction techniques, building regulations, and cultural environment?

These issues will be used as a base for an expert cross-impact survey, which will finally result as the prospective scenarios for wood products market.

Literature cited

1. Godet, M. 1987. Scenarios and Strategic Management. (Translation of: *Prospective et Planification Stratégique.*) Butterworth Scientific Ltd. 210 pp.
2. ________. 1993. From anticipation to action. A handbook of strategic prospective. UNESCO Publishing, Paris, France. 277 pp.
3. Meristö, T. 1991. Skenaariotyöskentely yrityksen johtamisessa (Scenarios in Strategic Management). Acta Futura Fennica 3. 187 pp.
4. ________. 1993. Yritysfuturologia: miksi, mitä, miten? (Corporate futurology: why, what, when?). Handout and lecture 3.6.1993.
5. ________. 1993. Skenaariotyöskentely strategisessa johtamisessa (Scenarios in strategic management). *In:* Vapaavuori, Matti (ed.). 1993. Miten tutkimme tulevaisuutta. VAPK, Helsinki. pp. 215-220.
6. Porter, M.E. 1985. Competitive Advantage. Creating and Sustaining Superior Performance. The Free Press. 557 pp.
7. Uotila, T. 1993. Skenaariot yrityksen toimintaympäristön kartoittamisessa (Scenarios in corporate environmental scanning). *In:* Vapaavuori, Matti (ed.). 1993. Miten tutkimme tulevaisuutta. VAPK, Helsinki. pp. 222-231.

Technical Forum Presentation:

Hardwood and softwood raw material: production and marketing opportunities within Argentina

Carl Jansen

Abstract

Argentina shares the "Southern Cone" of South America with Chile. Chile has 7.6 million ha of timberland with a population of 9 million. Argentina has 44.5 million ha of timberland with a population of 34 million. Most of the extreme south is covered with hardwood trees (lenga is the predominate species). The northern portion of Argentina is composed principally of plantation pine (40 years and younger). Argentina is a full democracy. Currently, its annual inflation rate is about 8.5 percent, although it was zero percent in August 1993 (Chile's inflation rate is over 13 percent). Argentina's lower inflation rate is due to the "new Argentina" free enterprise system and the discontinuation of big government control. These changes were introduced 2 years ago. A recent decree in Argentina clarifies existing laws concerning deregulation and economic reforms. Specifically, the decree authorizes foreign investors to repatriate their capital or liquid assets at any time and entitles foreign investors to invest in the country without any previous authorization and with the same capacity as local investors. Opportunities exist for foreign owners to purchase timberlands. Hardwood stands exceeding 55,000 ha are available and the estimated wood cost on the stump is approximately $10/m^3$. Softwood plantations exceeding 30,000 ha are available and can yield over 25 m^3/ha. per year compared to >12 m^3/ha. per year in the southern United States. There is a lack of advanced wood product manufacturing practices in Argentina. Current processing methods in Argentina are similar to those used in the United States 30 years ago. This presents a significant opportunity to U.S. and Canadian producers to transfer new technology and management methods. Opportunities also exist in expanding the marketing base of Argentina. In 1990, Argentina was a net importer of wood fiber. Opening the flow of processed wood fiber to world markets represents an opportunity for U.S. and Canadian firms. Plans have been implemented to export face-grade veneer and industrial shop lumber to U.S. customers. Argentina can become a significant supplier of wood fiber to world markets, partially in the form of logs, chips, veneer, and hardwood and softwood lumber.

(full paper not available)

The author is President, Search North America, Portland, Oreg.

Technical Forum Presentation:

Making more with less: the globalization of supply for materials used in wood veneer-wrapped millwork

Lee J. Jimerson

Imagine being able to increase the global supply of high grade wood used in the millwork industry by 42 times! This can be achieved by utilizing profile wrapping technology. Profile wrapping is the process of wrapping a thin veneer around a molded substrate.

The increasing scarcity globally of high grade wood is causing more value-added wood products to be made from composites rather than from a solid piece of wood. The veneers used for profile wrapping can be any species that we are accustomed to, but are finding harder to buy, and more expensive, in high grades. The substrates, which the veneers are wrapped on, can be a wide variety of wood species available globally, as well as non-wood substrates. The wood substrates can come in a number of forms including solid wood, finger-jointed, laminated, and medium density fiberboard. Nonwood substrates include aluminum, plastic, metal, fiberglass, and vinyl.

The possibilities are virtually endless for products using profile wrapping technology. Profile-wrapped products include household moldings, window parts, residential and office furniture parts, and kitchen and bath cabinets.

Species for finger-jointed substrate material

Domestic species (including Canada)

- Ponderosa Pine
- Spruce
- Hem/fir
- Eastern and western white pine
- Douglas-fir
- Southern yellow pine
- Poplar

Imported species

- Meranti (mixed light hardwoods) (Asia)
- Radiata pine (Chile, New Zealand, and Australia)
- Russian pine and larch (Siberia)
- Elliotis Pine (Latin America)
- Taieda Pine (Latin America)
- Jelutong (Indonesia and Malaysia)
- Agathis (Indonesia and Malaysia)

Wood-based composite substrates

Domestic and imported

- Fingerjointed solid wood
- Edge-glued solid wood
- Medium density fiberboard
- Particleboard
- Laminated veneer lumber
- Oriented strandboard

Nonwood-based substrates

- Aluminum
- Metal and alloys
- Plastic
- Polystyrene
- Fiberglass
- Vinyl
- Carbon

Wood veneer faces

Domestic (including Canada)

- Red and white oak

The author is Director, Lejé International, Portland, Oreg.

- White ash
- Maple
- Cherry
- Ponderosa pine
- Douglas-fir
- Hemlock
- Walnut
- Red and white birch

Imported

- Radiata pine (Chile, New Zealand, and Australia)
- Ramin (Indonesia and Malaysia)
- Agethis (Indonesia and Malaysia)
- Meranti (Asia)
- Genuine mahogany (Latin America)
- Teak (Asia)

End-use applications for profile-wrapped millwork

- Mouldings (jambs, casing, base, crown, etc.)
- Window and door parts
- Furniture parts (office and residential)
- Kitchen and bath cabinets

Global markets for U.S. manufactured profile-wrapped millwork

Asia/Pacific

- Japan, Taiwan, Korea, Australia, New Zealand

Western Europe

- Germany, Spain, France, Italy, The Netherlands, England

North America

- Canada and Mexico

Technical Forum Presentation:

Wood: the miracle resource

Robert F. Legg

The forest products industry has an amazing story to tell which can dramatically shift public opinion about forests, forestry, wood, and the forest products industry. In general, the industry has focused on process rather than product, on production rather than consumption, and has failed to position itself as a "provider" responding to a need which all people have, a need for environmentally sound building materials. While it is important to communicate with the public about stewardship, it is more important to make the public aware of their consumption options and the importance of their choices as responsible consumers. The industry has also failed to articulate a vision, hence the public assumes that the vision is simply to cut more trees and make more money. In fact, the industry is in a perfect position to talk about its contribution to a sustainable economy and society, thereby contributing to the "sustainable development" vision. This vision was presented in *Our Common Future,* the World Commission on Environment and Development's landmark report to the United Nations General Assembly in 1987.

The Temperate Forest Foundation has developed an exhibit titled *Wood: The Miracle Resource,* which was first shown at the Pacific Logging Congress (PLC) live-woods demo in Snoqualmie, Wash. in September 1993. The PLC officers agreed that the public needed to know something about wood before going out into the forest to see machines harvesting and processing the timber. The Foundation volunteered to produce an exhibit to get the job done, because we knew that some 2,500 middle school children, plus teachers and parents, would come to the show. We knew that if we did it right the public would have a much different perception of what was going on in the woods. We decided that there were four main objectives:

1. Present the forest as a factory where wood is manufactured through the remarkable process of photosynthesis. Also explain how trees capture carbon and return pure oxygen to the atmosphere. Because wood is produced in the forest, very little energy is required to convert it into useful products. A healthy forest is a productive forest.
2. Show all the ways we use wood in our everyday lives to include solid wood, engineered wood products, panels, and all kinds of products made from wood fiber and cellulose. Once people see this, they realize why an estimated 10,000 products from houses to milk shakes have wood in them.
3. Show how wood compares to plastics, steel, aluminum, concrete, and brick. Steel, as an example, requires three times as much energy, three times as much water, and produces three times as much CO_2. So the environmental burden is much higher for these alternative building materials.[1]
4. Show how wood contributes to a sustainable economy and a sustainable society. The world population is exploding and could double to 11 million people in the next 50 years at current growth rates. Doubling time is calculated by di-

The author is President and CEO, Temperate Forest Foundation, Beaverton, Oregon.

[1] This data is available from the CORRIM Report, in the proceedings of the Forest Product Society's 1991 Meeting on *Wood Product Demand and the Environment* in Vancouver, B.C., Canada, and in this proceedings. Also, Dr. Jim L. Bowyer of the University of Minnesota has done extensive work in this field and the Foundation has taped his presentation on *Materials and the Environment.*

viding the number 70 by the growth rate which is now about 1.4 percent on a global basis. Known oil reserves are all used up in 47 years at *current* consumption rates! If we were to make a massive shift to substitute building materials, it would create an unprecedented demand for additional energy and an environmental catastrophe. The only way out, without a dramatic drop in quality of life, is to accelerate our use of renewable resources, and to reduce our dependence on non-renewables. Not only can the forest products industry contribute to sustainable development, it must lead the charge. This is how the industry repositions itself in the public consciousness, as an industry dedicated to the welfare of current and future generations.

Everyone who went through our exhibit in Snoqualmie and then out to the woods, went home with a dramatically improved understanding of the role wood plays in their lives, and a dramatically improved perception of the folks who produce and process this resource for them. Every single person in society is a consumer and can make a difference through their choices. By involving them as consumers we initiate effective communications.

While it is important to do everything we can to promote the use of renewable resources to meet our needs, that is *not* enough. Our message is only credible if we also advocate that people reduce, reuse, and recycle. Responsible consumption is more important to the conservation of natural resources than responsible production, which is where we have focused all of our attention. In fact, responsible production and consumption go hand in hand.

All of the efforts to set aside forest reserves, and to practice "ecosystem management," will be in vain if we don't square up to the cumulative effects of population, consumption, and technology. We have to increase the earth's carrying capacity to support all the people that we are going to have. The only way we are going to do that is with renewable resources, and by using technology to produce more with less (i.e., less raw material, waste, and pollution of air, soil, and water.)

It is time for the industry to get serious and to elevate the level of debate. What a miracle it is that trees capture the sun's energy which is diffused all over the earth's surface. These trees produce a miracle resource called wood. It is a renewable resource and therefore an important part of the solution to the challenges we face. It is also energy efficient, biodegradable, recyclable, non-toxic, durable, versatile, and beautiful! It is *the* environmentally sound option.

Technical Forum Presentation:

Wood quality of plantation grown *Populus* × *Euramericana* hybrids

Ilona K. Peszlen

Abstract

Three clones of the hybrid *Populus* × *euramericana* (Dode) Guinier cv. — the Italian I-214 and the Hungarian Kopecky and Koltay — were investigated on two different sites. Six trees were sampled from each clone at breast height to examine patterns of variation for selected anatomical, physical, and mechanical properties.

Site and clonal effects were significant in terms of wood quality, but the differences were relatively small to be considered important from the utilization standpoint. Specific gravity did not prove to be the most important single factor influencing strength properties. No consistent relationship was found between growth rate and wood properties of the clones.

Background

Shortage of timber supply in many parts of the world has lead to the development of fast-growing hybrids of certain tree species. Members of the *Populus* genus have received special attention by tree breeders due to the relative ease of their hybridization and vegetative propagation.

In Europe hundreds of clones have been produced by hybridizing North American and European poplars. Some of these clones, such as the Italian bred I-214, have been planted all over Europe.

In Hungary, where only about 18 percent of the land area is forested, some 9 percent of all forests consist of hybrid poplars. Annual harvest of poplars is even higher, accounting for more than 20 percent of the overall cuttings. Research on wood properties of poplars provides essential information to both foresters in managing their tree improvement programs, and to the wood industries in applying new technologies for the best utilization of trees planted.

Objectives

The overall goal of this research was to enhance utilization potential of hybrid poplars. The specific objectives were to: 1) determine site and clonal effects on variation of selected wood properties and relationships to growth rate; and 2) establish bending, compression, and tensile strengths for each clone and relate these to specific gravity.

Materials and methods

Three *Populus* × *euramericana* (Dode) Guinier clones — the Italian I-214 and the Hungarian Kopecky and Koltay — were investigated on two different sites in Hungary. The Koltay clone was not available on Site 2. The plantation on Site 1 (Daka) was 15 years old, but on Site 2 (Zalavar) it was only 10 years old. Site 2 is considered to be a better site due to the more evenly distributed rainfall and reduced extremes in temperature.

Six trees were sampled from each clone on both sites. Measurements were taken at breast height within each tree from each growth ring. Anatomical properties were determined using an image analyzer, and specific gravity using the maximum moisture content method. Strength properties were measured on standard and micro specimens in three age groups along the radius on Site 1.

The author is Assistant Professor, Dept. of Wood Sci., Univ. of Sopron, Sopron, Bajcsy-Zs. u. 4., Hungary. The author acknowledges the Dept. of Wood Sci. and Forest Prod., Virginia Tech, Blacksburg, Va. for the financial support provided for this research. Special thanks are extended to F.M. Lamb for his valuable advice and to G. Ifju for constant care and encouragement.

Results

- Statistically significant differences existed among clones on two sites for most of the measured anatomical characteristics. However, the differences were small and negligible from a practical standpoint (Table 1).
- Cell wall material production of poplar clones was very low, less than one-third of the total cross-sectional area. Cross-sectional area of cell wall, vessel, and fiber lumen, and ray area exhibited only slight changes along the radius. Vessel and fiber lumen area increased approximately 8 percent from pith to bark (Fig. 1).
- Gelatinous fibers were detected in almost every sample as scattered cells or groups without specific pattern. Based on visual estimates, gelatinous

Table 1. — Mean anatomical properties and the coefficients of variation for the three clones on two sites.

Property	Site	Growth ring	Kopecky Mean[a]	Kopecky COV (%)	I-214 Mean[a]	I-214 COV (%)	Koltay Mean[a]	Koltay COV (%)
Vessel lumen diameter (μm)	Site 1	14	86.2 A	12.6	81.2 B	15.6	85.8	16.6
		10	84.8 A a	14.4	77.7 B a	16.8	82.2 AB	17.9
	Site 2	10	75.5 A b	13.3	77.3 A a	16.4	- -	- -
Vessel lumen area (%)	Site 1	14	34.3 A	10.3	32.8 B	13.8	34.7 A	14.2
		10	33.7 A a	11.2	31.6 B a	14.0	32.9 AB	13.5
	Site 2	10	28.7 A b	15.2	28.0 A b	17.8	- -	- -
Number of vessels per mm^2	Site 1	14	63.4 B	24.8	70.4 A	29.5	66.9 B	28.8
		10	64.4 B a	27.5	73.9 A	31.2	69.0 AB	30.3
	Site 2	10	67.6 A a	20.9	62.9 A b	28.4	- -	- -
Fiber lumen diameter (μm)	Site 1	14	16.2 B	11.8	16.4 B	8.9	17.0 A	9.0
		10	15.9 B a	12.9	16.0 B a	9.3	16.6 A	9.3
	Site 2	10	15.2 B a	12.3	15.9 A a	9.8	- -	- -
Fiber lumen area (%)	Site 1	14	51.2 C	7.2	54.1 B	6.3	56.1 A	5.0
		10	51.2 C b	7.0	53.7 B b	6.1	56.2 A	4.9
	Site 2	10	56.2 A a	10.4	57.0 A a	9.4	- -	- -
Fiber length (mm)	Site 1	14	1.10 B	16.2	1.12 AB	13.6	1.17 A	15.6
		10	1.03 B a	15.9	1.07 AB a	14.2	1.11 A	16.4
	Site 2	10	1.04 A a	16.6	1.06 A a	15.2	- -	- -
Gelatinous fibers (%)	Site 1	14	48.1 A	76.2	52.3 A	53.3	48.6 A	53.2
		10	54.8 A a	51.1	59.5 A a	46.2	57.5 A	42.9
	Site 2	10	56.5 A a	56.8	62.8 A a	41.3	- -	- -
Cell wall area (%)	Site 1	14	28.4 A	10.5	27.0 B	9.1	25.2 C	9.3
		10	28.6 A a	10.9	27.4 B b	9.0	25.8 C	8.9
	Site 2	10	27.4 A a	13.4	27.4 A a	14.7	- -	- -
Ray area (%)	Site 1	14	7.4 B	18.4	8.3 A	22.5	7.8 B	15.6
		10	7.8 B b	17.1	9.0 A a	17.8	8.2 B	14.0
	Site 2	10	8.7 A a	23.8	8.3 A b	19.2	- -	- -
Specific gravity	Site 1	14	0.343 B	7.2	0.304 C	7.2	0.353 A	6.8
		10	0.338 B a	7.3	0.305 C a	7.7	0.356 A	7.3
	Site 2	10	0.348 A b	9.6	0.307 B a	11.2	- -	- -

[a] Differences between clones within sites for each vessel parameter are denoted by uppercase letters. Differences between sites within clones for each vessel parameter are denoted by lowercase letters. Means with common letters are not significantly different at the 0.05 level as determined by Duncan's mean separation procedure.

fiber content varied among trees within clones and was higher near the pith than close to bark. It was not related to site, clone, or growth rate.

- The site did not affect specific gravity, however, significant differences existed between clonal means. The highest specific gravity was found for Koltay, the lowest for I-214 (Table 1).
- No significant clonal effect was found for the samples on Site 1 regarding modulus of rupture (MOR), crushing strength, maximum tensile strength (MTS), and modulus of elasticity (MOE) measured in tension. Strength values were low for each clone (Table 2).
- There were significant differences between age group means. The mechanical properties were higher near the bark than close to pith, except for ultimate tensile strength (Table 3).
- No significant correlations were found between

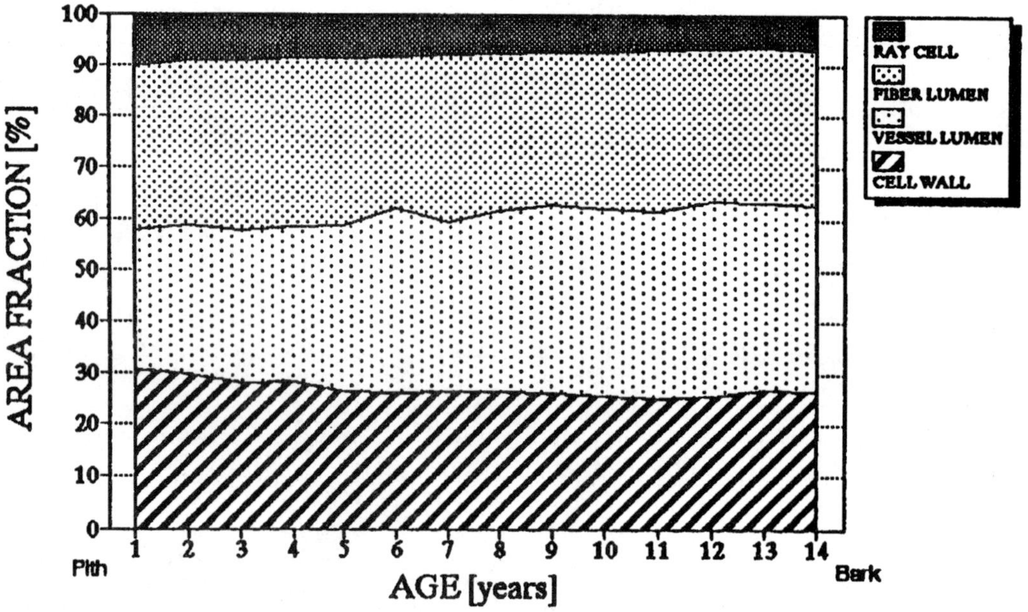

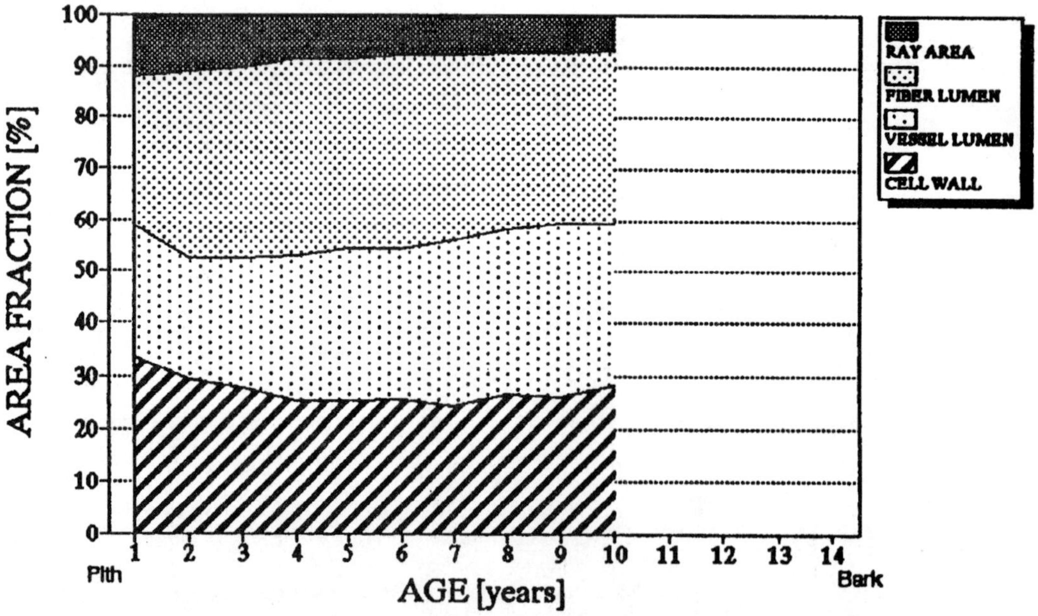

Figure 1. — Radial changes of distribution of anatomical elements. Top: based on three clones on Site 1 (15 years old). Bottom: based on two clones on Site 2 (10 years old).

Table 2. — Mean mechanical properties and specific gravities with the coefficients of variation for the three poplar clones.

	Clone					
	Kopecky		I-214		Koltay	
Property	Mean[a]	COV (%)	Mean[a]	COV (%)	Mean[a]	COV (%)
Specific gravity (bending, compression)	0.3358 B	16.0	0.3042 C	5.2	0.3517 A	5.5
MOR (kPa)	30,383 A	14.0	26,062 B	10.0	29,597 AB	10.3
Crushing strength (kPa)	11,709 A	14.7	11,903 A	13.6	12,337 A	11.1
MOE (compression) (MPa)	1,537 A	21.4	1,208 B	16.0	1,488 A	23.7
Specific gravity (tension)	0.3404 A	5.0	0.2932 B	8.1	0.3464 A	6.0
MTS (kPa)	33,170 A	40.8	27,148 A	23.9	30,102 A	42.9
MOE (tension) MPa)	1,774 A	20.5	1,527 A	18.0	1,746 A	23.9

[a] Differences between clones for each property are denoted by uppercase letters. Means with common letters are not significantly different at the 0.05 level as determined by Duncan's mean separation procedure.

Table 3. — Mean mechanical properties and specific gravities with the coefficients of variation for radial positions of the clones.

	Radial position from bark to pith					
	Age group 1 Ring no. 8-13 (bc), 11(t)		Age group 2 Ring no. 5-7 (bc), 7(t)		Age group 3 Ring no. 2-4 (bc), 3(t)	
Property	Mean[a]	COV (%)	Mean[a]	COV (%)	Mean[a]	COV (%)
Specific gravity (bending, compression)	0.3295 AB	7.8	0.3262 B	9.2	0.3361 A	10.5
MOR (b) (kPa)	33,437 A	12.5	28,432 B	15.8	24,174 C	11.2
Crushing strength (c) (kPa)	13,766 A	8.8	11,182 B	16.9	10,862 B	17.4
MOE (compression) (MPa)	1,649 A	23.4	1,307 B	22.9	1,267 B	25.9
Specific gravity (tension)	0.3347 A	7.1	0.3235 A	10.3	0.3217 A	11.4
MTS (t) (kPa)	32,295 A	41.1	30,142 A	33.8	27,984 A	31.1
MOE (tension) MPa)	2,199 A	21.5	1,645 B	20.8	1,204 C	23.5

[a] Differences between age groups for each property are denoted by uppercase letters. Means with common letters are not significantly different at the 0.05 level as determined by Duncan's mean separation procedure.

Table 4. — Pearson correlation coefficients between mechanical properties and specific gravities based on the total data set (bottom left) and separately for each clone.

Property	Clone	Specific gravity (bending, comp.)	MOR	Crushing strength	MOE (compression)	MTS	MOE (tension)	Specific gravity (tension)
Specific gravity (bending, compression)	Kopecky	1	0.357	0.663[a]	0.683[b]	- -	- -	- -
	I-214		–0.238	–0.122	–0.253			
	Koltay		–0.304	0.185	0.500			
MOR	Kopecky	0.235	1	0.597[b]	0.595[b]	–0.003	0.564	- -
	I-214			0.605[b]	0.876[b]	–0.102	0.607[a]	
	Koltay			0.215	0.058	0.056	0.542	
Crushing strength	Kopecky	0.223	0.480[b]	1	0.931[b]	0.108	0.510	- -
	I-214				0.769[b]	–0.229	0.051	
	Koltay				0.807[b]	0.491	0.472	
MOE (compression)	Kopecky	0.476[b]	0.551[b]	0.788[b]	1	0.040	0.434	- -
	I-214					–0.141	0.493	
	Koltay					0.398	0.183	
MTS	Kopecky	- -	0.048	0.145	0.225	1	0.635[a]	0.323
	I-214						0.368	–0.356
	Koltay						0.684[a]	0.092
MOE (tension)	Kopecky	- -	0.558[b]	0.361[a]	0.366[a]	0.628[b]	1	0.323
	I-214							0.107
	Koltay							–0.325
Specific gravity (tension)	Kopecky	- -	- -	- -	- -	0.100	0.134	1
	I-214							
	Koltay							

[a] Statistically significant at the 5 percent level.
[b] Statistically significant at the 1 percent level.

specific gravity and bending and tensile strengths of clones. However, the relationship was clonal specific in compression (Table 4).

- Sites, clones, and/or their interactions affected significantly one or more of the properties measured. However, most of the variation existed within trees from pith to bark. In other words, age had significant effect on growth ring width, anatomical characteristics, specific gravity, and mechanical properties.
- No consistent relationship was found between growth rate (expressed by either tree height, tree diameter, or growth ring width) and wood properties of the three clones from two sites.

Conclusions

The aging process caused larger within-tree variation in wood properties than the site and the genetic differences. This has to be considered in decisions on clonal selection and on rotation age. No consistent relationships was found between growth rate and wood properties, indicating that growth rate can be increased without influencing wood quality of poplars.

Utilization of this clonal material is limited. No structural applications are suggested because of their low strength properties. Poplars can be used as internal parts for furniture, as wood composites for different boards, and for pulp and paper production because of their relatively long fibers with thin cell walls.

These hybrid poplar clones with rapid growth and high productivity could be considered as an important fiber source, not for timber production, and only for short-rotation, intensively managed plantations.

Technical Forum Presentation:

Would radical carbon dioxide fees restructure future wood markets?

Heikki Seppälä

Background

The amount of living biomass in boreal forests has increased on average during recent decades and plays an increasingly important role in sequestering atmospheric carbon dioxide. Should forestry consequently be credited for CO_2 sequestration according to possible carbon taxes? We assume here that a marginal increase in the net fixation of CO_2 in biomass is equally valuable as a marginal reduction in the emission of CO_2 into the atmosphere, e.g., by reduced use of fossil fuels.

The forests of Nordic countries, especially Finland and Sweden, capture great amounts of CO_2 counteracting the greenhouse effect and the concomitant climate changes (Fig. 1). The volume of the growing stock of timber in Finland is currently increasing by about 30 million m^3 per year, or 35 to 40 percent of yearly gross increment. The fixation of CO_2 in the whole forest ecosystem in Finland is at about the same level as CO_2 emissions from fossil fuels. The concept of net national emissions suggests that carbon sequestration in forestry should be taken into account when countries buy CO_2 permits or pay CO_2 taxes.

This paper discusses some problems on inclusion of carbon and other nonmarket goods of forests in the national accounts. It also discusses the potential effects of CO_2 taxes (i.e., shadow price on CO_2) on timber markets.

National accounting and forestry

In Nordic countries, especially Norway and Sweden, there has been activities in extending the present national accounting of forestry towards a green national accounting system (1,2,4,7). Following Solberg and Svensrud (7) and Hoen (1), we make an effort to "green" the traditional National Accounting System in the Finnish forestry.

It has been the common practice in national accounts to focus on gross national product as a measure of welfare change. This can be justified when reproducible capital is the only stock that is considered and that stock can be assumed to be replaced in a constant manner. However, this practice is clearly not acceptable when the generation of gross income is combined with a continuous depletion of natural resources like forests (2). Therefore, changes in stocks should be valued by marginal net prices.

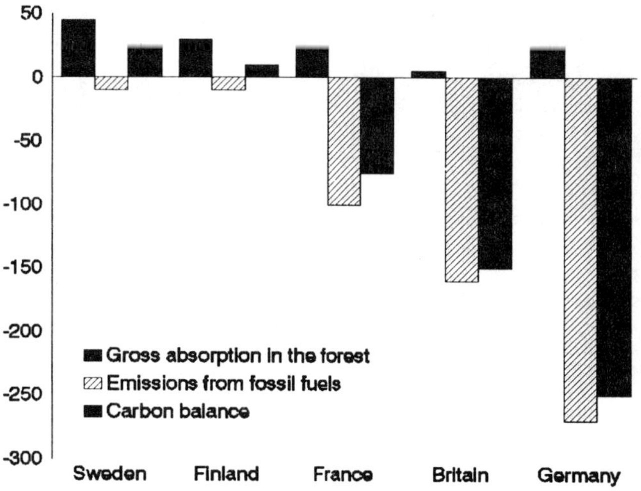

Figure 1. — Carbon flows in certain European countries (3).

The author is Senior Research Officer, Finnish Forest Res. Inst., Helsinki, Finland.

In addition to the sales value of timber extraction we have also estimated the value of net increase of timber stock, extraction of other than timber products, fixation of atmospheric carbon, and negative environmental effects from intensive forest management like clear cuttings (Table 1). Since we have a precise estimate only for contribution to net national product (NNP) of timber sales, we have for other items derived "guesstimates" based on "best judgment" from different sources (5,8). The focal point here is fixation of atmospheric carbon in net growth of timber stock compared to contribution to NNP of timber sales.

In Finland, carbon dioxide tax for fossil fuels is combined with other taxes. Therefore, to calculate the present discounted value of the CO_2 flow related to the marginal change in the growing stock, we used the Norwegian CO_2 tax for petrol of 3 NOK or about US $0.35 per gallon. This corresponds to a fee of US $30 per ton CO_2. We assumed that it takes 10 years before the timber is harvested (1) and wood-in-use decayed with average lifetime of 15 years (5) resulting in total average lifetime of 25 years. Using this decay pattern and a social rate of discount (SRD) of 5 percent per year, the net benefit of CO_2 sequestration in timber stock would be US $655 million or 47 percent of the net value of timber sales.

Table 1. — An extension of national accounts of Finnish forestry in 1992.

Timber extraction	(US $)
Gross value of sales	$1,913 million
Intermediate consumption	$80 million
Net depreciation of fixed capital	$448 million
Contribution to net national product	$1,385 million
Increase in timber stock	
Gross growth 80 million m^3	
Total drain 49 million m^3	
Net growth of stock 31 million m^3	
Value of growth (estimate) [excluded, because 4 > 2]	[$465 million]
Nontimber products	
Berries, mushrooms, reindeers, hunting, etc.	
Value of products (estimate)	$200 million
Fixation of atmospheric carbon	
Net growth of timber stock, 31 million m^3 sequestering 24 million tons CO_2. According to the "Norwegian" CO_2 tax ($30/ton CO_2), average lifetime of wood 25 years, and the social rate of discount 5 percent per year, the net benefit of CO_2 sequestration in timber stock [included, because 2 < 4]	$655 million
Negative environmental effects	
Willingness to pay for more "multiple use" oriented forestry instead of "intensive" forest management (estimate)	$100 million
TOTAL	$2,140 million

For market value of the net increase in timber stock (Table 1) we use a marginal price of US $15 per m^3 or about 55 percent of the average price derived from the unit value of NNP from timber extraction. To sequester carbon in net growth of stock we cannot harvest it, and vice versa. Since the value of fixation in atmospheric carbon is higher than that of the "market" value of net growth, we include the value of carbon fixation to our final calculation. The summary of our calculation in Table 1 is as follows:

	(US $)	
Timber extraction (NNP)	$1,385 million	65%
Nontimber products	$200 million	9%
Fixation of atmospheric carbon	$655 million	31%
Negative environmental effects	(–$100 million)	–5%
TOTAL	$2,140 million	100%

In addition to the level of CO_2 tax used as a shadow price, the value of growing stock change depends on the social rate of discount and the decay patterns of timber. Using another, more reasonable tax level of US $5/ton of carbon equivalent to US $18.3/ton of CO_2, the impact on NNP would be US $400 million in 1992 or 29 percent compared to that on NNP of timber sales assuming again the 5 percent SRD and average total lifetime of 25 years. These levels of SRD and decay patterns reduce the value of an increase in the growing stock by about 70 percent compared to that without considering SRD or lifetime of timber.

Concluding remarks

The attempt for extending national accounts presented above is preliminary in nature. The calculations, however, indicate that potential large-scale CO_2 subsidies paid to forest owners according to carbon sequestration in forests with CO_2 taxes would most probably impact the timber market. High shadow prices derived from high CO_2 taxes would obviously change allocations in both the market sector and between the market and the environmental (nonmarket) sectors. The use of high shadow price also has impact on marginal timber prices used in valuation of growing stock changes in extended national accounts.

Calculations are sensitive to the levels of CO_2 taxes used as shadow prices as well as to the social rate of discount and the decay patterns. The recent research of product life cycles and carbon in wood-in-use (5) has, however, produced more reliable information on decay patterns.

Literature cited

1. Hoen, H.F. 1993. Forestry and national accounting — a comment. *In:* Valuing Biodiversity. M. Lindel and A. Naskali, eds. Scandinavian Forest Economics 34:136-146.
2. Hultkranz, L. 1992. National accounts of timber and forest environmental resources in Sweden. Environmental and Resource Economics 3(2):283-305.
3. Kauppi, P.E. and E. Tomppo. 1993. Impact of forests on net national emissions of carbon dioxide in West Europe. Water, Air, and Soil Pollution 70:187-196.
4. Mäler, K.G. 1991. National account of timber and forest environmental resources. Environmental and Resource Economics 1(1):1-15.
5. Seppälä, H. and V. Siekkinen. 1993. Wood consumption and carbon balance. A study on the impact of wood consumption on carbon flow in Finland 1990. Res. Pap. 473. Finnish Forest Res. Inst.
6. ________ and ________. 1993. Accounting system for wood in use as a carbon sink. *In:* The Globalization of Wood: Supply, Processes, Products, and Markets. Forest Prod. Soc., Madison, Wis. pp. _____.
7. Solberg, B. and A. Svensrud. 1992. Environmental factors and national account of forestry — some findings from Norway. Paper for IUFRO Centennial Meeting.
8. Yearbook of Forest Statistics. 1992. Official Statistics of Finland.

Technical Forum Presentation:

Accounting system for wood-in-use as a carbon sink

Heikki Seppälä and Virpi Siekkinen

Background

Carbon dioxide is the most important of heat-trapping gases in the predicted global climate change. Forests, both in their role as contributors to global warming through deforestation, and as contributors to the solution to the problem by reforestation and adequate management of natural forests, are subjects of frequent discussion. Carbon budgeting of forest ecosystems promises to be one the most popular topics in 1990s forestry research.

The role of wood-in-use has also been mentioned as a potential balancing factor of the global carbon budget. To what extent has been questionable — mostly because of lack of relevant information.

In this paper, an accounting system for carbon in wood-in-use in the Finnish forest sector is introduced (5). The system is compatible with national accounts. Carbon balance calculations are made for 1990. Apart from these, estimates are made of carbon emissions to the year 2040 from 1990 timber harvests and wood use. The estimates are based on product life-cycle analyses and trade statistics.

Carbon accounting

Carbon accounting is based on wood material accounting in Finland (4). Wood material accounting describes changes in the growing stock of Finland's forests, and the flows of wood material in different sectors of the national economy. The accounting contains forestry activities, other industrial and end-use activities, and an assortment of some 220 raw materials, intermediate goods, commodities, and wastes. All industrial activities and end-use activities are consistent with the national accounts.

Wood material statistics are based on the data and statistics of Finnish Forest Research Institute, industrial statistics of Statistics Finland, statistics made by forest industries, foreign trade statistics, and various studies on forests and the use of wood. Carbon accounting is refined from wood material accounting by using conversion factors of wood dry weight and carbon content by tree species.

Structure of the carbon accounting

The framework and structure of carbon accounting follows the one in wood material accounting, which is a Finnish application of the Norwegian model (2). The system includes three main accounts: wood balance, use account, and mass account.

The wood balance describes the annual change in the growing stock of wood due to biological growth, natural losses, and fellings (Table 1). Growth

Table 1. — Carbon in forest balance 1990, 1000 metric tons of carbon.

	Pine	Spruce	Broad-leaved	Total
Stock 1.1.1990	183,582	137,421	85,559	406,562
Annual increment	7,642	5,223	3,694	16,559
Total drain	4,470	4,472	2,896	11,838
Natural losses	134	74	66	274
Total removals	4,337	4,373	2,830	11,540
Silvicultural waste	287	247	549	1,083
NET removals	4,050	4,126	2,281	10,457
Sawlogs	1,938	2,103	384	4,425
Pulpwood	1,998	1,916	1,332	5,246
Fuelwood	115	106	564	785
Stock 31.12.1990	186,754	138,197	86,357	411,308

The primary author is Senior Research Officer, Finnish Forest Res. Inst., Helsinki, Finland.

Table 2. — Carbon in use account 1990, 1000 metric tons of carbon.

	Sawlogs	Pulpwood	Fuelwood	Residues	Sawn wood	Plywood veneers	Particleboard	Fiberboard	Mechanical pulp	Chemical pulp	Waste liquors	Paperboard
Removals	4542	5229	721	- -	- -	- -	- -	- -	- -	- -	- -	- -
Imports	73	1224	6	60	12	4	5	5	4	6	- -	38
Exports	–60	–75	- -	- -	–870	–220	–64	–16	–12	–302	- -	–1665
Inventories	–21	215	- -	- -	–11	–1	–1	–1	- -	–2	- -	4
Primary Supply	4576	6592	728	60	–895	–214	–57	–8	–8	–298	- -	–1612
Sawing and planing												
In	–3619	–127	–101	- -	–71	- -	- -	- -	- -	- -	- -	- -
Out	4	18	2	2102	1492	- -	- -	- -	- -	- -	- -	- -
Plywood and veneers												
In	–525	- -	–24	- -	–16	–6	–0	- -	- -	- -	- -	- -
Out	- -	- -	2	235	9	258	3	- -	- -	- -	- -	–0
Particleboard production												
In	- -	- -	–2	–198	–1	- -	–0	- -	- -	- -	- -	–1
Out	- -	- -	- -	2	3	2	113	- -	- -	- -	- -	- -
Fiberboard production												
In	- -	–4	- -	–52	- -	- -	- -	–2	- -	- -	- -	- -
Out	- -	- -	- -	–0	- -	- -	- -	1	21	- -	- -	- -
Mechanical pulp production												
In	–189	–1719	- -	–276	- -	- -	- -	- -	- -	- -	- -	–12
Out	- -	- -	- -	81	- -	- -	- -	- -	801	- -	- -	- -
Chemical pulp production												
In	–76	–4398	- -	–1120	- -	- -	- -	- -	- -	- -	–1767	- -
Out	- -	- -	- -	403	- -	- -	- -	- -	- -	1109	1860	- -
Paper and board production												
In	- -	- -	- -	- -	- -	- -	- -	- -	–810	–723	- -	–73
Out	- -	- -	- -	- -	- -	- -	- -	- -	- -	- -	- -	1926
Other supply	- -	- -	73	2	3	- -	- -	- -	- -	- -	- -	- -
Used as fuel	- -	- -	1113	–1118	- -	- -	- -	- -	- -	- -	- -	- -
Losses, statistical errors	–56	62	96	–223	–85	–15	–13	–5	- -	- -	–152	15
End use	126	15	1880	23	523	25	94	24	8	61	3	413
Agriculture	- -	- -	148	- -	- -	- -	- -	- -	- -	- -	- -	- -
Prefabricated buildings	75	8	9	14	95	1	6	2	- -	- -	- -	- -
Building materials	27	3	8	3	95	6	18	2	- -	- -	- -	0
Furniture, fixtures	3	- -	12	- -	23	4	40	3	- -	- -	- -	0
Paper goods, packaging	- -	- -	- -	- -	- -	- -	- -	- -	8	20	- -	188
Printing, publishing	- -	- -	- -	- -	- -	- -	- -	- -	- -	- -	- -	202
Other manufacturing	1	- -	9	6	13	2	2	- -	- -	41	3	23
Electricity, gas, water	- -	- -	944	- -	- -	- -	- -	- -	- -	- -	- -	- -
Building, construction	20	5	- -	- -	296	11	28	16	- -	- -	- -	- -
Commerce, transportation	- -	- -	6	- -	- -	- -	- -	- -	- -	- -	- -	- -
Public administration	- -	- -	34	- -	- -	- -	- -	- -	- -	- -	- -	- -
Private households	- -	- -	711	- -	- -	- -	- -	- -	- -	- -	- -	- -

Table 3. — Carbon in mass account 1990, 1000 metric tons of carbon.

	Sawlogs	Pulpwood	Sold fuel	Residues	Main product	Used as fuel	By-products	Wood wastes	Losses, error
Sawing and planing									
In	3619	127	- -	- -	73	- -	- -	- -	- -
Out	4	18	497	1361	1572	108	- -	184	65
Particleboard production									
In	525	- -	- -	- -	6	- -	16	- -	- -
Out	- -	- -	71	160	258	24	12	- -	21
Plywood and veneers									
In	- -	- -	- -	182	- -	- -	2	- -	- -
Out	- -	- -	2	- -	146	2	8	- -	27
Fiberboard production									
In	- -	4	- -	49	2	- -	- -	- -	-
Out	- -	- -	- -	- -	44	- -	2	2	6
Mechanical pulp production									
In	189	1719	- -	276	- -	- -	22	- -	- -
Out	- -	- -	- -	78	1934	- -	- -	58	135
Chemical pulp production									
In	189	1719	- -	276	- -	- -	22	- -	- -
Out	- -	- -	- -	78	1934	- -	- -	58	135
Paper and board production									
In	- -	- -	- -	3247	131	- -	- -	- -	- -
Out	- -	- -	- -	- -	3370	- -	- -	21	–12

and total drain are followed by tree species; fellings are followed by roundwood assortments.

The use account follows the processing and use of harvested roundwood as intermediate goods, end products, and waste. The use balance contains removals of timber, imports and exports, and inventories of timber and wood products. The sectors of economy are 10 for forest industry activities, and 40 for other industrial and end-use activities. Roundwood, intermediate goods, commodities, and wastes are aggregated to 12 classes.

For forest industry activities, both inputs and outputs of carbon in wood and wood products are accomplished, and only inputs for other industry and end-use activities.

The mass account is accomplished only for forest industry activities. The mass account indicates the amount of carbon contained annually in forest industry roundwood, intermediate products, and wood containing waste.

Flow of carbon in wood-in-use

In addition to carbon balance calculations for 1990, estimates were made of carbon emissions from 1990 timber harvest and use in Finland to 2040. These were based primarily on product life-cycle analyses and trade statistics. Lifetime information in product life-cycle analyses were derived from Aalstadt (1).

The results of the analyses indicate that even though the lifetime especially of wooden houses and wood-based constructions is very long, the average circulation time of wood harvested and processed in Finland is relatively short (Fig. 1). This is because of the production structure of the forest products industry is focused on pulp and paper products due to massive exports of these products. Over 70 percent of the total value of industrial production are exported.

In 5 years, or by 1995, almost 80 percent of timber harvested in 1990 for wood consumption would be emitted back to the atmosphere as CO_2. After that, however, the emission rate would be very slow.

Roughly half of the carbon would be emitted in Finland and the other half in export markets of the products, mainly the European Community region. Of the wood used for energy, 70 percent was combusted as waste liquors in pulp industry, almost 15 percent in sawmilling, and only less than 15 percent for heating in households and farming.

In 1990, wood from sawlogs was combusted almost as much as it was allocated for domestic consumption of sawn wood (Fig. 2). Also taking into account chips and sawdust from sawmills combusted in pulpmills, sawlogs were used by a quarter more for energy than was bound in exports of sawn wood.

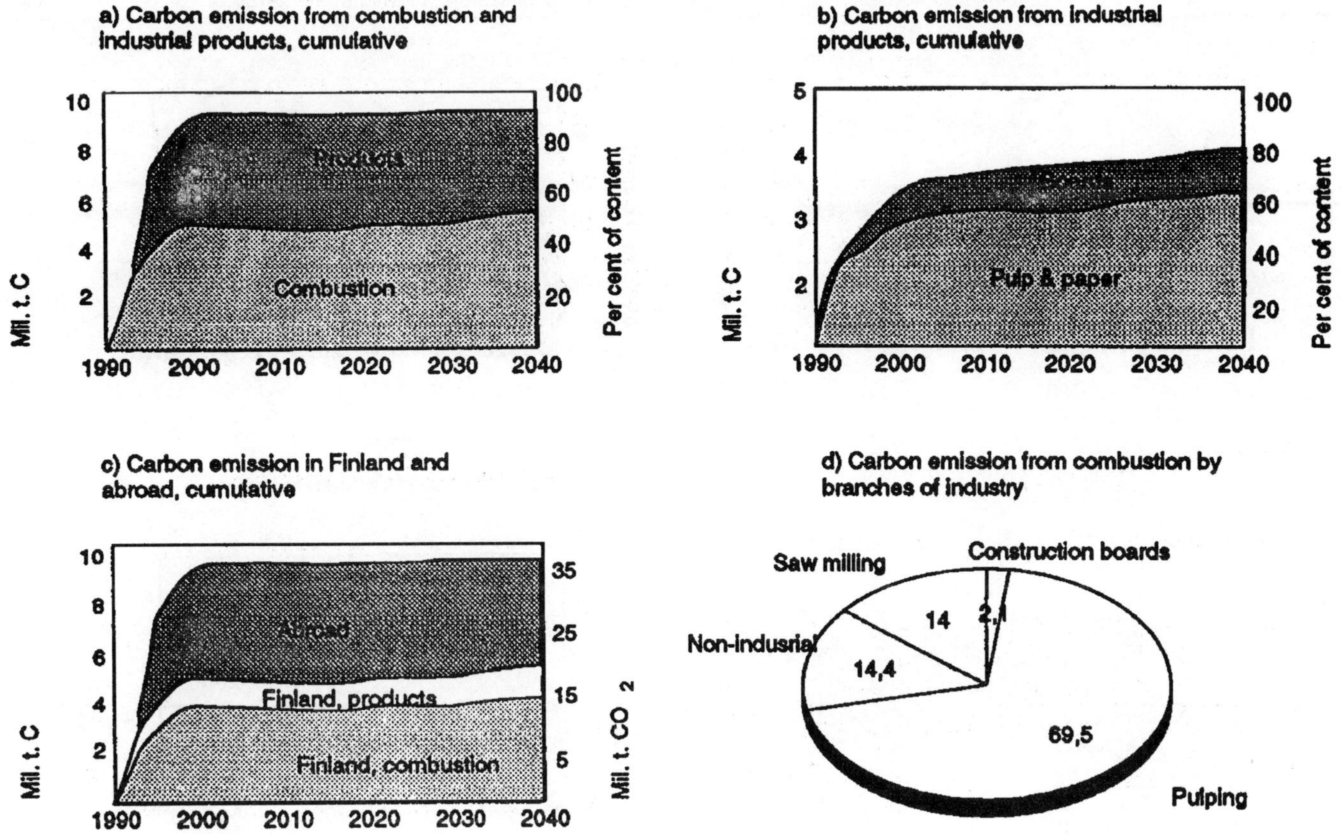

Figure 1. — Estimate for carbon emission from wood used in Finland, 1990.

However, almost half of woody material combusted was bark.

Even though the life cycle of sawn wood is over 150 years (after Aalstad 1990), more than half of carbon in sawlogs used for sawmilling would be emitted back to the atmosphere as CO_2. Carbon would discharge first by combustion and there after from pulp and paper products made of sawdust and chips. Emission from domestic consumption is faster than that from exports during the early years, exceeding them by about a third.

Deposition of wastes in pulping has increased along with intensified combustion of wastes for energy. This has resulted in a shorter lifetime of wood in pulp and paper industry, but also diminished use of fossil fuels for energy.

The lifetime of pulp and paper products is usually less than 5 years. About 40 percent of wood raw material allocated to pulping will be combusted for energy, and consequently ever shortening the lifetime. This is why emission in Finland is faster during the early years than abroad. It also follows that in spite of export-orientated production, only half of carbon in wood for pulping is finally emitted abroad.

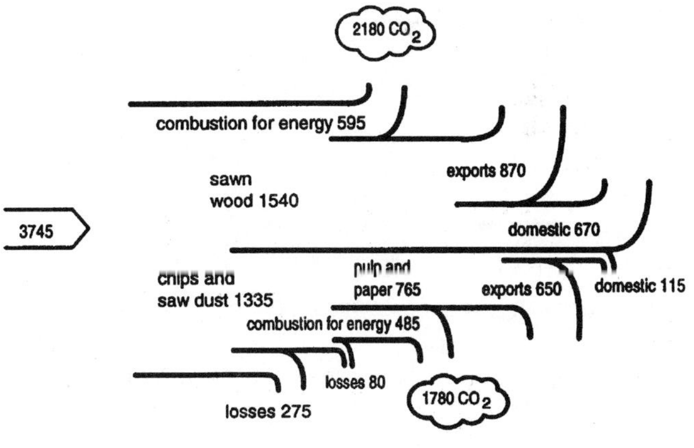

Figure 2. — Flow of carbon in sawlogs for sawmilling in Finland, 1990.

Plywood products practically reach the end of their lifetime by the year 2040. Because of a relatively small share of combustion of wood and a fairly long antropogen lifetime of products the life-cycle curve is kinked. By 2015 almost all carbon released originates from combustion in Finland. After that

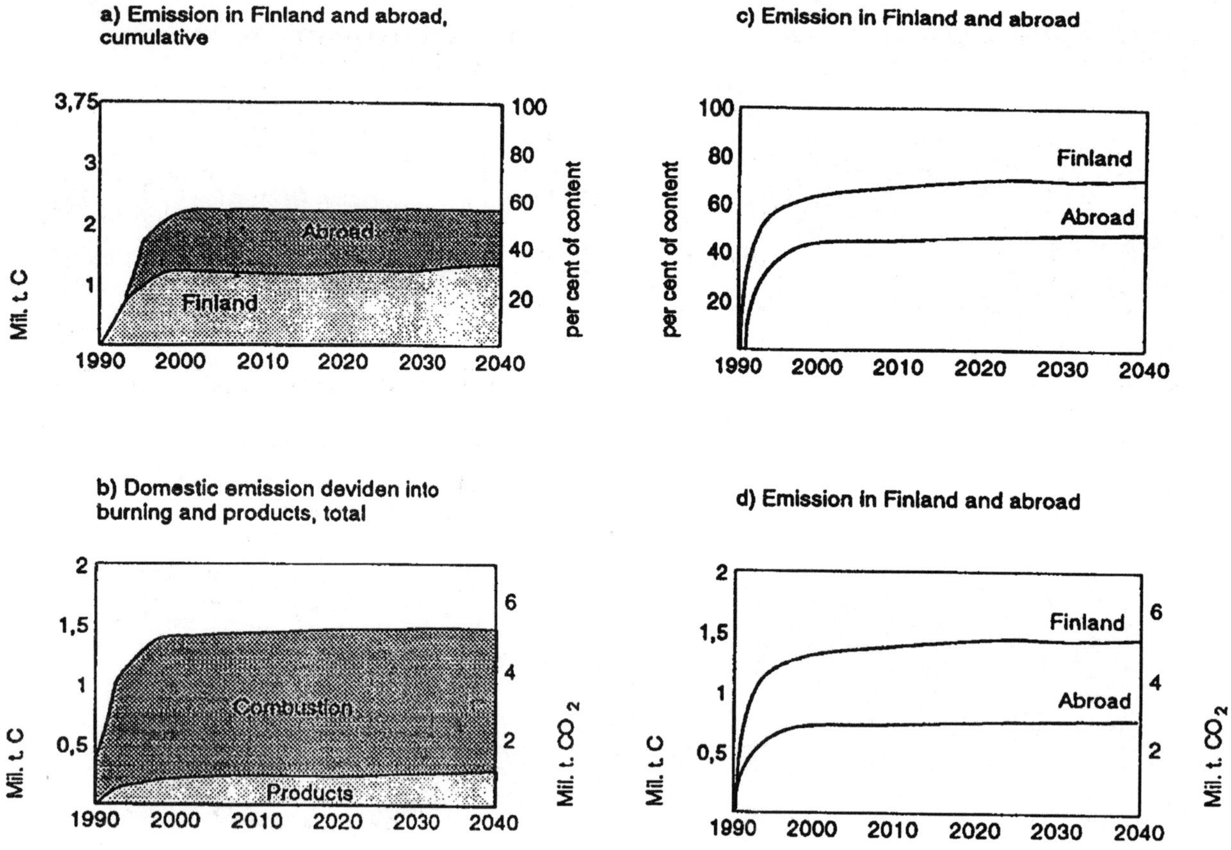

Figure 3. — Estimate for carbon emission from sawlogs for sawmilling in Finland, 1990.

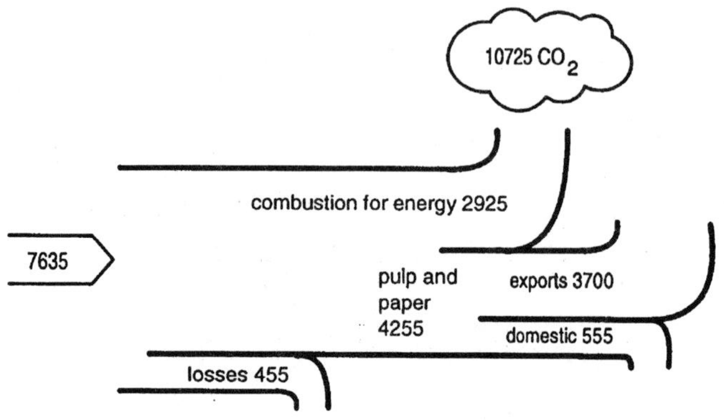

Figure 4. — Flow of carbon in wood for pulping in Finland, 1990.

the rates of emission are at the same level both in Finland and export markets.

Carbon emissions from fiberboard and particleboard closely follow that of plywood products. The production is, however, more directed in domestic markets. Production quantities have been on the decline during the 1980s and in 1990 only 1.2 million m^3 of wood, equivalent to 0.25 million tons of carbon, were used as raw material in these industries.

Discussion

Carbon accounting for wood in use describes the annual changes of carbon sequestration in the growing stock of wood due to biological growth, natural losses, fellings, and the processing and use of harvested roundwood as intermediate goods, end products, and waste. The accounting system, being part of account of natural resources and compatible with national economy statistics, provides a possibility to assess impacts, both environmental and economic, on national economy of wood-in-use.

The research proceeds by constructing a formal computational model tracing detailed impacts on carbon cycle of harvesting and use of wood in Finland. The impacts on carbon cycle of alternative scenarios for harvests, processing technologies, use of wood, life cycles, and other factors will be easy to simulate with the model.

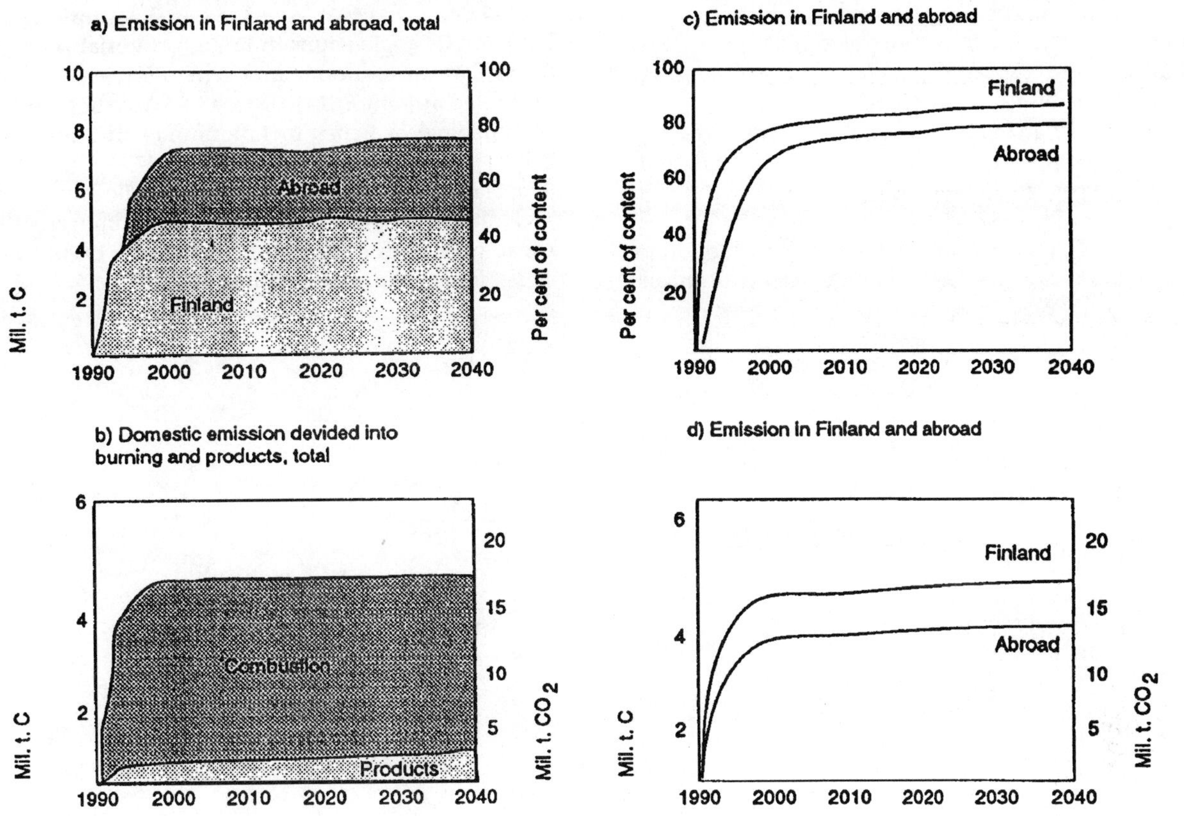

Figure 5. — Estimate for carbon emission from wood raw material for pulping, Finland 1990.

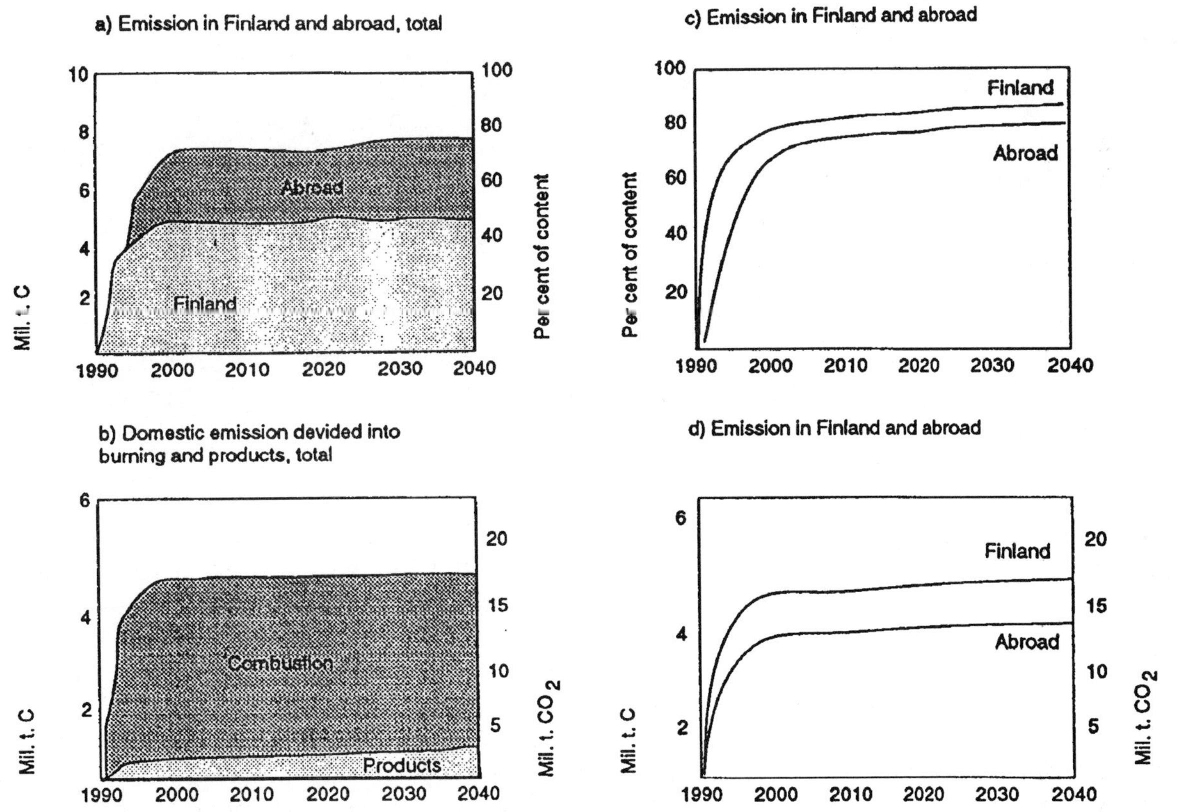

Figure 6. — Estimate for carbon emission from wood raw material for plywood manufacturing, Finland 1990.

Literature cited

1. Aalstadt, T. 1991. Consumption of timber in Norway: what are the end-uses and long lifetime does the product have? Dept. of Forestry, AUN. AS. 15 pp.
2. Alfsen, K., T. Bye, and L. Lorentsen. 1987. Natural resource accounting and analyses: The Norwegian experience 1978-1986. Sociale og Okonomiske Studier 65. Statistisk Sentralbyrå. Oslo, Norway.
3. Ayres, R. 1978. Resources, Environment, and Economics. Applications of the Materials/Energy Balance Principle. John Wiley & Sons, New York.
4. Muukkonen, J. 1993. Wood material accounting in Finland and its applications in linking materials/energy balances data to economic and environmental data. Paper presented at joint ECE/EUROSTAT Workshop on Specific Methodological Issues in Environmental Statistics, Bratislava, Slovakia.
5. Seppälä, H. and V. Siekkinen. 1993. Wood consumption and carbon balance. A study on the impact of wood consumption on carbon flow in Finland 1990. Res. Pap. 473. Finnish Forest Res. Inst.

Technical Forum Presentation:

Establishing allowable design values for structural lumber of foreign species for use in the United States

Bradley E. Shelley and David W. Green

Abstract

Historically, market needs for structural lumber in the United States have been met from domestic and Canadian timber resources. In recent years, changing government policy related to the industrial use of federal timber has caused a major decline in commercially available timber, particularly in the northwestern United States. To fill this shortfall, manufacturers and consumers have begun to look to offshore sources. The procedures used for assigning allowable design values to structural lumber in the United States are not generally well understood outside of the regional lumber grading rules writing agencies. In addition, there have been many recent changes in the procedures as a result of the North American in-grade testing program. The purpose of this paper is to present a summary of the current procedures used to establish design values for structural lumber in the United States with special emphasis on species of foreign origin.

Introduction

Historically, market needs for structural lumber in the United States have been met from domestic and Canadian timber resources. In the 1980s, the growing public debate over the preferred use of publicly owned timberland in the northwest began to diminish the land base designated for commercial timber production and harvest. By the early 1990s, the continued contentious litigation over the northern spotted owl, and other species designated as threatened or endangered, has virtually halted all timber sales from federally owned timber in the Pacific Northwestern states of Washington, Oregon, and California.

Because the U.S. federal government owns approximately 32 percent of the unreserved timberland in the State of Washington, and 51 percent in Oregon, the disruptions in timber supply will continue at least until the federal government develops new management plans for federal land. It is further likely that even after new management plans are developed, the harvest levels will be substantially below historic levels. As a result, manufacturers in the Pacific Northwest have begun to look to offshore sources to meet some of the demand for timber. In order to import lumber or timber from foreign sources for structural purposes, the material must be approved for import by the USDA Animal and Plant Health Inspection Service, and have approved allowable design values assigned. The purpose of this paper is to present a brief summary of the procedures for obtaining approval of design values for use in the United States.

Model building codes

The United States has a rather unique system of approval for building components and systems. Approval of buildings, systems, and components are governed by local building departments which in turn subscribe to one of three regional model building codes (7,12,19). Some building jurisdictions choose

The authors are, respectively, Executive Vice President, West Coast Lumber Inspection Bureau, Portland, Oreg.; and General Engineer, USDA Forest Serv., Forest Prod. Lab., Madison, Wis.

to develop their own building codes. Often these are based on one of the major model codes.

While the local building officials are technically responsible for the approval of buildings and components, they have of necessity come to rely on the model codes to specify approved materials and construction practices. New, unique, or proprietary systems and components are evaluated by the model code organizations which issue research reports on their findings and approved uses.

Structural lumber, too, has a rather unique relationship with the model codes. The lumber industry has a standardization program which sets grade standards, evaluates third party inspection performance, and approves design values. This national system has been evolving over the last 70 years. The current system is described in Product Standard PS20, the American Softwood Lumber Standard (17). It is probably better known as the American Lumber Standards (ALS). The building codes and building officials have come to rely on the ALS and its certification and accreditation program to ensure the reliability and product performance of structural lumber.

American Lumber Standards

The American Lumber Standards Committee (ALSC) established Product Standard PS20-70 to develop standard grades and sizes of dimension structural lumber (nominal dimensions of 2 to 4 in. in thickness, 2 in. and wider), accredit grading and inspection agencies, monitor the performance of accredited agencies, and certify grading rules and assigned design values. The ALS system (Fig. 1) as established by PS20-70 is a consensus standards process with representative participation by all interested groups, producers, users, and general interest. Under this standard an independent Board of Review, elected by the ALSC and concurred in by the Secretary of Commerce, is established to carry out the accrediting and monitoring of grading agencies and certification of grading rules and design values.

Design values certified by the Board of Review, while published by a single rules writing agency, are non-proprietary. The design values apply to structural grades of all lumber produced of that certified species which is graded under the supervision of any ALS-certified grading or inspection agency. The U.S. Department of Commerce acts as the secretariat and publisher of Product Standard PS20-70. For a more detailed discussion of the structure and function of the American Lumber Standards Committee, the reader is referred to PS20-70. Copies of PS20-70 are available from the American Lumber Standards at the following address: American Lumber Stand-

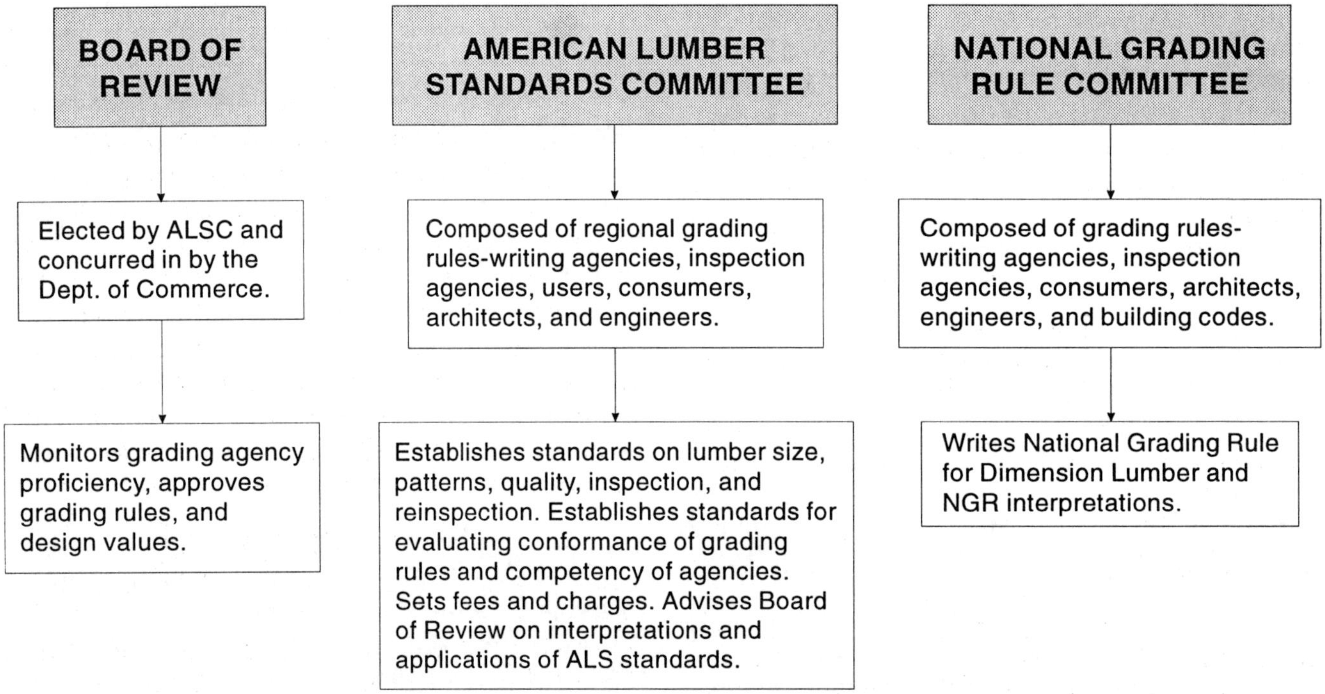

Figure 1. — American Softwood Lumber Standard PS20-70.

ards, P.O. Box 210, Germantown, MD 20875; phone 301-972-1700.

Approval procedure for design values

It is the responsibility of the regional grading rules writing agencies (14-16,18,20,21,23) to develop appropriate design values for the species and grades covered in their respective rule books. For new domestic species or foreign species, the rules writing agency must first seek authorization from the Board of Review of the American Lumber Standards Committee to include these species in their rule book. If authorization is granted, the agency may then develop and submit proposed design values for certification.

The design values which the grading agency develops are also submitted to the American Lumber Standards Board of Review for certification. The specifics of developing these proposed design values are discussed in greater detail by the West Coast Lumber Inspection Bureau (WCLIB) (22). The purpose of this section is to outline the general approval process.

When proposed design values are submitted for certification, the Board of Review refers the proposal to the National Institute for Standards and Technology (NIST) for review. Section 5.3.1 of PS20-70 requires that all standards or "other appropriate criteria" used in the development of proposed design values must be found to be appropriate by NIST. NIST makes its determination with the advice and counsel of the USDA Forest Products Laboratory. In addition, the Board of Review must determine that the requirements of the standards and "other criteria" were faithfully followed in preparation of the submission. The Board of Review utilizes the services of the USDA Forest Products Laboratory (FPL) in Madison, Wis. to perform the detailed review of the technical content of the submission. Upon completion of its review, the FPL issues a written report of its findings to the Board of Review.

Public notice is made of all proposals presented to the ALS Board of Review. This public notification permits other interested parties the opportunity to comment on the submission to the Board of Review. If the Board of Review finds that the proposed values were developed in accordance with the provisions of PS20-70 and the ALSC policies, the values are certified for publication by the submitting rules writing agency.

Upon approval the design values are published in the proponent's grading rules, and subsequently the National Design Specification Supplement (1). The approved values are also then submitted to the model codes for reference in the building codes.

Table 1. — Design value approval process under PS20-70.

1. Agency seeks approval to include species in grade rule book.
2. Agency develops sampling and testing plan with advice of ALS and USDA FPL.
3. Lumber is sampled and tested in accordance with approved sampling and testing plan.
4. Agency analyzes data by procedures of ASTM D 1990 and other appropriate standards, if needed.
5. Agency submits proposed design values to ALS Board of Review.
6. Submission is reviewed by ALS staff and USDA FPL.
7. Submission is available for comment by other agencies and any other interested parties.
8. ALS Board of Review approves (or disapproves) design values with modifications (if needed) based on all available information.
9. Agency publishes new design values for the species.

Acceptance of foreign species

Prior to 1992, U.S. lumber design values were derived using the clearwood values of ASTM D 2555 (6) and the grade derivation of design values in ASTM D 245 (3). Because the process relied on clear wood data which is generally not available for foreign species, new procedures had to be developed. In 1975 Galligan and Ethington (8) developed procedures based on ASTM D 2915 (5), for establishing design values for foreign species for use in the United States, known by its initials as T/I/QC. This procedure required extensive testing of lumber properties and ongoing quality control. The program was accepted by the Board of Review, but no foreign lumber was ever commercially imported under the program.

In the late 1970s, the United States and Canada began a massive testing program to evaluate the mechanical properties of the major commercial softwood species used in the two countries. The program became known as the In-grade Testing Program (11). From the late 1970s to the mid-1980s, a total of some 70,000 pieces of lumber were tested, most to destruction. The results of this testing program were used as the basis to develop a new ASTM Standard D 1990 (4). ASTM D 1990 provides a standardized means of assigning allowable design values to dimension lumber based on tests of full-size specimens. After this standard was approved in 1991, new design values were developed by the regional grading rules writing agencies and approved by the ALS Board of Review.

ASTM D 1990 also provided a new method of acceptance of foreign lumber for structural use. Proposed new procedures for acceptance of foreign lumber were developed by Green and Shelley (9,10) in two papers which were submitted to the ALS Board of Review for approval. The first paper (9) discusses how to adapt ASTM D 1990 procedures for determining design values of foreign lumber that has been visually graded to standard ALS grades. The second paper (10) provides acceptance procedures for machine stress-rated lumber (MSR). These two procedures have been approved for use by the ALS Board of Review.

The general procedures for visually graded lumber covered by Green and Shelley were further explained in a publication of the WCLIB (22). The WCLIB publication included proposed sampling and testing schemes together with estimates of the costs of testing required to develop data for establishing allowable properties.

The procedures for MSR lumber are designed to verify the applicability of the U.S. grading model. Specifically, the procedures establish a standardized methodology for evaluating the relationships of mechanical properties as they relate to machine grading, and segregation by modulus of elasticity.

Conclusions

The new procedures approved to establish design values for foreign lumber provide a standardized basis whereby foreign species can enter the U.S. structural market and be treated the same as domestic species. To the present no foreign species have been approved under these new procedures. It is anticipated that when the economic conditions are favorable for the importation of foreign material, these procedures for their acceptance will be utilized.

Literature cited

1. American Forest and Paper Association. 1991. National Design Specification. AFPA, Washington, D.C.
2. American Society for Testing and Materials. 1992. Annual Book of Standards, Vol. 04.09, Wood. ASTM, Philadelphia, Pa.
3. ________. 1988. ASTM D 245. Standard practice for establishing structural grades and related allowable properties for visually graded lumber. ASTM, Philadelphia, Pa.
4. ________. 1991. ASTM D 1990. Standard practice for establishing allowable properties for visually graded dimension lumber from in-grade tests of full-size specimens. ASTM, Philadelphia, Pa.
5. ________. 1990. ASTM D 2915. Standard practice for evaluating allowable properties for grades of structural lumber. ASTM, Philadelphia, Pa.
6. ________. 1988. ASTM D 2555. Standard test methods for establishing clear wood strength values. ASTM, Philadelphia, Pa.
7. Building Officials Code Administrators International. 1993. National Building Code. BOCAI, Country Club Hills, Ill.
8. Galligan, W.L. and R.L. Ethington. 1975. Philosophies of testing/inference where the goal is to judge whether an imported grade meets or exceeds an ALS approved grade of dimension lumber. A report to the Board of Review of the American Lumber Standards Committee.
8. Green, D.W. and B.E. Shelley. 1991. Guidelines for assigning allowable properties to visually graded foreign species based on test data from full sized specimens. Report to the Board of Review of the American Lumber Standards Committee.
9. ________ and ________. 1993. Guidelines for assigning allowable properties to mechanically graded foreign lumber. Report to the Board of Review of the American Lumber Standards Committee.
10. In-Grade Testing of Structural Lumber. 1989. Proceedings 47363. Forest Prod. Res. Soc., Madison, Wis.
11. International Conference of Building Officials. 1991. Uniform Building Code. ICBO, Whittier, Calif.
12. ________. 1991. Uniform Building Code Standards. ICBO, Whittier, Calif.
13. National Lumber Grades Authority. 1991. Standard Grading Rules for Canadian Lumber. NLGA, Vancouver, B.C., Canada.
14. Northeastern Lumber Manufacturers Association. 1991. Standard Grading Rules for Northeastern Lumber. NLMA, Cumberland Center, Maine.
15. Northern Softwood Lumber Bureau. 1991. NSLB, Cumberland Center, Maine.
16. Product Standard 20-70. Amended 1986. American Softwood Lumber Standard. U.S. Dept. of Commerce, Washington, D.C.
17. Redwood Inspection Service. 1992. Standard Specifications for Grades of California Redwood Lumber. RIS, Novato, Calif.
18. Southern Building Code Congress International. 1991. Standard Building Code. SBCCI, Birmingham, Ala.
19. Southern Pine Inspection Bureau. 1991. Standard Grading Rules for Southern Pine Lumber. SPIB, Pensacola, Fla.
20. West Coast Lumber Inspection Bureau. 1991. Standard Grading Rules for West Coast Lumber No. 17. Revised 1993. WCLIB, Portland, Oreg.
21. ________. 1993. Acceptance Procedures for Foreign Species Imported into the United States. WCLIB, Portland, Oreg.
22. Western Wood Products Association. 1991. Western Lumber Grading Rules 91. Western Wood Prod. Assoc., Portland, Oreg.

Technical Forum Presentation:

Chilean lumber export update: a need for increased drying capacity

H. Peter Steinhagen, Arnaldo Jelvez, and Nazhla I. Mohor

Introduction

Chile's forest products output is targeted to the world market (5). The present study examines Chilean lumber export targets vis-à-vis installed kiln-drying capacity, as pressures to kiln dry export lumber are building up.

Lumber exports

At present, Chilean lumber exports per year total 1.1 million m^3 (4), mostly of radiata pine (Fig. 1). Lumber production per year is 3.3 million m^3 (4). Both production and export are expected to increase steadily, since most Chilean radiata pine plantations are presently less than 15 years old (3) and will enter sawlog production at age 20. Chile's lumber export market is dominated by eight companies whose market share is over 70 percent (Table 1). Most lumber is presently shipped green after preservative treatment with penta salt, which is leachable and extremely toxic to fish. There are international marketplace pressures to phase out the use of penta and kiln dry the lumber instead (1).

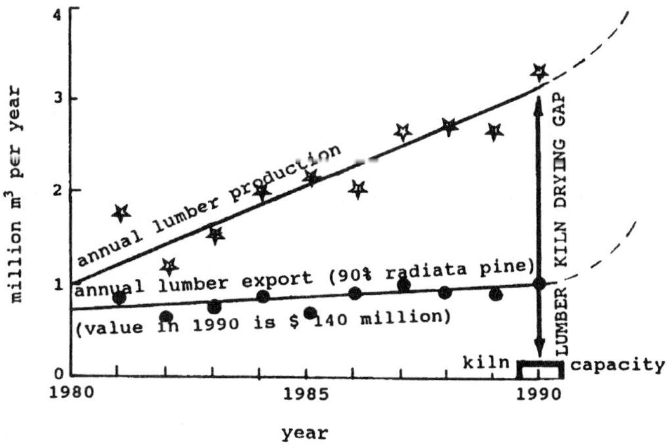

Figure 1. — Chile's lumber export 1980-1990 (4).

Lumber drying

Chile's annual kiln-drying capacity is only 0.2 million m^3 (4), and yard drying is no alternative because of potential problems with blue stain. Chile, therefore, can kiln dry no more than 18 percent of the lumber volume exported and 6 percent of the lumber volume produced. This means that Chile

Table 1. — Chile's lumber exporters, 1990 (4).

Number of sawmills	Mill size	Annual production	Annual exports	Annual kiln capacity
		(million m^3/year)		
1,600	All sizes	3.3	1.1	0.2
8	Large mills (>50,000 m^3/year/mill)	1.0	0.8	<0.2

Table 2. — Chile's lumber dry kilns, 1990 (2).

Kiln type	Number of kilns	Combined kiln size	Average kiln size
		(m^3)	(m^3/kiln)
Conventional steam kilns	35	5,400	155
High-temperature steam kilns	4	740	185
Other kilns	15	830	55

The authors are, respectively, Assoc. Professor, Univ. of Idaho, Moscow, Idaho; Assoc. Professor, Univ. of Bio-Bio, Concepcion, Chile; and former Graduate Student, Professional Institute Diego Portales, Concepcion, Chile.

appears about 1 to 3 million m^3 short of annual kiln-drying capacity (Fig. 1).

Conventional steam kilns are presently the preferred kilns in Chile (Table 2). Target moisture content is often less than 10 percent of the ovendry weight, and the lumber is steam-conditioned after drying to relieve residual stresses.

Using drying cost figures by Neild (6), it is estimated that the Chilean kiln capacity gap translates into investment needs for kilns and boilers in the amount of $20 to 70 million, or $2 to 7 million per year, assuming a 10-year, straight-line depreciation. Also, additional power plants (cogeneration plants) will be needed to satisfy electricity demands.

Conclusions

There are opportunites for investors and manufacturers in the lumber drying area in Chile. Chilean investment needs for dry kilns and boilers appear to be $20 to 70 million, to maintain present lumber export and production levels, and may be greater once plantations mature.

Literature cited

1. Chapotelle, D.N. 1990. International marketplace pressures to kiln dry lumber. Proc. Western Dry Kiln Assoc., Forest Res. Lab., Oregon State Univ., Corvallis, Oreg. p. 1.
2. Corporacion del Fomento (CORFO). 1991. Secado de pino a alta temperatura. FDP 89/91, Santiago, Chile.
3. Instituto Forestal (INFOR). 1991. Estadisticas Forestales 1990. Boletin Estadistico No. 21, Santiago, Chile.
4. ________. 1991. La industria del aserrio 1990 (Censo). Boletin Estadistico No. 23, Santiago, Chile.
5. Jelvez, A., K.A. Blatner, R.L. Govett, and H.P. Steinhagen. 1989. Chile's evolving forest products industry. Part 1. Its role in international markets. Forest Prod. J. 39(10):63-67.
6. Neild, P. 1982. A technical/economic evaluation of alternative lumber dry kiln heating/drying methods. Proc. 6th Inter. FPRS Industrial Wood Energy Forum '82, Vol. 2, Forest Prod. Res. Soc., Madison, Wis. p. 371.

Technical Forum Presentation:

Reinforced glulam: improved wood utilization and product performance

Daniel A. Tingley and Robert J. Leichti

Abstract

The technology for glued-laminated timber (glulam) relies on E-rated laminae to provide strength and stiffness to the engineered composite member. However, the required grades for tensile laminations are becoming difficult to obtain and more expensive. A consortium of industry and academia has been working on substituting the high-grade tensile laminations with a fiber-reinforced-plastic pultrusion comprised of a high-strength fiber, in this case Kevlar® 49, embedded in a plastic matrix. The plastic pultrusion has been dubbed "high-strength, fiber-reinforced plastic," or FiRP. The test results have demonstrated substantial gains in design capacities and stiffness for glulam reinforced with FiRP. The product is economically viable because the reinforcement leads to reduced section dimensions, lumber grade reductions, and reduced chemical treatment and transportation costs. A design method has been devised and verified. Several structures using the new technology are now ready for installation.

Project overview

Traditional glulam technology relies on E-rated laminations to provide strength and stiffness to flexural members. The most important of these laminations is the tensile laminae, which are often required to be L1 or L2 grades. Given the current forestry enigma, the availability of these grades is often limited while the cost is escalating. The concept of tensile reinforcement in glulam has been investigated primarily to reduce the demand for the high cost and difficult to procure lumber grades.

Testing and theoretical analysis has been conducted over the past 10 years on the use of high-strength, fiber-reinforced plastic (FiRP) to be used as tensile reinforcement in wood composites. During the last year, research in this area was accelerated by a team comprised of a glulam manufacturer (American Laminators), a timber engineering firm (Western Wood Structures), and the Departments of Forest Products and Civil Engineering at Oregon State University. Full-scale beams have been manufactured and tested in a range of sizes and degrees of reinforcement and a design method has been verified.

The testing program included conventional 24F Douglas-fir beams as controls and reinforced members made from lower grades of ponderosa pine. Reinforcement materials have included plastics with both Kevlar® and carbon fibers. The test results indicate that reinforced members are less variable in strength, stiffness is enhanced from 11 to 50 percent, and moment capacity may be increased by 85 to 115 percent. In addition, the reinforced members require smaller sections for most design applications.

Ultimately, the reinforcement of glulam with high-strength fiber reduces the demand for the expensive and difficult to procure laminating grades of Douglas-fir. The reduced wood and grade requirements coupled with secondary cost factors such as transportation and chemical treatment lead to a cost competitive product.

In 1993, four commercial structures using the new reinforcement technology were scheduled for construction. These are located in Wisconsin, Oregon, Washington, and California. Other commercial building structures are in the planning and design stage.

The authors are, respectively, Graduate Research Assistant and Assoc. Professor, Oregon State Univ., Dept. of Forest Prod., Corvallis, Oreg.

Manufacturing the FiRP/wood composite

The FiRP is produced in various thicknesses, widths, and lengths. Regular quality control sampling is used to establish the strength criterion of the FiRP prior to use in the structural application.

The FiRP is placed in the glulam beams using the same process as the wood laminae. The FiRP/wood composites production process allows the manufacturer to continue using the resorcinol adhesives and current manufacturing practices with very few changes. For example, the manufacturer cuts the FiRP to desired length and introduces the FiRP layers in the specified locations and thicknesses during the lay-up process. If desired, the FiRP can pass through the same glue spreader as the wood laminae (although some modifications may be necessary). The same lay-up pressures are needed, but some additional time in the racks is required for beams wider than 6-3/4 inches.

The FiRP-reinforced beams are planed in the same manner as the regular wood products, and on-site fabrication and pressure treatments are still possible

Table 1. — Selected properties of materials.

Property	Wood	FiRP	Steel
Material character	Orthotropic, planes of symmetry	Orthotropic, unidirectional continuous fibers in a plastic matrix (fibers: Kevlar® 49)	Isotropic (based on ASTM A 529)
Stiffness (psi)	1.8×10^6 (D-f at 12% MC)	KRP, 11×10^6	30×10^6
Tensile strength (psi)	15,000 (D-f at 12% MC)	KRP, 200,000	60,000
Cyclic load resistance	Excellent	KRP, excellent	Poor
Long-term load resistance	Poor	KRP, excellent (92% creep to failure rate)	Good to excellent
Strain at yield point (%)	D-f (compression, 12% MC), 0.8; D-f (tensile, 12% MC), 0.9	KRP, 2.8	2.0
Poission's ratio	Average softwoods, 0.033 to 0.47	KRP, 0.12 to 0.36	0.30
Weight density (lb./ft.^{3})	35 to 40 (D-f, 12% MC)	KRP, 80 to 90	460
Coefficient of thermal expansion (in./in./°F)	0.0000025 (softwoods)	0.0000029 (KRP, longitudinal)	0.000025
Thermal conductivity (Btu-in./hr.-ft.^{2}-°F)	0.85 (D-f, specific gravity=0.5, OD)	KRP, 0.0858	310
Thermal diffusivity (in.^{2}/sec.)	0.00025 (D-f, specific gravity=0.5, OD)	KRP, 0.000030	0.02
M ıchinability	Excellent plant and site machinability	Good to excellent plant and site machinability	Specialized equipment for machining
Strength-to-weight ratio	Very high	KRP has the best of all high-strength fibers	Very low
Moisture sensitivity	Hygroscopic (shrinks and swells); strength and stiffness dependent on moisture content.	KRP, low hygroscopicity, moisture gain of 6 to 8% with an 8% loss in strength	No appreciable effect
Environmental	Generally good resistance to chemicals, susceptible to fungal decay	Excellent resistance to a broad range of chemicals	Poor resistance to chemicals and susceptible to oxidation in moisture environments
Fire resistance	Excellent resistance in sizes greater than 6- × 6-inch sizes, loss of strength begins at 150$F, slow charring occurs in the range 212 to 482°F, active charring occurs in the range 536 to 932°F.	KRP begins loss of strength properties at 300°F, maintains 50% of tensile strength for 7 days at 492°F.	Strength loss begins at 200°F, substantial loss beginning at 350°F.

with conventional methods. Machining experience with FiRP-reinforced materials has shown that carbide-tipped tools have fewer wear problems than high-speed steel tools.

Experimental beams

The ponderosa pine used to produce these series of test beams had no knot grade rule and was E-rated prior to manufacturing the glulams. The E-values ranged from 1,800,000 to 600,000 psi. Three basic groups of glulams were produced, and each group had the FiRP reinforcement located between the bottom two laminations.

- E-stacked — stacked according to E-value, lumber with lowest E-value on the bottom
- Random E-stack — stacked without respect to E-value of the lumber
- No fingerjoints — joints within laminations were butt joints with no glue

Results of testing

For the three types of glulam, several trends can be identified. The ponderosa pine glulam with the E-stacked layups were shown to be equal in moment capacity to V4-2400 Douglas-fir beams, although the stiffness of the FiRP-reinforced glulam with one layer of KRP was equal to a V3-1600 Douglas-fir layup. In general, the E-stacked layups with minimum reinforcement exhibit dramatic design capacity increases, stiffness improvements from 11 to 50 percent, and reduced variation of ultimate capacity.

Results of other testing with Douglas-fir members has been used to develop and verify a design method. The reinforcement technology offers an economic advantage of 25 to 35 percent depending on the member size, degree of reinforcement, chemical treatments, and transportation.

The future of FiRP and composite design

Currently, Western Wood Structures of Tualatin, Oreg. has implemented the technology for bridges and some other structures. Necessary testing for code compliance is underway. The technology is exclusively licensed by Elk River Enterprises, Inc., Drain, Oreg. Negotiations are underway to sublicense the technology to others in the industry.

The first commercial structures are due to be constructed in Wisconsin and Oregon late in 1993. Other structures are in the planning and design stages.

Technical Forum Presentation:

Current status of U.S. hardwood chip export trade to Asia

Cynthia D. West

Abstract

Exports of U.S. hardwood pulp chips to Asia have increased dramatically since the mid-1980s. The majority of this material is exported to Japan. In 1992, 84 percent of U.S. hardwood pulp chips exports were to Japan. The remaining 16 percent was almost equally divided between Korea and Taiwan.

Increasing exports to Japan have been driven by an increase in domestic consumption of paper and paperboard products. However, following Japanese pulping capacity expansion in the late 1980s, consumption flattened and declined slightly. 1992 U.S. hardwood chip exports to Japan grew by only 12 percent compared to an average annual growth rate of over 70 percent between 1988 and 1991. In the first half of 1993 exports to Japan remained flat.

Table 1. — U.S. hardwood chip exports (revised 9/20/93).

Region	1992	1991	1990	1989	1988
	(metric tons, green weight)				
West Coast[a]					
Japan	1,355,974	1,413,484	1,185,109	974,199	722,352
Korea	0	0	0	0	0
Taiwan	0	0	0	0	0
Gulf Coast					
Japan	1,688,530	1,193,691	755,134	437,843	218,685
Korea	260,797	33,909	74,794	26,350	0
Taiwan	206,138	128,427	54,596	56,767	41,548
East Coast					
Japan	708,774	736,461	288,605	208,597	67,002
Korea	100,196	38,136	31,217	63,781	0
Taiwan	191,196	305,255	160,429	76,964	172,214
TOTAL					
Japan	3,753,278	3,343,636	2,228,848	1,620,639	1,008,039
Korea	361,473	72,045	106,011	90,131	0
Taiwan	397,334	433,682	215,025	133,731	213,762

[a] PIERS estimates do not include chips not specified as hardwood or softwood shipped from West Coast ports.

The author is Research Assistant, Northeastern Forest Expt. Sta., USDA Forest Serv., Princeton, W. Va.

The U.S. currently holds 28 percent of the Japanese market share of imported hardwood pulp chips. Australia leads with 32 percent and Chile rounds out the leaders with 21 percent. The remaining 19 percent is predominantly shared by South Africa, Indonesia, and China. In 1992, U.S. hardwood fiber was the costliest of these six leading suppliers to Japan.

The rapid increase in the demand for hardwood fiber by the Japanese pulp and paper industry between 1985 and 1991 has slowed. This industry has been successful in diversifying their supply sources during this time period through strategic foreign investment. It appears that Japanese buyers have regained an advantage in the market and are not expected to substantially increase demand for U.S. hardwood fiber, although there may continue to be regional shifts in U.S. supply.

Currently, the Gulf Coast ports lead U.S. chip exports to Asia. In 1992, these ports shipped an estimated 45 percent of U.S. hardwood chip exports to Japan, 72 percent of exports to Korea, and 52 percent of exports to Taiwan. Gulf Coast shipments of hardwood fiber, principally through the Mobile Export District, have experienced the greatest growth of the three regions due to an extensive water transportation system which accesses hardwood fiber from northern Mississippi, northern Alabama, and Tennessee. This region is expected to dominate hardwood chip exports to Asia in the short term.